FOODOPOLY

FOODOPOLY

The Battle Over the Future of
Food and Farming in America

Wenonah Hauter

Requests for permission to reproduce selections from this book should be mailed to:
Permissions Department, The New Press, 38 Greene Street, New York, NY 10013.

Chapter 3 draws from the 2012 Food & Water Watch report "Why Walmart Can't Fix
   the Food System."
Chapter 4 draws from the 2011 Food & Water Watch report "A Decade of Dangerous
   Food Imports from China."
Chapters 8, 9, and 10 draw from the 2010 Food & Water Watch report "Factory Farm Nation:
   How America Turned Its Livestock Farms into Factories."
Chapter 9 draws from the 2008 Food & Water Watch report "The Trouble with Smithfield:
   A Company Profile."
Chapter 13 draws from the 2012 Food & Water Watch report "Genetically Engineered Foods:
   An Overview."
Chapter 16 draws from the 2012 Food & Water Watch report "Farm Bill 101."

Published in the United States by The New Press, New York, 2012
Distributed by Perseus Distribution

LIBRARY OF CONGRESS CATALOGING-IN-PUBLICATION DATA
Hauter, Wenonah.
   Foodopoly : the battle over the future of food and farming in America / Wenonah Hauter.
      p.   cm.
   Includes bibliographical references and index.
   ISBN 978-1-59558-790-9 (hc. : alk. paper)
   ISBN 978-1-59558-794-7 (e-book)
   1. Food supply—United States.   2. Agricultural industries—United States.   3. Agriculture—
Economic aspects—United States.   I. Title.   II. Title: Battle over the future of food and farming
in America.
   HD9005.H358 2012
   338.10973—dc23   2012025605

Now in its twentieth year, The New Press publishes books that promote and enrich public
discussion and understanding of the issues vital to our democracy and to a more equitable world.
These books are made possible by the enthusiasm of our readers; the support of a committed group
of donors, large and small; the collaboration of our many partners in the independent media and
the not-for-profit sector; booksellers, who often hand-sell New Press books; librarians; and above
all by our authors.

www.thenewpress.com

Composition by dix!
This book was set in Minion

Printed in the United States of America

10   9   8   7   6   5   4   3   2   1

*This book is dedicated to
family farm defenders who steward the land and fight for justice.*

CONTENTS

ACKNOWLEDGMENTS

So many people have helped make this book possible.

I could never have completed *Foodopoly* without the help of my really smart, talented, and wonderful research assistant, Lily Boyce. She was always cheerful and efficient, spent months creating the charts and graphics featured in this book, and helped tirelessly with research. Lily is a star!

I want to acknowledge and thank the extraordinary staff of Food & Water Watch. I owe an intellectual debt to Patrick Woodall, our brilliant and talented research director, for his deep thinking, number crunching, and research, and for being a patient and supportive sounding board as I struggled through the complex web of issues that have created the dysfunctional food system. A special thanks to Patty Lovera, director of the food program, who is incredibly knowledgeable on a broad range of issues, and who helped tremendously with many aspects of this project. Many thanks to Darcey Rakestraw, communications director, who was enormously supportive in so many ways during this project, including helping with editing. I am grateful to Lisa Mastny, who was extremely helpful with editing and with suggestions for clarity, making this dense material more readable. My colleague Lane Brooks, the chief operating officer, took over many of my duties and responsibilities as I wrote this book. I am forever thankful to him for his good judgment and for being a calm, dependable, and good-humored partner in running Food & Water Watch.

I greatly appreciate the wonderful Food & Water Watch staff, who provided research and technical support, covered for me during this long project, and offered endless moral support: Sarah Alexander, Dave Andrews, Sarah Borron, Royelen Boykie, Jon Brown, Tony Corbo, Zach Corrigan, Scott Edwards, Noelle Ferdon, Clay Gatewood, Anna Ghosh, Kim Girton, Mitch Jones, Doug Lakey, Michele Merkel, Eve Mitchell, Rachel Nissley, Darcey O'Callaghan, Matt Ohloff, Genna Reed, Mark Schlosberg, Ben Schumin, Tim Schwab, Adam Scow, Tyler Shannon, Elanor Starmer, Yi Wang, Anna Witowaska, Emily Wurth, Gabriella Zanzanaini, and Ron Zucker.

I am deeply grateful to Helaine and Sid Lerner for their confidence in and encouragement and support for this project. I can truly say that without their dedication to the creation of a better food system, this book would never have been written and so much important work would never be done. A special thanks to GRACE Communications Foundation for ongoing support and assistance with so many areas of my work: Scott Cullen; Leslie Hatfield; Lisa Kleger; Destin Layne; and the staff of the Eat Well Guide, the Meatrix, and Sustainable Table. I am so appreciative of the support and encouragement from Joan and Bob Rechnitz, who have had the great foresight to understand that nature should not be financialized and that extreme forms of energy threaten our food and water.

This book would not have been possible without the knowledge and perspectives of the many people that I interviewed and provided material for the book. I am indebted to the following people for taking the time to speak with me and provide valuable insight and information: Mark Arax, John Bunting, Ben Burkett, Mike Callicrate, Lloyd Carter, Dale Coke, Joaquin Contente, Roberta Cook, Agatha D'Esterhazy, Cap Dierks, Diane Endicott, Hugh Espey, Larry Ginter, Joel Greeno, Andrew Gunther, Sean Hallahan, Kyal Hamilton, John Hansen, Michael Hansen, Diane Hatz, Gary Hoskey, Frederick Kaufman, Kurt Kelsey, Robby Kenner, Kendra Kimbirauskas, John Kinsman, Garry Klicker, Judy Labelle, Anna Lappé, Dr. Robert Lawrence, Ray Leon, Scott Marlow, Michael Masters, Mas Masumoto, Larry Mitchell, Carole Morrison, Dr. Keeve Nachmann, George Naylor, Dr. Marion Nestle, Felicia Nestor, Harvey Nijjer, Kathy Ozer, Stan Painter, Rhonda Perry, Michael Pertschuk, Chris Peterson, Dr. Daryll Ray, Matt Rogers, Valerie Ruddle, Rebecca Spector, Steven Stoll, Dr. Robert Taylor, Warren Taylor, Bruce von Stein, Lori Wallach, Dr. David Wallinga, Mike Weaver, Tom Willey, Brad Wilson, Donna Winburn, and Mark Winne.

I am extremely grateful to Marc Favreau, my editor at The New Press, for having the confidence in this project and for helping me every step of the way. I am so very fortunate to have had Marc shepherding this project. His patience and understanding have made this experience a pleasure. Thank you also to my production editor, Sarah Fan, who along with Marc provided an experienced eye with editing and made this a much better book than it would have been otherwise. I am grateful to Rachel Burd for the thorough and painstaking copyediting. Thank you also to Azzurra Cox and all of the other helpful staff of The New Press.

Many thanks to my dear friend and colleague Maude Barlow, who saw the value of this book when it was just an idea and who helped make it possible with important introductions and ongoing moral support when the going got tough. I appreciate the advice, positive reinforcement, and camaraderie of my colleague and dear friend Lisa Shubert, who was always available with a kind word.

I am fortunate to have been trained by the brilliant organizers of the Midwest Academy, who taught me that the only way we can bring about long-term progressive change is by building political power, and that it takes a long-term grassroots organizing strategy. I owe Steve Max, Jackie Kendall, and David Hunt a deep debt of gratitude for years of organizing mentoring.

I spent almost a decade working for Public Citizen, a group that Ralph Nader founded forty years ago. I am grateful to Ralph for helping to shape my worldview and for his decades of fighting the foodopoly, among so many other injustices. He also had the wisdom and foresight to help support and publish the landmark book by the late Al Krebs, *The Corporate Reapers: The Book of Agribusiness*. In writing *Foodopoly*, I drew freely from Al's work. Those of us in the fair food and farm movement miss him deeply. His knowledge and insight on agriculture and farm policy—from the 1770s to 1990—was unmatched.

I also owe an intellectual debt to social scientists Dr. Bill Heffernan and Dr. Mary Hendrickson, who were pioneers in documenting the foodopoly. Their excellent research on consolidation in the food system and the impact this has had on farmers and consumers laid the groundwork for this book.

I want to acknowledge Tim Wise, director of research and policy at the Global Development and Environment Institute at Tufts University, whose important work on farm income and rural development provided the underpinning for many of my arguments on agriculture policy. Likewise, I would like to recognize Dr. Daryll E. Ray, director of the Agricultural Policy Analysis Center at the University of Tennessee, for his decades of work on farm policy, which provided the basis for my analysis on the negative role of overproduction on our food system and recommendations for changes in policy.

I am very appreciative of Harriet Barlow, who emanates wisdom "on the commons" and helped facilitate my opportunity to spend a delightful and idyllic month at the Blue Mountain Center, during which time several chapters of this book were written. The center's staff—Ben Strader, Alice Gordon, Sis Eldridge, Diane McCane, Nico Horvitz, and Jamie Barret Riley—made the writer's retreat an enjoyable and productive experience, as did all of the interesting people who were part of the residency program.

With all my heart I am so thankful to have loving and supportive friends and family, who have been patient and enormously understanding over the last year while I was completely absorbed in this project. My beloved children,

Adrina Miller and Che Miller, always my most enthusiastic cheerleaders, provided constant love, care, and encouragement. I am so lucky that our family bonds are deepened by mutual respect and friendship.

My dear stepdaughter and friend, Christy Nichols, texted or e-mailed me almost daily with uplifting messages to spur me on and sent me flowers to brighten the day. I so deeply appreciate her kindness and positive reinforcement. My cherished friend of four decades Sue Hays and her husband, Tom Hays, never ceased to offer support, care, and delicious meals. My extended family and friends—Erin Dougherty, Alton Dulaney, Debbie and Wayne Hauter, Kelsie Kerr, Pat Lewis, Kathy and Chip Reid, Mary Ricci, Leo and Jan Scolforo, and Kelly Wolf—were all helpful and supportive through the different phases of this project. I am grateful for the young people in my life, who inspire me to keep on fighting for a better world: Mark and L.J. Hilberath; Tyler, Christian, and Bennett Nichols; and Jackson Wolf.

Lastly, I must thank the farmer in my life, my loving husband, Leigh, who tutored me in politics and has been my friend, companion, and comrade. We have shared almost three decades in the struggles not only for a fair food system but also for social, economic, and environmental justice.

INTRODUCTION

In 1963 my dad bought a ramshackle farm with rich but extremely rocky soil in the rural Bull Run Mountains of Virginia, forty miles southwest of Washington, D.C. Today it is on the verge of suburbia.

He grew up in Oklahoma during the Dust Bowl, rode the rails, and eventually, in his late fifties, found his way "back to the land." So we moved to what was then a very rural landscape—a place culturally a world away from the nation's capital and physically linked only indirectly by two-lane roads. Our old farmhouse, with a mile-long rutted driveway accessible only by four-wheel drive, was off another dirt road and had no electricity or plumbing. Eventually my dad did manage to get the local rural electricity co-op to put in poles and hook up power, but he never did get around to installing indoor plumbing.

He was an unusual man—a religious iconoclast and an organic gardener at a time when few people knew the term. He was considered a crank and a hobby farmer, if you can call it that, growing a few vegetables and keeping bees. His wild-blossom honey was the only vaguely successful part of his farming venture. My dad, who died in 1991 at the age of eighty-one, would be shocked now to see both his farm and the massive development around it.

Today the hundred acres of mostly wooded land is bordered by a mega-mansion subdivision on one side and an expensive "gated community" a mile away as the crow flies. Thousands of town houses and new subdivisions have cropped up where once there were fields dotted with cows. This has brought on the box stores, including Walmart and fast-food joints—blights on the once bucolic rural landscape. A major highway, I-66, recently engineered to be either six or eight lanes depending on the location, means we can zip into the nation's capital during the rare times that commuters are not clogging the road.

Since 1997, my husband has run the farm as a community-supported agriculture (CSA) program, feeding five hundred families each season with subscription vegetables grown using organic practices. It's a successful family business that suits my activist husband, who taught high school and college and worked for public interest organizations, but who really prefers the challenge of farming without chemicals. It works financially, because we own the land outright, and because we live near a major metropolitan area where urban and suburban residents are seeking greater authenticity in the food they eat. They want their children to see where food comes from and to learn that chickens enjoy living together in a pasture. We often joke that for most people the CSA is more about having a farm to visit than the vegetables.

As a healthy-food advocate, I feel privileged to have grown up a rural person and to have had the real-life experience of pulling weeds, squishing potato bugs, canning vegetables, gutting a chicken, baking bread, and chopping wood for the cookstove. As a teenager I felt deprived, but as an adult I am grateful to know where food comes from and how much work it takes to produce it. My family is also extremely lucky that my dad bought almost worthless land in the 1960s that today is located near a major metropolitan area populated by a largely affluent and educated population. But most farmers, or people aspiring to be farmers, aren't so lucky. Fortunately, farmers' markets and similar venues help capture the excitement and nostalgia for farming, and for a simpler and healthier lifestyle, and they are delightful for the customers and can be profitable for farmers.

But despite my firsthand knowledge of and appreciation for the immense benefits of CSAs and farmers' markets, they are only a small part of the fix for our dysfunctional food system. Food hubs, which aggregate and distribute local food, are beneficial for participating farmers and the purchasing food establishment. But, so far, they must be subsidized by nonprofits or local governments because they are not self-sustaining. We must delve deeper into the history of the food system to have the knowledge to fix it. I decided to write this book because understanding the heartbreaking story of how we got here is not only fascinating but necessary for creating the road map for changing the way we eat.

The food system is in a crisis because of the way that food is produced and the consolidation and organization of the industry itself. Solving it means we must move beyond the focus on consumer choice to examine the corporate, scientific, industrial, and political structures that support an unhealthy system. Combating this is going to take more than personal choice and voting with our forks—it's going to take old-fashioned political activism. This book aims to show what the problem is and why we must do much more than create food hubs or find more opportunities for farmers to sell directly to consumers. We must address head-on the "foodopoly"—the handful of corporations that control our food system from seeds to dinner plates.

While the rhetoric in our nation is all about competition and the free market, public policy is geared toward enabling a small cabal of companies to control every aspect of our food system. Today, twenty food corporations produce most of the food eaten by Americans, even organic brands. Four large chains, including Walmart, control more than half of all grocery store sales. One company dominates the organic grocery industry, and one distribution company has a stranglehold on getting organic products into communities around the country.

Further, science has been allowed to run amok; the biotechnology industry has become so powerful that it can literally buy public policy. Scientists have been allowed to move forward without adequate regulation, and they are now manipulating the genomes of all living things—microorganisms, seeds, fish, and animals. This has enabled corporations to gain control over the basic building blocks of life, threatening the integrity of our global genetic commons and our collective food security. Biotechnology has moved into the world of science fiction, as scientists actually seek to create life-forms and commercialize them. Reining in and regulating the biotechnology industry is critical to reforming the dysfunctional food system.

These structural flaws are often overlooked by the good-food movement, which focuses on creating an alternative model from the ground up that will eventually overtake the dysfunctional system. However, this approach raises the question: for whom and how many? A look at the most recent statistics on local food illustrates this point. A November 2011 study by the U.S. Department of Agriculture's Economic Research Service, using 2008 data (the most recent available), found: "Despite increased production and consumer interest, locally grown food accounts for a small segment of U.S. agriculture. For local foods production to continue to grow, marketing channels and supply chain infrastructure must deepen." [1]

The study found that levels of direct marketing to consumers are highest in the Northeast, on the West Coast, and in a few isolated urban areas outside these regions. Direct marketing of local foods to consumers at farmers' markets and CSAs, along with local food sales to grocery stores and restaurants, generated $4.8 billion in sales in 2008. [2] This figure is infinitesimal in comparison to the $1,229 trillion in overall sales from grocery stores, convenience stores, food service companies, and restaurants.

According to the USDA, only 5 percent of the farms selling into the local food marketplace are large farms (with over $250,000 in annual sales), but these large farms provided 93 percent of the "local foods" in supermarkets and restaurants. Eighty-one percent of farms selling local food are small, with under $50,000 in annual sales, and 14 percent of farms selling local foods are medium-sized, with $50,000 to $250,000 in sales. The small and medium-sized farms sell nearly three quarters of the direct-to-consumer local foods (both CSAs and farmers' markets) but only 7 percent of the local foods in supermarkets and restaurants. Although the 5,300 large farms averaged $772,000 in local food sales, small farms sold only $7,800 and medium-sized farms sold only $70,000 local foods on average.[3]

Of special significance is the finding that over half of all farms that sell locally are located near metropolitan counties, compared to only a third of all U.S. farms. This illustrates the difficulty that farmers who grow corn, soy, wheat, and other feed or cereal grains for commodity markets have in converting their farming operations to direct sales to consumers. These farmers sell crops that reenter the food system as a component of another food—as a sweetener, an oil, a starch, or as feed for animals. The lack of a local market, a distribution network, or in many cases the infrastructure needed to harvest, aggregate, or process local foods is also a tremendous hindrance to creating an alternative food system.

Look at a map of the large agricultural middle of this nation to understand that the few remaining farmers who grow the millions of acres of corn and soybeans, fencerow to fencerow, do not live where they can sell directly to the consumer. Most farmers don't have nearby affluent urban areas to which to market their crops. They can't switch from commodities to vegetables and fruit even if they had a market, because they have invested in the equipment needed to plant and harvest corn and soy, not lettuce, broccoli, or tomatoes.

Overly simplistic solutions are often put forward by some leaders in the good-food movement that take the focus away from the root causes of the food crisis—deregulation, consolidation, and control of the food supply by a few powerful companies. One of the most prevalent policy solutions put forward as a fix for the dysfunctional system is the elimination of farm subsidies. This silver bullet prescription implies that a few greedy farmers have engineered a farm policy that allows them to live high off the hog on government payments, while small farms languish with no support. Proponents of this response say that if we remove these misapplied subsidies to these few large farms, the system will right itself.

Unfortunately, the good-food movement has been taken in by an oversimplified and distorted analysis of farm data. It is based on a misinterpretation of misleading U.S. Department of Agriculture statistics that greatly exaggerate the number of full-time family farm operations. A close look at the USDA's *Census of Agriculture* shows that one third of the 2.2 million entities counted as farms by the agency have sales of under $1,000 and almost two thirds earn under $10,000 a year. These small business ventures are counted even though they are far from being full-time farming operations. In most cases these are rural residences, not farms, and the owners are retired or have significant off-farm income. They have a part-time agriculture-based business as part of their rural lifestyle—anything from having a vineyard to growing flowers or mushrooms.

Counting these small ventures as farms not only skews the statistics on the number of farms in the United States; it also makes it appear that only a small percentage receive government payments. In reality, we have under a million full-time farms left, and almost all of them, small and large, receive government subsidies. This is not to say that the subsidy system is good policy. Rather that it is a symptom of a broken food production system, not the cause of the problems. If we penalize farmers for policies that the powerful grain traders, food processors, and meat industry have lobbied for, we will never create a sustainable food system. We need midsize farming operations to survive and to be transitioned into a sustainable food system.

Midsize family farmers have an average income of $19,277—a figure that includes a government subsidy.[4] The cost of seeds, fertilizers, fuel, and other inputs is continuing to rise as these industries become more monopolized. Most farmers are scratching by, trying to hold on to their land and eke out a living. We are losing these farms at a rapid rate, resulting in the consolidation of smaller farms into huge corporate-run industrial operations with full-time managers and contract labor. Telling these farmers that all they have to do once the subsidies are taken away is grow vegetables for the local farmers' market is not a real solution for them or their communities. Rural communities are seeing the wealth and the profit from agriculture sucked into the bottom lines of the largest food corporations in the world.

Economically viable farms are the lifeblood of rural areas. Their earnings generate an economic multiplier effect when supplies are bought locally, and the money stays within the community. The loss of nearly 1.4 million cattle, hog, and dairy farms over the past thirty years has drained not only the economic base from America's rural communities, but their vitality. These areas have become impoverished and abandoned, and the only hopes for jobs are from extractive industries such as hydraulic fracturing or from building and staffing prisons.

Something is fundamentally amiss in a society that does not value or cherish authentic food that is grown full time on appropriate-size family farms. The benefits of farmers—rather than corporate managers—tending crops and the land are many. Fred Kirschenmann, a North Dakota farmer and a leader in the sustainable agriculture movement, along with his colleagues at the Agriculture in the Middle project write extensively on this point and poignantly outline the benefits of these vulnerable midsize farms in today's economic landscape. They fall between the large, vertically operated commodity operations and the small-scale ones that sell directly to consumers. Farms in the middle also provide wildlife habitats, open spaces, diverse landscapes, soils that hold rainwater for aquifers, perennials that reduce greenhouse gases by removing carbon from the atmosphere, and crop and pastureland that reduce erosion and flooding.[5]

These are the farms that could be changed to provide sustainably grown organic food for the long term. Many are located in the Midwest and South, where there is no large population to buy directly from them, but they have the capacity to produce food for the majority of Americans—if given a chance.

Changing farm policy to provide that chance is key to preventing our nation's rural areas from becoming industrial sites and to truly remaking the food system for all Americans. We must address the major structural problems that have created the dysfunction—from the failure to enforce antitrust laws and regulate genetically modified food to the manipulation of nutrition standards and the marketing of junk food to children. We need to move beyond stereotypes and simplistic solutions if we are to build a movement that is broad-based enough to drive policy changes.

Most people are several generations away from the experience of producing their own food. This leads to many misconceptions—from over-romanticizing its hard, backbreaking work to the dismissal of farmers as greedy, ignorant, and selfish "welfare queens." Understanding the difficult challenges they face is critical to developing the policy solutions necessary for saving family farms and moving into a sustainable future. We need to develop a rural economic development plan that enables farmers to make a living while at the same time providing healthy, affordable food choices for all Americans.

We have the opportunity, before it is too late, to change the course of our food system's development away from factory farms and laboratories and toward a system that is ecologically and economically sound. We can challenge the monopoly control by fighting for the reinstatement of antitrust laws and enforcement of them. We have the land and the human capacity to grow real food—healthy food—but it will take a wholesale effort that includes restructuring how food is grown, sold, and distributed. It means organizing a movement to hold our policy makers accountable, so that food and farm policy is transformed and environmental, health, and safety laws are obeyed.

It will require a massive grassroots mobilization to challenge the multinational corporations that profit from holding consumers and farmers hostage and, more important, to hold our elected officials accountable for the policies that are making us sick and fat. We must comprehend the complexity of the problem to advocate for the solutions. We cannot shop our way out of this mess. The local-food movement is uplifting and inspiring and represents positive steps in the right direction. But now it's time for us to marshal our forces and do more than vote with our forks. Changing our food system is a political act.

We must build the political power to do so. It is a matter of survival.

Note: Full-color versions of the graphics that appear throughout the book are available at www.foodandwaterwatch.org/foodopoly-infographics/.

PART I

Farm and Food Policy Run Amok

The dysfunctional food system that we suffer from today is the result of long-standing farm and food policies that were first proposed by some of the most powerful men in the country shortly after World War II. These men envisioned a future in which most young rural men would supply cheap labor for manufacturing in the industrial North rather than continuing to farm, and in which a small number of large industrialized farms would supply the necessary food. They foresaw a future in which food production would be globalized for economic efficiency and the "free market" would create the cheap inputs necessary for processed food. The visions that these powerful men had in the late 1940s and early 1950s were eventually enshrined in federal farm policy and in global trade agreements.

1

GET THOSE BOYS OFF THE FARM!

Burn down your cities and leave our farms, and your cities will spring up again as if by magic; but destroy our farms and the grass will grow in the streets of every city in the country.
—William Jennings Bryan, "Cross of Gold" speech, July 9, 1896

Although most consumers—eaters—view food first and foremost as the sustenance necessary for life, Big Business thinks of our kitchens and stomachs as profit centers. The unwavering determination by the leaders of a handful of powerful multinational corporations to concentrate ownership and control of the food production and delivery systems has created unprecedented consolidation down the entire food chain. Food and agricultural products have been reduced to a form of currency on income statements that cause a rise or fall of quarterly profits. The worth of these products is measured on the return on investment, or as an opportunity for mergers or acquisitions, that drive the strategy of the parent company. Their value is described in a Wall Street–speak of deals, synergies, diversification, and "blockbuster game changers."

Even hedge funds, those poorly regulated firms that played a role in causing the recent financial crisis, have become some of the largest investors in food companies, farmland, and agricultural products. These firms invest the money of high-wealth individuals and institutions into broad segments of the economy—including food and agriculture. They have speculated in food commodity markets (contributing to price spikes in corn and soybeans) and bought restaurant chains (Dunkin' Donuts), and are buying up farmland in the United States and the developing world. A private investment company even owns Niman Ranch, the firm that pioneered producing pork more sustainably.[1]

Hedge funds have been big proponents of grabbing land—they have bought farmland worldwide—to capitalize on expectations of profitability from the catastrophic impacts of climate change on agriculture. The dramatic increase in the price of land in the U.S. Midwest over the past few years has led the president of the Federal Reserve Bank of Kansas City to warn about the crash that could result from a farmland bubble. The U.S. Senate's Agriculture Committee warns that "distortions in financial markets" will catch the country by surprise again.

This financialization of food and farming has wreaked havoc on the natural world. The long list of the consequences of industrialized agriculture includes the polluting of lakes, rivers, streams, and marine ecosystems with agrochemicals, excess fertilizer, and animal waste. Nutrient runoff (nitrogen and phosphorus) from row crops and animal factory farms, one of the foremost causes of the conditions that starve waterways and the ocean of oxygen, is creating massive dead areas of the ocean, such as one at the mouth of the Mississippi River the size of the state of New Jersey. Planting and irrigating row crops has caused serious erosion, as irrigation and rainwater wash the topsoil away at the rate of 1.3 billion tons per year. And as soil scientists are fond of saying, "No soil, no life."

The relentless drive for profit by agribusiness has had long-lasting and negative effects on all aspects of society. Public health has been sacrificed on a diet of heavily advertised processed foods that are high in calories and low in nutrients, resulting in consumers who are overweight and poorly nourished. Obesity affects 35 percent of adults and 17 percent of children in the United States, and causes a range of health problems from heart disease to diabetes. And while many Americans are overfed and dieting, one in six Americans frequently goes hungry.

No segment of society has been more affected by agribusiness and its allies in government over the past sixty years than farmers. After World War II, farmers became the target of subtle but ruthless policies aimed at reducing their numbers, thereby creating a large and cheap labor pool. In more recent times, federal policy has been focused on reducing the number of farms as labor has been replaced by capital and technology. In 1935, 54 percent of the population lived on 6.8 million farms; between 1950 and 1970, farm populations declined by more than one-half. Today under a million farms produce the bulk of the food produced in the United States, and farmers are less than 1 percent of the nation's population.

The struggle to eke out a living has intensified each decade since 1950, because farmers have been locked into a system of low crop prices, borrowed capital, large debt, high land prices, and a weak safety net. Unchecked corporate mergers and acquisitions have increased the economic pressure, since fewer firms are competing to sell the seeds, equipment, and supplies that farmers use every day. At the same time, they have few choices where to sell their products. A handful of agribusiness and food industry multinational corporations stand between the farmers who produce the food and the more than 300 million people who consume it in the United States.

Consolidation at the top of the food chain has affected every segment below, including farming. Large-scale industrial operations comprising only 12 percent of U.S. farms make up 88 percent of the value of farm production. Family farming stands on the edge of extinction; most small and medium-size farms are dependent on off-farm income for survival. Although crop prices have been higher since 2008, the increased income has been gobbled up by higher costs for seeds, chemicals, fertilizers, fuel, and feed.[2]

The loss of farms has caused a rural bloodletting, leaving rural towns and counties forlorn, boarded-up, and in some cases completely gone. A *Los Angeles Times* analysis of census data from fourteen hundred rural counties in the U.S. heartland, the region between the Mississippi River and the Rocky Mountains, found that rural areas are sparsely populated and continuing to lose people.[3] When farms go out of business, the local businesses that depend on them also disappear: the implement dealers and farm supply companies and all of the stores and service providers. Hard times also mean that rural youth disappear to urban areas in search of jobs—even those who would prefer to farm and live a rural lifestyle.

Farmers have fought back against the rural exodus that has stretched over more than a century. Activists have long been engaged in a struggle with banks, railroads, and business interests over their inequitable position within the economic system. The nineteenth and twentieth centuries were marked by populist uprisings against the unfair economic policies that threatened farm family livelihoods. They banded together to form organizations: as part of the Grange, the Farm Alliance, and the National Farmers Union, they organized, ran candidates, and joined with progressive allies in labor and social justice movements. Most of this story has been erased from public consciousness, especially the history of the post–World War II farm movement. Farmers were still a large and vital political force that had to be reckoned with in the 1950s. And they were willing to take militant action to protect their families and communities.

The National Farmers Organization (NFO) organized in 1955 to protest a move to reduce crop prices that was being perpetrated by President

Eisenhower's secretary of agriculture, Ezra Benson. Benson was set on destroying the New Deal program for agriculture—measures that had been designed to ensure fair farm prices. The large and powerful grain-trading, food-processing, banking, and industrial giants had been conspiring to cut the cost of grains and to drastically reduce the number of farm families. Farmers were considered "excess labor" by the captains of industry—workers who should be shifted into factories, while large, highly capitalized farms produced all the foods needed for domestic consumption and for the global trade they envisioned.

In 1942, several businessmen and an advertising executive had created an organization that was to have a powerful role in shaping the post–World War II economy and society—an influence that continues to this day. They aimed to make the Committee for Economic Development (CED) a place where leaders of business could hammer out their differences on economic policy, and then use the new technique of public relations to promote their agreed-upon agenda. Among the founders were Paul Hoffman, president of Studebaker; William Benton, the inventor of modern consumer research and polling; and Marion Folsom, an Eastman Kodak executive.

All three eventually were placed in high government positions. Hoffman was appointed by President Truman to administer the Marshall Plan, the large-scale economic aid program designed to rebuild war-torn Europe and to combat communism. Later, as president of the Ford Foundation and administrator of the United Nations Development Programme, he became one of the architects of the "Green Revolution."

Benton eventually left public relations and was instrumental in organizing the United Nations. He published the *Encyclopedia Britannica* and became a senator representing Connecticut. Folsom staffed the U.S. House Special Committee on Postwar Economic Policy and Planning. He was instrumental in developing the first tax law revision since 1874 as Eisenhower's undersecretary of the Treasury Department in 1953 and was later appointed by Eisenhower as secretary of health, education, and welfare.

In the early 1960s the very influential CED, at that time a think tank headed by men representing Ford Motor Company and Sears, had released a report declaring that there were too many farmers. The corporate solution: get farm boys off the farm and into vocational training for industrial skills and relocated to where their labor was needed.[4]

So, in August 1962, when twenty thousand farmers convened for the annual NFO convention in Des Moines, Iowa, they were fighting mad. The CED report had only added insult to injury. Agribusiness, the food-processing industry, and the nation's banks had been lining up over the previous decade to depress farm prices.

The release of the CED's screed against farmers during the summer of 1962 stirred the NFO to organize "catalog marches" in seven cities, where protesters dumped Sears catalogs in front of their stores. Long caravans of Ford cars and trucks drove in circles around Ford establishments in several cities. Shortly thereafter, both companies disavowed the report, and hearings were held in the U.S. Senate and House agriculture committees to discredit the proposed solution to the so-called farm problem that the CED had been peddling.

The CED, operating in a quasi-public sphere, represented the most powerful economic interests in the nation. Its members called "for action by government working with the free market, not against it."[5] During its first fifteen years of existence, thirty-eight of its trustees held public office and two served as presidents of the Federal Reserve Bank. The organization maintained strong relationships with the Truman, Eisenhower, and Kennedy administrations, helping to direct government research dollars as well as to provide funding for academic research. The strong ties to academia resulted in policy prescriptions shrouded in sophisticated economic rhetoric and focused on weakening the reform-liberalism of the New Deal. They couched their proclamations on shrinking the farm population as moving "labor and capital where they will be most productive."[6]

A demonstration of the group's power took place in 1962, when a conflicted President Kennedy was debating with his staff the merits of a massive tax cut. Kennedy was influenced to support the tax cut by a CED report that called for "a prompt, substantial and permanent reduction" that the White House legislative liaison's office distributed to members of Congress.[7] The CED then helped organize the Business Committee for Tax Reduction, endorsed by Kennedy, which actively lobbied Congress, eventually resulting in the passage of legislation in 1964 cutting individual tax rates by 20 percent across the board and reducing corporate tax rates.

CED members viewed the organization as a merchant of ideas. Its leadership had strong media connections that enabled it to publicize and popularize policy recommendations with elected officials and the public. Its information committee included members of several advertising agencies, the editors of the *Atlanta Constitution* and *Look*, the publisher of the *Washington Post*, the head of the Book-of-the-Month Club, the board chairman of Curtis

Publishing, and the presidents of *Time-Life* and the Columbia Broadcasting System (CBS). When the CED spoke, its propagandists wrote: a 1958 pamphlet, "Defense Against Inflation," was discussed in 354 papers and magazines, reaching 31 million people.

Immediately after its formation, the CED began mapping a postwar program to expand chemical-intensive agriculture and to grant industrial and financial interests more control over it. It worked to create a postwar economy built on massive and profitable industrial growth in the North, which would require an enormous pool of cheap labor. Their first report on agriculture was published in 1945, at a time when farmers were doing very well by feeding a war-ravaged world. Farmers flourished even with higher postwar production costs due to New Deal farm measures that ensured that farm income would keep up with the cost of farming—an important policy known as parity. The CED opposed continuation of these programs, which had been created by the Agricultural Adjustment Act of 1933 to help farmers receive prices for their products that were on par with the rest of the economy—much like a livable wage.

Among the programs created by the legislation to achieve parity were acreage reduction and land set-asides, which were both focused on reducing the bane of agriculture: overproduction. The Commodity Credit Corporation (CCC) established a price floor by making loans to farmers when the food processors or grain corporations refused to pay farmers a price that covered the cost of production. Farmers pledged their crops to the government as collateral against the loans, effectively ensuring that they were paid a fair price. The loan rate, set by the CCC and based on parity, acted as a price floor, because a farmer could sell to a national grain reserve that was established as a last-resort market.

The grain reserve was filled when crops were abundant and prices were low; grain was released when crops were scarce. In this way the reserve prevented crop prices from skyrocketing during times of drought or low production. Since this policy stopped products from reaching the market if the price was not fair, prices inevitably returned to a normal level, and farmers could pay off their loans. Together these policies helped keep overproduction in check and reduced commodity price volatility. This meant farmers could make a living *without* subsidies.

The parity programs worked so well that there was real prosperity in rural areas during World War II and that postwar period. This was strikingly different from the post–World War I era when, without supply management, farm prices collapsed. The programs also worked for Main Street by reducing price volatility, and the grain reserve actually made a profit of $13 million over twenty years as the crops were sold on the commodity market. Meanwhile, the food-processing and grain industries preferred overproduction, because it led to cheap prices for the products they needed. Still, today, they continue to wage a propaganda war against any policy that gives farmers a shot at fair prices.

The CED carried on a campaign against these programs for political reasons, beyond the desire for cheap commodities and an increased cheap industrial labor pool. These interests feared the political power of farmers, who since the Civil War had been on the vanguard of populism, protesting against abuses by the railroads, banks, and grain merchants, among other monied interests.

Farmers hard hit by the depression of the 1870s had reacted desperately to a tight money supply and to the high shipping rates charged by railroads, and they organized political groups, including the Grange and the Farmers' Alliance. The populist agrarian revolt, which lasted from 1860 through the early twentieth century, was spurred by the incongruity of farmers, who were central to the nation's well-being, suffering from poverty and bankruptcy. As a result of these hardships, the portion of farmers in the country's labor pool dropped from 58 percent to 38 percent during this period. In 1850, farmers owned almost 75 percent of U.S. wealth, but by 1890 this had plummeted to 25 percent.

Farmers on an economic roller coaster were a dynamic political force through the end of the nineteenth century: they worked for candidates; pushed to regulate the railroads, grain elevators, and meatpackers; and joined with labor unions to demand an end to their economic plight. They sought alternative structures to improve their standard of living, forming cooperatives, founding banks, and pushing to end the gold standard and to use silver coinage to lessen the control of the banking interests.

By the turn of the twentieth century, as the industrial revolution transformed the country, the wealthy concentrated in urban centers, and the income gap grew for rural Americans, whether they were sharecroppers in the South or grain farmers in the Great Plains. The economic upheaval forced a combination of major industries that created large corporate bodies known as trusts that relied on price fixing to avoid competition and charged high prices for necessities. Low prices devastated cotton and tobacco farmers early in the century. In response, farmers organized actions to withhold their products from the market in an attempt to boost prices that generated violent responses from the industries dependent on their crops. In 1902, the National Farmers Union, which continues to be a political force today, was formed in Rains County, Texas, to advocate for a family farm system of agriculture.

The first two decades of the new century were a prosperous time for many farmers, as prices increased and the number of farms declined. Many farmers still felt victimized by the banks, railroads, and grain companies. The Non-Partisan League (NPL) was formed in North Dakota to advocate for state-owned grain elevators and flour mills. By working in a farmer-labor coalition NPL candidates won most of the elected offices in North Dakota in 1916 and initiated reforms, including state inspection of grain elevators, regulation of railroad shipping rates, and a reassessment of land taxes for farmers.

Yet the continuing problem of the monopoly control of crucial industries and consumer items caused hardships for both rural and urban Americans, creating the political momentum for the Progressive era. Theodore Roosevelt, elected on a trust-busting platform, underscored the need for fair government regulation of "special interests." His administration's first target was J.P. Morgan, a financier who controlled Northern Securities, a railroad that monopolized freight movement across the northern United States. Roosevelt's attorney general filed suit to break up the railroad, and Morgan lost on appeal at the U.S. Supreme Court, in a vote of 5 to 4. At the end of his term, Roosevelt said that he had given Americans a "square deal."

Government policies designed to increase prices encouraged farmers to put out a staggering amount of products during World War I. But prices crashed in 1920 when commodity markets shrank after battlefields in Europe returned to farm fields and U.S. crops were no longer necessary, wartime price supports were eliminated, and President Wilson lowered tariffs to encourage imports. The world prices for grain and cotton collapsed, leading to bankruptcy for farms across the nation. While the twenties were "roaring" for some, the Depression came early to rural farming communities, where prices decreased on average to half of wartime levels.

It was during this time that the American Farm Bureau Federation, an organization that continues to represent agribusiness today, was organized to counteract the strong farmer-labor organizing that was becoming a formidable political force. To this day, the Farm Bureau has acted in concert with the U.S. Chamber of Commerce, together protecting the economic interests of industry and agribusiness and serving as an instrument to blunt the power of a progressive coalition and to divide farmers. At its inception, the group worked with members of Congress who feared the Progressive movement.

Called the "farm bloc," these farm-state legislators wanted to address their constituents' concerns without changing the structure of the economic system. At that time only five corporations controlled more than 60 percent of the meat industry—an industry that is even more consolidated today.

Despite their efforts, in 1921 Congress passed the Packers and Stockyards Act, a hard-fought yet poorly designed bill that sought to restore competition to the meatpacking industry. The legislation vested "expansive rule-making authority" in the secretary of agriculture to develop a set of "market facilitating regulations" that would address the unfair practices of the industry that prevent livestock growers from receiving a fair price for their animals. The law was also supposed to enhance existing antitrust law by providing "a second layer of comparable law that the Secretary would enforce through cease and desist orders." Although the law worked pretty well to curb excesses through the 1970s, neither of these mandates has been seriously enforced since the 1980s.

The Obama administration of 2008 failed to keep a campaign promise to write regulations curtailing the abuses of the meat-industry monopoly. This was after a yearlong investigation and hearings held around the country during which farmers testified about abusive practices. In the fall of 2011, Republicans who controlled the House of Representatives ensured that there would be no funding at the Departments of Agriculture or Justice to pursue regulations. So the 1921 legislation continues to be ineffective—as was intended by many of those who voted for it in 1921.[8]

Similarly, the post–World War I crash of farm prices led populists to call for the abolition of futures trading, because the speculation and price manipulation involved in it worsened the situation. President Harding, seeking to head off legislation to end trading, proposed legislation to regulate it. The Futures Trading Act of 1921 mandated that trades not conducted on an exchange licensed by the federal government would be taxed. It was overturned by the U.S. Supreme Court in 1922, and was reintroduced after manipulation caused wheat prices to collapse. The new bill, the nearly identical Grain Futures Act of 1922, used the commerce powers granted to Congress under the Constitution to tax futures trading. Speculation continued to plague farmers during the period leading up to the 1929 stock market crash.[9]

One sixth of U.S. farms were lost to bankruptcy, foreclosure, or delinquent-tax sales between 1930 and 1935. Thousands of other families left their farms voluntarily or took on debt-induced mortgages. Franklin Roosevelt feared that the growing alignment of farmers, the unemployed, and labor unions could become a political force that could organize a socialist revolution, so he set about undercutting the coalition that had formed and won third-party electoral victories in several states. Two weeks after his election, as part of his program to save capitalism, Roosevelt announced a "new means" for rescuing agriculture. Central to the plan were the policies enacted in the Agricultural Adjustment Act of 1933, which established parity for farmers, and the Commodity Exchange Act of 1936, which aimed to halt rampant speculation in farm products.

The backlash against agrarian revolt—from the last half of the nineteenth century to the Progressive era and the Depression of the 1930s—was powerful. Corporate leaders feared a reorganization of the economy and a more socially, economically, and racially egalitarian society, and so they took measures to disrupt the growing coalition of the economically disadvantaged. This response took many forms, from red-baiting and jailing dissenters to organizing disinformation campaigns and creating groups like the CED.

The CED's members were afraid of the potential political power of farmers and sought to reduce this population by creating agricultural policies that would shrink rural numbers and solve the problem of a rural insurrection that could withhold food from urban areas. The disparate economic interests represented in the group were united in viewing the world with an urban lens and as one reinvented by technology. Reforming agriculture meant substituting capital for farm labor and replacing small farms with large ones that would be vertically integrated into the companies that needed the raw materials to standardize and mechanize the food system.

The CED's strategy for agriculture culminated in the 1962 publication of "An Adaptive Program for Agriculture," the aforementioned radical document that laid out a plan to drastically reduce the number of farmers, creating a labor pool for industry and a vision for the globalization of food production, through a free trade agenda. The report, which was prepared by fifty influential business leaders and eighteen economists from leading universities, declared that "agriculture's chief need is a reduction of the number of people." This would be accomplished "by getting a large number of people out of agriculture before they are committed to it as a career."

The report recommended stopping the promotion of agriculture in vocational training, reeducating rural young men for jobs in industry, and providing help to relocate them to the places where their new skills were needed. It recommended removing parity price supports after five years so that "farm prices would be governed by free Market forces," and went on to conclude that these suggested programs "would result in fewer workers in agriculture, working a smaller number of farms of greater average size."

The CED's views on trade, which were printed in its 1945 pamphlet "International Trade, Foreign Investment and Domestic Employment," have become the official policy of the United States today. The CED said, "Restrictions to world trade prevent the free flow of goods, services and capital from where they are available to where they are needed. This obstruction prevents efficiency in the use of the world's human and material resources." [10]

The 1962 agricultural report laid out one of the fundamental pieces of the Agreement on Agriculture that is currently overseen by the World Trade Organization (WTO). The CED notes, "In an efficient organization of the world economy, the United States would make much larger exports of farm commodities to Europe" and that the removal of restrictive quotas in Europe "should be a cardinal point of United States trade policy." A 1964 CED report, "Trade Negotiations for a Better Free World Economy," became a road map in the trade negotiations taking place at that time and in the future. The manifesto justified trade liberalization as "economic efficiency" and argued that it must include the "reduction of impediments to trade of agricultural goods."

The emphasis at the U.S. Department of Agriculture on the alleged economic efficiency of larger farms capable of investing in new equipment and using new technologies began during the aftermath of World War II, at the same time that the CED was formed. A 1943 USDA-sponsored report said, "[O]ur advocacy of a public policy which favors the family farm does not mean that we favor the retention of all small farms." The report noted that of the existing 6 million farms, 2.5 million did not meet the agency's production criteria. It stated that 84 percent of farm products came from one third of U.S. farms and that the goal was "to direct the surplus manpower into productive nonagricultural activities" [11]

The continuation of New Deal programs that had established parity for farmers created a temporary rural prosperity during the decade following the war. In addition, the demand for U.S. farm products increased due to the devastation of Europe, the disintegration of colonial empires, assorted weather catastrophes around the world, and the adoption of the Marshall Plan for Europe. These good times were short-lived.

In 1953, the first stage of the propaganda war was won with the appointment of Ezra Benson, a far-right ideologue trained as an agricultural economist, as secretary of agriculture. During his eight years in President Eisenhower's administration, Benson served simultaneously in the governing hierarchy of the Mormon church—a multibillion-dollar agribusiness operation to this day—as one of the Quorum of the Twelve Apostles. He eventually became the thirteenth prophet and president of the Church of Jesus Christ of Latter-day Saints.

Under Benson's management, the dismantlement of the 1930s parity legislation begaᵤ and prices for farmers began to fall, beginning the ongoing modern farm crisis. Benson vehemently denounced the New Deal farm programs as "socialism." Although Eisenhower campaigned on a platform of parity for farmers, Benson's goal was to eviscerate it, and he successfully pushed the concept of flexible price supports, giving the secretary of agriculture the ability to lower the prices farmers received for goods. The phrase was doublespeak for making cheap grain available and getting government out of agriculture.

Benson specifically labeled supply management as "socialist" and worked with the CED to recruit academics into a wholesale propaganda effort against existing farm programs. Numerous organizations beyond the CED, including the U.S. Chamber of Commerce and the American Bankers Association, engaged in a full-court press to destroy parity for farmers, including producing research, publishing reports, and lobbying all branches of government intensely.[12]

John Davis, Eisenhower's assistant secretary of agriculture, went on to head a business and agriculture program at Harvard funded by the Corn Products Refining Corporation. He wrote in the *Harvard Business Review* that vertical integration was the best alternative to "big government programs." He called this new type of farming "agribusiness," a term that came into common usage during the Eisenhower years.[13]

The thirty-year erosion of the New Deal programs, which had enabled farmers to earn an income on par with urban workers, forced farmers either to leave farming or borrow heavily. As a result, the farm population declined 30 percent between 1950 and 1960, and by another 26 percent between 1960 and 1970—the precise outcome the industrial tycoons had plotted to facilitate. But as the farm crisis worsened, a new protest movement grew.

No one had a larger impact on agriculture during the second half of the twentieth century than Benson protégé and Cornell University graduate Earl Butz. Cornell had been the mouthpiece for farm policy under both the Eisenhower and Kennedy administrations. Alan Emory, a reporter writing in February 1954 in the *Watertown Times*, an upstate New York paper, observed that the "boys from Ithaca" have a "preference for big agriculture over the individual farmer" and "appear more interested in low prices for raw materials than in increased purchasing power for the farmer."

Butz, as President Nixon's second secretary of agriculture, facilitated the agribusiness agenda, becoming infamous for slogans like "agriculture is big business" and for saying that farmers must "adapt or die." As prices continued to drop in the 1960s and 1970s, the Farm Bureau and an agribusiness coalition challenged an opposing coalition of twenty-five farmers' organizations in the battle over the farm bill in 1970, a reauthorization that takes place approximately every five years. The legislation ended up reducing price supports in an attempt to encourage exports. The Nixon administration set in motion an economic policy that relied heavily on the export of grain, arguing that the United States had a "comparative advantage" in producing capital-intensive crops, while the developing world was suited to growing labor-intensive fruits and vegetables.

Crucial to the expansion of food trade were the negotiations around the General Agreement on Tariffs and Trade, or GATT, the forum that served multilateral institutions until the WTO replaced it in 1995. During the early 1960s, the CED and like-minded corporate-sponsored organizations geared up to push what became the 1995 WTO treaty, which included the Agreement on Agriculture. While the purported function of this treaty is to remove trade barriers and increase competitiveness, its purpose is to allow the largest food companies and grain traders to source crops where they can be obtained at the lowest cost.

The dismantling of parity in the 1950s, the farmer unrest it garnered, and a background of grain and food industry demands for cheap commodities resulted in a new farm program in the early 1970s and set the stage for the farm crisis of the 1980s. The new program created a target price for commodities, and if the price fell below the rate set by the USDA, then U.S. taxpayers, rather than the food and grain companies, would have to pay the cost of production. Participating farmers received an annual "deficiency check" to make up for the dismantling of parity programs. Food companies such as Kellogg and General Mills, grain traders such as Cargill, and their trade associations, using their political clout, had managed to increase their profits by waging a multi-decade campaign that not only lambasted farmers for inefficiency but increasingly accused them of being on the public dole.

Further, in the summer of 1972, hoping to bolster farm incomes and dissuade farmers from voting for George McGovern, Agriculture Secretary Butz blessed Cargill and four other grain traders' secret negotiations with Russia, which was suffering from a bad harvest. The grain cartel made a $1.5 billion deal to send a quarter of the U.S. grain harvest overseas—a move that had the effect of increasing food prices in the United States.

Butz, an ever-popular farm-circuit speaker, traveled the country telling farmers to "get big or get out" and to "plant fencerow to fencerow" to meet the global demand. In 1972, to boost production of grains, Butz removed 25 million acres from the New Deal set-aside program, shrinking it to 7.4 million acres in 1973. Under his guidance, the 1973 Farm Bill reduced parity for crops to 50 percent or less. But many farmers followed Butz's advice and began investing heavily in expanding operations. As a result, by 1978, just 19 percent of U.S. farms were producing 78 percent of the country's crops.

The election of Jimmy Carter marked the worsening of the most severe farm crisis since the 1930s. Because of energy supply shortages in the United States and abroad, the cost of petroleum-based fertilizer, farm equipment, and diesel fuel had skyrocketed, while the price for crops continued to drop. With land prices also increasing, the only choice for many farmers was to sell or borrow—farm debt ballooned by 400 percent between 1960 and 1977.

In the fall of 1977, the American Agriculture Movement was born when a group of farmers meeting in Springfield, Colorado, called for a national farmers' strike. They were desperate after the passage of yet another Farm Bill that failed to address their continuing loss of equity. Farmers from across the nation drove tractors to Pueblo, Colorado, and demanded that the secretary of agriculture take action to preserve family farms.

Long before the Internet, news of the strike spread by word of mouth, and on December 10, 1977, tractorcades took place in many state capitals. In January 1978, fifty thousand farmers rallied in Washington, D.C., and farmers continued to organize for a stop to foreclosures and to reinstitute parity. In January 1979, a large tractorcade blocked traffic in the nation's capital, but when a blizzard hit the city during the protest, farmers used their tractors to clear roads and provide transportation for emergency services. The protest, which received national coverage, raised public awareness, but policy makers did not take the actions necessary to provide relief.

The crisis snowballed in the 1980s under President Ronald Reagan. Crop prices continued to drop because of overproduction, and farm debt increased dramatically. Farmers had to borrow against their land to survive and keep farming, often using all of their equity until the bank foreclosed. Thousands of farm families faced the loss of their homes and land. The overseas demand for grain that Butz had promised never materialized, and the land boom crashed, dramatically reducing the value of farm property. The farmers who survived often had to depend on off-farm income. Meanwhile, cheap grain allowed the expansion of factory farms, and the food processors and grain traders enjoyed record profits from the cheap commodities that received a taxpayer subsidy.

· · ·

In April 1982, longtime farm activist Merle Hansen organized a meeting in Des Moines, Iowa, and more than fifty farm, labor, and community organizations gathered to oppose Reagan-era policies and to demand parity prices. He became the president of the North America Farm Alliance, a new coalition that brought together many groups that are still collaborating today. He was emblematic of family farm leaders who have put their lives on the line throughout our nation's history, and he once observed of agribusiness, "They are not only setting farm, trade and food policies for the U.S.; they set them for the world." [14]

Hansen was born in 1919, and as a teenager he worked with his father at the militant Farmer's Holiday Association, an organization formed in the Midwest during the 1930s, organizing "penny auctions" at farm sales: neighbors of foreclosed farms would bid only pennies on each item offered for sale and return it to the family. Hansen was deeply affected by the experience, and after serving in the Pacific during World War II, he became a field organizer, first for the South Dakota Farmers Union (SDFU) and later for the Iowa Farmers Union (IFU). He worked closely with African American activist Edna Griffin, the Iowa organizer of one of the first desegregation campaigns in the nation.

Hansen returned to his family's Nebraska farm in the 1950s, when the Korean War caused a bitter split in the IFU. There he became vice president of the antiwar U.S. Farmers Association and was active in the local chapter of the NFO. Hansen viewed farm policy through a social justice lens and believed deeply in broad-based coalitions. In 1970, he became president of Nebraskans for Peace, and later in the decade became a state officer in the American Agriculture Movement. He eventually became Reverend Jesse Jackson's chief adviser on agriculture policy, and he nominated Jackson for president at the 1984 Democratic National Convention.

Hansen continued to act as an articulate spokesperson for the farmer during the devastating crisis of the 1980s and helped organize the benefit concert Farm Aid; in addition to founding the North America Farmers Alliance, he started the National Family Farm Coalition (NFFC).

George Naylor, an Iowa farmer and former president of the NFFC, worked closely with Hansen during the Reagan era. He reminisces about Hansen, who died in 2009: "He could give a wonderful speech, so that people's eyes were filled with tears at the end of his speeches. He was just a tremendous person." Naylor says Hansen had the ability to bring together people with different political beliefs to talk and take action. Merle Hansen originally brought many of the groups now working as part of NFFC together. [15]

When he was in his eighties, Hansen commented:

There was a time when I was an outcast in my own community, but that is
not true today. When I opposed the Vietnam War, people wrote me letters
threatening to kill my cattle and saying that I was a terrible, unpatriotic
person. My kids paid dearly for it. Our car was vandalized, and in school
our kids were taunted and called Communists. My son John was starting
to wear long hair and, boy, were they out to get him. . . . Today, when he
comes to Newman Grove and speaks as president of the Nebraska Farmers Union, some of these very same people say he's really good.[16]

John Hansen, the sixth generation in his family to farm, followed in his
father's footsteps, also becoming an activist in the early 1970s. He worked tirelessly for passage of the 1982 law banning corporate farming in Nebraska—the
strongest law in the nation until it was overturned by a judge on a technical-
ity in 1982. John has continued his father's work at the Nebraska Farmers
Union where he has been elected president ten times since 1989. The younger
Hansen says that the ongoing fight for the family farm raises fundamental
questions about what kind of society we are going to have.

Hansen charges: "The American families who produce our food and fiber
are hemorrhaging. The pressure from one-sided and unfair farm and trade
policies is taking a tragic toll on farm families, farm businesses, rural communities and the soul of America."[17]

Naylor, whose grandfather acquired his family farm in 1919, agrees with
this assessment. He only manages to survive in Iowa as an independent family
farmer by living very modestly, repairing his own equipment, and keeping
costs very low.

In August 1999, the heartland was still reeling from the elimination of all
of the New Deal programs and safety net. Naylor testified at the U.S. Senate
Democratic policy hearing held by Senator Tom Harkin in Bondurant, Iowa,
about the "boarded-up small towns" and the "anguish of farm families" as a
result of the legislation.

In 2012, Naylor reflected, "No betrayal was more galling, or the effects
more devastating for farmers and eaters, than Bill Clinton's single-minded
pursuit of free trade and his support for the 1996 'Freedom to Farm' bill."[18]

Fifty years after the CED had developed its plan to remove "excess labor"
out of farming, Clinton had facilitated the final blow. The number of U.S.
farmers had been reduced by more than two-thirds, and long after its original
founders were dead, CED's plan for eliminating barriers to free trade and

submitting agriculture to market forces had become U.S. law. It took the wholehearted support of President Clinton, a center-right Democrat, to fully accomplish the radical agenda laid out by the CED in the 1940s—a goal sought for decades by the most powerful economic interests on Earth.

During the twenty years leading up to the free trade policies and the 1996 Freedom to Farm bill, a range of corporate interests—banking, manufacturing, energy, pharmaceuticals, and agribusiness—had coalesced to control the rules that would govern the global economy. Sharing the philosophy that commercial interests trump all else, they used campaign contributions, influence peddling, and raw political power to subordinate domestic health, safety, and environmental regulations. Through a complex web of corporate-funded trade associations, nonprofit think tanks, public foundations, private clubs, and other institutions, they plotted the content of international trade rules and U.S. farm policy to maximize profitability—at the expense of humankind and the environment.

Clinton signed the North American Free Trade Agreement (NAFTA) on September 13, 1993, committing his administration to fight for congressional approval of the measure. The heavily contested legislation divided and weakened the Democratic Party and led to the devastating 1994 midterm elections that ended forty years of Democratic rule in Congress. Unfortunately, the Congress under Democratic leadership had not been an advocate for family farmers during this period, and the same was true of President Clinton. As he promised, NAFTA passed shortly after the 1994 elections—but it did not bring the prosperity the president envisioned. Since that time, according to the Economic Policy Institute, NAFTA has been responsible for the loss of seven hundred thousand U.S. jobs. And 2 million Mexican peasant farmers, unable to compete with U.S. corn, have been forced off their land.

Clinton led the battle to join the controversial World Trade Organization. He fought successfully for WTO-specific "fast track" authority, an undemocratic procedure that authorizes the president to negotiate trade agreements that allows Congress to vote only up or down on them, without the possibility of amendment or parliamentary challenge.

U.S. entry into the WTO was approved under the fast-track procedure, during a lame-duck session of Congress, on December 1, 1994. Countries that join the WTO yield their authority for deciding if agriculture programs, food safety rules, environmental regulations, and worker safety laws are illegal trade barriers. The WTO's structure promotes commerce at the expense of other societal goals. As the final arbiter of trade disputes, the WTO almost always rules in favor of business interests, restraining member countries from having health, safety, or environmental rules that companies contend impede trade.

In one of its latest actions against consumer interests, the WTO ruled in November 2011 that the U.S. requirement for mandatory country-of-origin labeling (COOL) for meat is a violation of international trade law. Canada and Mexico had challenged it, which the United States had passed in 2008 with the aim of providing consumers with information vital to making informed food choices.

Lori Wallach, director of Public Citizen's Global Trade Watch, says the COOL ruling makes very clear that "these so-called trade pacts have little to do with trade between countries and everything to do with major agribusiness corporations selling mystery meat in the United States." According to Wallach, this was the third WTO ruling in 2011 against popular U.S. consumer policies. She says, "The ban on candy and clove-flavored cigarettes commonly used to hook teenagers, and the dolphin-safe tuna labels instrumental in reducing fishing fleets' killing of dolphins, have also been declared WTO-illegal."

Wallach, a Harvard-trained lawyer, was lobbying Congress on food-safety issues in the late 1980s when she learned about the impact of the ongoing trade negotiations on domestic laws. At hearings on various food-safety improvements, corporate lobbyists declared that the trade pacts then under negotiation would not permit them. Wallach obtained an early draft of the WTO text in 1991 and realized that the WTO itself—and not just Congress and U.S. agencies— had to be engaged in the battle to ensure health and safety protections.

Wallach has been fighting the sneak-attack domestic regulations since that time, including playing a major role in organizing the 1999 protests against the WTO in Seattle. She confirms that Daniel Amstutz, a former executive of both agribusiness giant Cargill and Goldman Sachs, authored the WTO's rules on agriculture. Under the Reagan administration, Amstutz was the chief negotiator for agriculture during the Uruguay Round of trade talks on GATT. He went on to be the executive director of the North American Export Grain Association, an industry trade association that lobbied for the rules and represents the interests of companies such as Cargill and Archer Daniels Midland (ADM).

The CEO of ADM at the time, Dwayne Andreas, who had transformed the company into an agribusiness powerhouse, was also instrumental in lobbying U.S. officials for the WTO. While Andreas was building ADM into the "supermarket to the world," according to its slogan, several top executives went to jail for price fixing and the company was assessed the largest fine in history for an antitrust violation.

Andreas always hedged his bets by financing an eclectic mix of powerful presidential candidates, including Humphrey, Nixon, Carter, Reagan, Bush, Clinton, and Dole. And he funded both sides of the political spectrum, from Jesse Jackson to Newt Gingrich. A believer in realpolitik, he once said: "There is not one grain of anything in the world that is sold in the free market. Not one. The only place you see a free market is in the speeches of politicians." [19]

The heavily contested Agriculture Agreement that was crafted by ADM, Cargill, and other multinational food companies was the centerpiece of the negotiations leading up to the establishment of the WTO. On behalf of the largest global corporations, the AA requires countries to enable "market access" by making import bans, import quotas, or quantity restrictions on agricultural imports illegal and by phasing out tariffs (taxes on imports), policies that were traditionally used to control the quality and volume of imported goods. AA rules allow the largest agribusiness companies to move operations overseas to where production is cheapest, to compete unfairly with local producers, and to pursue a global race to the bottom for farm prices. Policies that ensure fair prices for farmers are considered trade barriers.

The WTO also administers and enforces several other agreements that affect agriculture—all of them favoring large, vertically integrated global corporations. Countries cannot base laws or regulations on value judgments or social priorities (such as promoting eco-labeling or banning Internet gambling), and use of the "precautionary principle" (the idea that it's better to be safe than sorry) is illegal. The WTO prescribes the use of expensive and impractical risk-assessment procedures to consider hazards, and this has had a chilling effect on a range of safety regulations, including those for pesticides, bacterial contamination, and hormones. Trade rules have been used to force genetically modified crops and meat produced with hormones on countries that have passed laws prohibiting them.

WTO rules have been designed to help agribusiness produce crops where labor is cheapest, environmental laws are weakest, and costs are lowest, so food production is increasingly moving to the developing world. Newly impoverished people have surged into urban slums. This has created a disincentive for domestic food production in poor nations, and many of them have allowed multinational companies to drive peasant farmers off their land so export crops can be grown. Since the adoption of the AA, the decline in domestic production has increased food insecurity around the world.

Wallach notes, "A dramatic decline in farm income in developed and developing countries alike has been the norm under the WTO, causing indebtedness and foreclosures in rich countries and loss of livelihoods and hunger in poor countries."

Regulatory Changes Affecting Agriculture and Food Policy

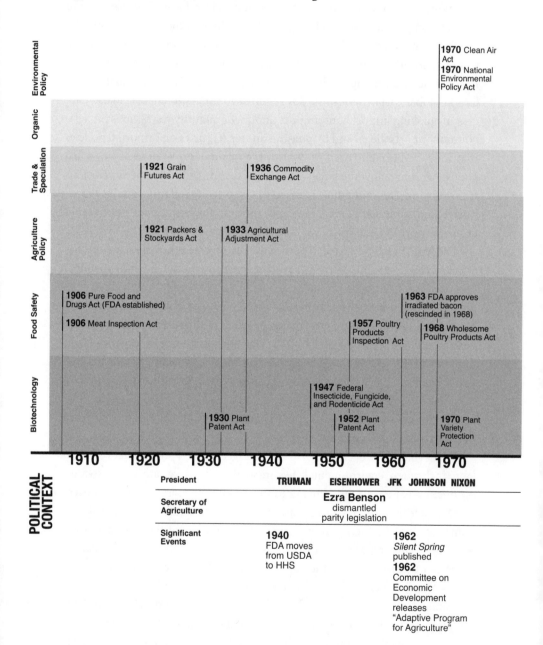

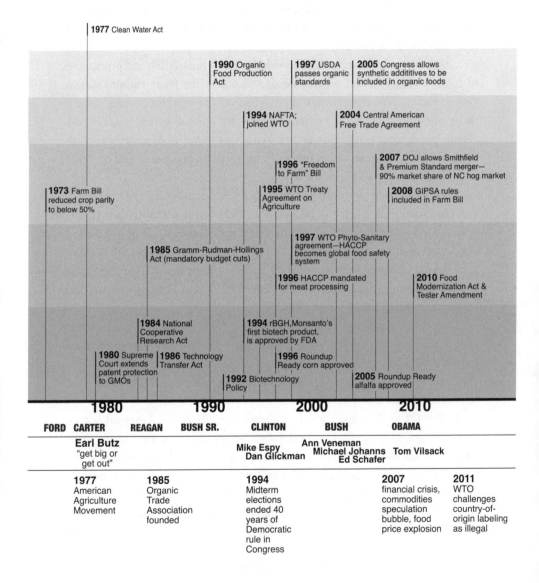

1977 Clean Water Act

1990 Organic Food Production Act

1997 USDA passes organic standards

2005 Congress allows synthetic addititives to be included in organic foods

1994 NAFTA; joined WTO

2004 Central American Free Trade Agreement

1996 "Freedom to Farm" Bill

2007 DOJ allows Smithfield & Premium Standard merger— 90% market share of NC hog market

1973 Farm Bill reduced crop parity to below 50%

1995 WTO Treaty Agreement on Agriculture

2008 GIPSA rules included in Farm Bill

1985 Gramm-Rudman-Hollings Act (mandatory budget cuts)

1997 WTO Phyto-Sanitary agreement—HACCP becomes global food safety system

1996 HACCP mandated for meat processing

2010 Food Modernization Act & Tester Amendment

1984 National Cooperative Research Act

1994 rBGH, Monsanto's first biotech product, is approved by FDA

1980 Supreme Court extends patent protection to GMOs

1986 Technology Transfer Act

1996 Roundup Ready corn approved

1992 Biotechnology Policy

2005 Roundup Ready alfalfa approved

1980	1990	2000	2010
FORD CARTER	REAGAN BUSH SR.	CLINTON	BUSH	OBAMA

Earl Butz "get big or get out"

Mike Espy
Dan Glickman

Ann Veneman
Michael Johanns
Ed Schafer

Tom Vilsack

1977 American Agriculture Movement

1985 Organic Trade Association founded

1994 Midterm elections ended 40 years of Democratic rule in Congress

2007 financial crisis, commodities speculation bubble, food price explosion

2011 WTO challenges country-of-origin labeling as illegal

As previously noted, Cargill is among the companies pushing for the policies that have caused so much suffering. It is the largest privately held company in the United States and the largest grain trader in the world. Although invisible to most consumers, it operates in every segment of the food industry. Most Americans eat something that Cargill has produced every single day. The company also manufactures fertilizer and feeds, provides loans to farmers and buys their crops, operates the grain terminals where farmers eventually deliver their crops, and provides marketing advisory services to guide them through this entire process.

Cargill has used its status as the world leader in the trading and processing of corn, wheat, soy, and other oilseeds to employ the trade rules to gain a global reach in procuring commodities wherever they are cheapest and selling them where they are most profitable. The company has played a significant role in promoting genetically engineered crops. Since it operates the grain elevators, it can choose which types of crops will be purchased and contract with farmers to grow specific varieties.

Cargill has been promoting trade policies for decades. In 1971, a Cargill vice chairman took a leave to become President Nixon's deputy special representative for trade, enabling him to shape agricultural trade policy during the decade that saw the opening of China to world markets and the Russian grain crisis. A former president of Cargill Investor Services was the chief agricultural trade negotiator from 1987 to 1989 for the GATT, the agreement that ultimately created the WTO, in 1995.

Unsurprisingly, Cargill had a relationship with the Clinton administration, since they were in agreement on trade rules and the deregulation of federal agriculture policy. Clinton appointed Ernest Micek, CEO of Cargill at the time, to his export council and traveled with him to Africa on a mission to encourage free trade.

In 1994, Cargill helped form a powerful coalition that emerged with a common goal: lining up U.S. agriculture policy with the new trade rules. The Coalition for a Competitive Food and Agricultural System included the Chicago Board of Trade, General Mills, Tyson Foods, Kraft Foods, Procter & Gamble, Union Pacific Railroad, the Fertilizer Institute, and more than one hundred other corporations. A U.S. Chamber of Commerce staffer said it was "the first time in history that a broad-based group in the food industry had gotten together." [20]

Clinton and the free traders of both parties were happy to oblige. First on the chopping block were the last vestiges of the New Deal programs to protect farmers and provide food security. In 1995, against the backdrop of

Newt Gingrich's Contract with America, a small group of farmer activists and critics of corporate globalization faced off against the largest corporations in the country.

Kathy Ozer of the NFFC has worked on every farm bill since 1987. She says the so-called Freedom to Farm scheme in the 1996 Farm Bill was hatched by the same corporations that dictated the WTO rules on agriculture: "Their goal was cheap grains for overproduction. So they attacked the 'supply management' policies that helped maintain a better balance between supply and demand. Freedom to Farm reduced the conservation land set-aside and grain reserve programs that prevented overproduction, low prices and volatility. By mirroring the WTO rules, it destroyed the policies that guaranteed the cost of production for farmers—similar to a minimum wage."

Ozer, who grew up in suburban Washington, D.C., never dreamed she would spend more than twenty years fighting for family farmers. She went to work at NFFC during the farm crisis because of her expertise on credit issues and became so committed to the issues that she never left. Ozer says, "The 1996 legislation introduced the concept of 'decoupling'—delinking farm payments from production or price. Transition payments were to be made over seven years to farmers and zeroed out after that. Companies like Cargill and ADM had lobbied for 'decoupled' payments to be designated as 'non-distorting' by the WTO. Before this deregulation, farmers didn't need to rely on the government for a fair income—the market provided."

Larry Mitchell, now administrator of the Grain Inspection, Packers & Stockyards Administration (GIPSA) at the USDA, concurs. He was a lobbyist for the National Farmers Union in 1996 and therefore a frontline soldier in the epic battle over agriculture. Mitchell, an expert on farm policy who has been working on these issues since the 1970s, says that Gingrich first attached Freedom to Farm to the Balanced Budget Act of 1995—legislation that Clinton vetoed during the fight over budget cuts in December of that year. Mitchell said that no one had seen this version of the bill until it was attached to the budget bill late one night. But eventually, since they were basically in agreement on farm policy, Gingrich and Clinton came to an understanding. Clinton signed Freedom to Farm in April 1996, marking the culmination of the deregulatory era.

Senator Paul Wellstone, the late progressive from Minnesota, voted against the bill, declaring: "In the long run, it says, you're on your own with Cargill. You're on your own with the Chicago Board of Trade."[21] Which, Mitchell says, is exactly what happened. He explains that the legislation phased out all government intervention in commodity markets, eliminated programs that prevented overproduction, and dismantled policies that maintained fair prices, leaving farmers to the vagaries of the market and speculative shenanigans.

In 1997, after working on Clinton's reelection campaign in 1996, Mitchell was appointed by the president to be the USDA deputy administrator of the Farm Service Administration (FSA). But in retrospect, he says, Clinton and the Republican candidate, Bob Dole, differed little on trade, banking, or agribusiness. While at FSA Mitchell argued for having its state staff come to Washington and testify at congressional hearings on the impact of deregulated agriculture, and he was told no go by the Clinton White House.

Mitchell was originally a farm boy from Texas and an activist in the American Agriculture Movement, and he says that too many policy makers, farm organizations, and farmers have been indoctrinated over the years that trade will ultimately fix the rural economy. He says, "Trade has only been an engine for agriculture three times—during World Wars I and II and the Soviet grain deals in the 1970s. But we've bet our entire farm policy on something that only worked three times in a century."

The failure of this strategy for American farmers is apparent from the lack of growth in exports. Dr. Daryll Ray, a well-respected agricultural economist from the University of Tennessee, says that export values were expected to soar when prices were lowered by deregulation, and that this would "propel U.S. agriculture to the promised land of accelerating export growth and financial prosperity." He testified to Congress in 2010: "While the export-centric narrative was successful in moving farm policies to the check-writing payment programs of today, the grain-export promises failed to occur. . . . Amazingly and contrary to general belief, the U.S. is now exporting a smaller proportion of its combined production of corn, wheat, and soybeans than in 1980—45 percent in 1980 and 25 percent in 2009."[22]

Rather than boosting prosperity from trade, the result of deregulation was a massive increase in the production of commodity crops, causing prices to plunge for most of the past fifteen years. By 1999, the real price of corn was 50 percent below 1996 levels, and the price of soybeans was down 41 percent. To quell the anger and political heat when crop prices collapsed in 1998, Congress first authorized "emergency payments" to farmers—essentially, grants to keep farmers afloat. Meanwhile, the meat industry, grain traders, and food processors enjoyed low-priced commodities at the expense of taxpayers for more than a decade.

According to Tim Wise at Tufts University, "The biggest winner from cheap commodity crops has been the industrialized meat industry"—the country's largest purchaser of corn and soybeans. Wise estimates that when crop prices plummeted, the cost to industry of producing meat dropped between 7 and 10 percent. Industrial producers pocketed the savings, enabling them to greatly expand their operations during the years of extremely low commodity prices.[23]

Wise's research demonstrates that small livestock producers are further disadvantaged when they pay the full cost of growing the grain they produce while corporate buyers pay below the cost of production. Diversified, smaller livestock growers that use hay, pasture, and grains require more labor and are more vulnerable to low prices. Wise says that "industrialized livestock operations drive down the price of livestock, further squeezing diversified farmers out of animal production and into bulk row crops." When farmers are forced out of small-scale livestock production and into commodity crops, overproduction snowballs even more.[24]

This fact has been apparent since 1997. But because of the political power of agribusiness, no attempt has been made to reintroduce supply-management policies or to shore up prices and increase food security with a grain reserve. Even with 1998's "emergency payments," net farm income declined 16.5 percent during the first five years after the passage of the legislation now known to many farmers as the Freedom to Fail Act.

Rather than address the primary cause of low prices, Congress set the treadmill on high speed and quelled the political fallout by making emergency payments permanent in the 2002 Farm Bill. As a result, the subsidy system we know today was born.

It should be no surprise, considering the history of farm policy, that farmers are blamed for the subsidy system—rather than the architects of the broken food system. Subsidies are an easy target, whereas understanding the cause of the dysfunctional system—a combination of deregulation and the dismantling of complex farm policies—is complicated. Oversimplifying the convoluted set of conditions that have landed us here, from the failure to enforce antitrust law to the system of legalized bribery that corrupts the political system, does a disservice to farmers and eaters.

Ozer, with the National Family Farm Coalition, says: "Unfortunately, it is widely believed that the biggest problem with the food system is subsidies, and that farmers are greedy. Blaming the victims of a system that is rigged to keep them overproducing and dependent on Monsanto and DuPont for overpriced and dangerous chemicals and seeds is unfair and counterproductive. Subsidies are a symptom of a broken system—not the cause."

Longtime Iowan farm activist Brad Wilson says that progressives are "unknowingly siding with agribusiness" in the discussion over subsidies. He laments in his blog, "To merely remove subsidies would drive the remaining diversified family farmers out of business." He says that farmers can "help progressives meet a wide range of policy goals, as well as key movement strategy goals in this time of budget cutting and deregulation." Wilson admonishes, "Don't destroy them! Enough already!"[25]

The real question for the sustainable-food movement should be, How can we advocate effectively to save family-owned farms? Working for the survival of midsize farms, which have a significant land base and where farming is the primary occupation for the owner, is crucial for the transition to a sustainable future. These family operations, barely hanging on even with a government payment, are vulnerable to becoming part of larger industrial farms or being transformed into yet another housing development or shopping center.

Fred Kirschenmann, former director of the Leopold Center for Sustainable Agriculture, has advocated for decades to save farms in the middle. He says: "Over 80 percent of farmland in the U.S. is managed by farmers whose operations fall between small-scale direct markets and large, consolidated firms. These farmers are increasingly left out of our food system. If present trends continue, these farms, together with the social and environmental benefits they provide, will likely disappear in the next decade or two. The 'public good' that these farms have provided, in the form of land stewardship and community social capital, will disappear with them."

A close look at the statistics shows that midsize family farms are barely surviving and depend heavily on subsidies. Eliminating subsidies without a major reform of the system will lead to an even more industrialized food system. Research by Dr. Daryll Ray shows that removing subsidies without instituting policies to prevent oversupply would decrease farm income by another 25 percent to 30 percent, driving more farmers out of business.

Regrettably, the data on subsidies used by the media and many organizations are based on deceptive USDA statistics. The Environmental Working Group's farm subsidy database, widely used by the media and activists to dramatize the problems with subsidies, relies on USDA's misleading estimate of 2.1 million farms. A large pool of farms makes it appear that a smaller segment of farmers receives government support, whereas in reality 82 percent of midsize farmers depend on some support to stay afloat. Tim Wise affirms: "It is false to suggest that the vast majority of full-time family farmers are excluded from federal farm programs. A significant majority receive such benefits."[26]

Wise explains that the misinformation is based on the USDA's use of an inflated number of farms. In its calculation of 2.1 million farms, the agency includes "residential/lifestyles" and "retirement" farms that averaged $100,000 in off-farm income in 2009. While these 1.4 million farms make up roughly two thirds of the entities categorized as farms, farming is not the primary occupation of their owners, and virtually all income is from off-farm sources. A more accurate number of farms includes only those where the owner is a full-time farmer—fewer than a million in the United States.[27]

Most farmers can hardly be said to be living high on the hog, especially by the standards of Washington, D.C., where the subsidies are routinely decried. The midsize farms—those grossing between $100,000 and $250,000—averaged a net income of approximately $19,270 in 2009, including government payments. Even those operations designated by the USDA as "large industrial farms" (making a gross income of between $250,000 and $500,000 in 2009) netted only $52,000 on average, including $17,000 in government payments.[28]

And, fulfilling the prophesies the CED made in the 1940s, the largest commercial enterprises now dominate sales of farm goods, accounting for 73 percent of U.S. farm income in 2010. These 115,000 industrial operations grossed over half a million dollars and had an average net income of $264,000.

Even though the price of crops has increased since 2006 for a combination of reasons, including ADM's success in lobbying for corn ethanol incentives, speculation, and drought in many regions of the world, net farm income has failed to keep up with the cost of inputs. Agribusiness consolidation has inflated the cost of seeds, fuel, fertilizer, feed, and other inputs, while higher commodity prices have meant lower government payments.

Wise says, "So much for a boom for family farmers. High prices for their products have been gobbled up by rising expenses, government payments have fallen, and, more recently, the recession has significantly contracted the off-farm income that these farm families depend on to make ends meet."[29]

PART II

Consolidating Every Link in the Food Chain

One of the Reagan administration's lasting legacies is the dismantling of the regulatory system for ensuring fair and competitive markets. Since that time the failure to stop massive consolidation has allowed a handful of companies to control the entire food chain—from seeds, fertilizer, and implements to processing, distribution, and retail grocery chains. The largest twenty food companies exert tremendous control over food and farming, as both buyers of ingredients and sellers of product. In every subsector—from dairy products and beef to potato chips, soup, and canned vegetables—a small number of companies completely dominates the marketplace. The retail sector is even more amalgamated, with Walmart and three other large retail chains controlling 70 percent to 90 percent of the market in many regions of the country. Reversing this corporate tyranny and concentration is critical for creating a fair and sustainable food system.

Top 20 U.S. Food Companies and Their Brands

PEPSI 1

Pepsi, Mountain Dew, Aquafina, Sierra Mist, Tazo, Sobe, Slice, Lipton, Propel, Gatorade, Tropicana, Naked Juice, Cap'n Crunch, Aunt Jemima, Near East, Rice-A-Roni, Pasta-Roni, Puffed Wheat, Harvest Crunch, Quaker, Quisp, King Vitamin, Mother's, Lay's, Maui Style, Ruffles, Doritos, Funyuns, Cheetos, Rold Gold, Sun Chips, Sabritones, Cracker Jack, Chester's, Grandma's, Munchos, Smartfood, Baken-ets, Matador, Hickory Sticks, Hostess, Miss Vickie's, Munchies, True North, Starbucks Frappuccino*, Starbuck's Doubleshot*, Seattle's Best Coffee*, Dole Juice*

Nestlé 2

Carnation, Coffee-Mate, Nescafe, Nespresso, Nestea, Juicy Juice, Kit-Kat, Butterfinger, Toll House, Crunch, Wonka, Dreyer's, Edy's, Haagen-Dasz*, Smarties, Power Bar, Cookie Crisp, Hot Pockets, Lean Cuisine, Jenny Craig, Chef-Mate, Stouffer's, Buitoni, Gerber, Tombstone, DiGiorno, California Pizza Kitchen*, Perrier, Poland Spring, San Pellegrino, Ozarka, Deer Park

3 KRAFT

Oscar Mayer, Lunchables, Boca, Claussen, Jell-O, Cool Whip, Miracle Whip, Jet-Puffed, Good Seasons, A1, Bull's Eye, Grey Poupon, Baker's, Shake 'n' Bake, Stove-Top, Louis Rich, Philadelphia, Athenos, Polly-O, Velveeta, Cheez Whiz, Breakstone's, Knudsen, Maxwell House, Capri Sun, Kool-Aid, Crystal Light, Tazo*, Starbucks*, Seattle's Best Coffee*, Oreo, Chips Ahoy, Newtons, Nilla, Nutter Butter, Ritz, Premium, Triscuit, Snack Wells, Wheat Thins, Cheese Nips, Honeymaid, Planters, Teddy Grahams, Back to Nature, Toblerone, Trident, Halls, Dentyne, Sour Patch Kids, Swedish Fish, Chiclets, Milka

Tyson 4

Cobb-Vantress, Bonici, Lady Aster, Wright, ITC, Mexican Original, Russer, Jordan's, Iowa Ham

5 JBS

Pilgrim's Pride, Swift Premium, G.F. Swift 1855, Swift Natural, La Herencia Natural Pork, American Reserve, Cedar River Farms, Showcase Supreme, Packerland, Moyer, ClearRiver Farms, Liberty Bell, Blue Ribbon Beef, Certified Angus Beef, Steakhouse Classic

* licensed

6 General Mills

Cheerios, Chex, Count Chocula, Honey Nut Clusters, Kix, Cookie Crisp, Basic 4, Oatmeal Crisp, Boo Berry, Frankenberry, Cinnamon Toast Crunch, Trix, Raisin Nut Bran, Lucky Charms, Golden Grahams, Total, Wheaties, Cocoa Puffs, Fiber One, Betty Crocker, Bisquick, Gold Medal Flour, Pillsbury, Muir Glen, Larabar, Bugles, Fruit Snacks, Gardetto's, Nature Valley, Progresso, Mountain High Yoghurt, Yoplait, Cascadian Farm, Green Giant, Totino's, Hamburger Helper, Old El Paso, V. Pearl, Wanchai Ferry, Suddenly Salad, Valley Selections, Diablitos, Haagen-dazs (U.S.), Reese's Puffs*, Hershey's*, Weight Watchers*, Macaroni Grill*, Good Earth*, Sunkist*, Cinnabon*, Bailey's*

Dean 7

Alta Dena, Arctic Splash, Atlanta Dairies, Barbers, Barbe's, Berkeley Farms, Brown's Dairy, Bud's Ice Cream, Broughton, Chug, Country Charm, Country Churn, Country Delite, Country Fresh, Country Love, Creamland, Dairy Fresh, Dean's, Dipzz, Fieldcrest, Friendship, Gandy's, Garelick Farms, Hygeia, Jilbert, Lehigh Valley Dairy Farms, Liberty, Louis Trauth Dairy Inc., Maplehurst, Mayfield, McArthur, Meadow Brook, Meadow Gold, Mile High Ice Cream, Model Dairy, Nature's Pride, Nurture, Nutty Buddy, Oak Farms, Over The Moon, Price's, Purity, Reiter, Robinson, Saunders, Schenkel's All-Star, Stroh's, Swiss Dairy, Schepps, Shenandoah's Pride, Swiss Premium, Trumoo, T.G. Lee, Tuscan, Turtle Tracks, Verifine, Viva, Horizon Organic, Silk, The Organic Cow, Alpro, Borden*, Land O' Lakes*, Foremost Farms*, Provamel, International Delight, Fruit2Day, Hershey's*, Knudsen*

Mars

3 Musketeers, Bounty, Celebrations, Combos, Dove, Galazy, Kudos, M&M's, Milky Way, Munch, Snickers, Twix, Abou Sioiuf Rice, Castellari, Dolmio, Masterfoods, Royco, Seeds of Change, Suzi Wan, Uncle Ben's, Pedigree, Royal Canin, Whiskas, KiteKat, Banfield Pet Hospital, Cesar, Nutro, Sheba, Chappi, Greenies, The Goodlife Recipe, 5, Wintermint, Boomer, Spearmint, Juicy Fruit, Altoids, Lifesavers, Orbit, Skittles, Big Red, Crave, Starburst, Airwaves, Eclipse, Excel, Extra, Freedent, Creme Savers, Hubba Bubba, Doublemint

8

Smithfield 9

Cumberland Gap, Olde Kentucky, Genuine Smithfield Ham, Basse's Choice, Smithfield Marketplace, Eckrich, Armour, Margherita, Healthy Ones, Carando, Curly's, Patrick Cudahy, Higueral, Realean, El Mino, La Abuelita, Riojano, Pavone, Stefano's, Ember Farms, Farmland, Kretschmar Deli, Gwaltney, Weight Watchers*, Peanut Shop of Williamsburg, Paula Deen Collection*

* licensed

Kellogg's 10

Kellogg's, All-Bran, Apple Jacks, Bran Buds, Cocoa Crispies, Corn Flakes, Complete, Carr's, Corn Pops, Cracklin' Oat Bran, Crispix, Crunchy Nut, Eggo, Cruncheroos, Crunchmania, Crunchy Nut, FiberPlus, Froot Loops, Frosted Flakes, Frosted Mini-Wheats, Fruit Harvest, Just Right, Mueslix, Pops, Product 19, Raisin Bran, Rice Krispies, Smacks, Smart Start, Special K, Kashi, Bear Naked, Morningstar Farms, Loma Linda, Natural Touch, Gardenburger, Worthington, Keebler, Austin, Cheez-It, Famous Amos, Hi-Ho, Gripz, Jack's, Krispy, Mother's, Murray, Ready Crust, Right Bites, Sandies, Stretch Island, Sunshine, Toasteds, Town House, Vienna Creams, Wheatables, Zesta

Coca-Cola 11

Coca-Cola, Diet Coke, Sprite, Fanta, Squirt, Canada Dry, Dasani, Thums Up, Fresca, Barq's, Cappy, Burn, Nos, Full Throttle, Mello Yello, A&W, Five Alive, Mr. Pibb, Tab, Vault, Minute Maid, Hi-C, Simply, Nestea Teas, VitaminWater, Lift, Fuze, Powerade, Aquarius, Schwepppes, Bonaqua, Cristal, Honest Tea, Gold Peak, Odwalla

13 Hormel

Hormel, Stagg, Spam, Jennie-O Turkey Store, Lloyd's BBQ, Dinty Moore, Valley Fresh, Farmer John, Di Lusso, Don Miguel, Embasa, House of Tsang, Herdex, La Victoria, Marrakesh Express, Peloponnese, Saag's, Herb-Ox, Chi-Chi's

12 ConAgra

Alexia Foods, Banquet, Kid Cuisine, Dennison's, Marie Callender's, Healthy Choice, Lamb Weston, Chef Boyardee, La Choy, Manwich, Hunt's, Wolf Brand Chili, Libby's, Peter Pan, Ro*Tel, Ranch Style Beans, Van Camp's Beans, Rosarita, Act II, Crunch 'N Munch, David, Hebrew National, Jiffy Pop, Orville Redenbacher's, Slim Jim, Snack Pack, Swiss Miss, Egg Beaters, Reddi-Wip, Wesson Oil, Parkay, Pam, Blue Bonnet

Cargill 14

Consumer Brands: Liza, Naturefresh Oil, Diamond Crystal Salt, Truvia, Peter's Chocolate, Wilbur's Chocolate, Honeysuckle White Turkey, Shady Brook Farms, Sterling Silver Meats, Rumba Meats, Tender Choice Pork, Good Nature Pork
Animal Nutrition Brands: Purina Feed, Sportsman's Choice, Nutrena, Acco Feeds, Right Now, Loyall

* licensed

Sara Lee — 16

Ball Park, Bryan, Deli D'Italia, Deli Perfect, Galileo, Gallo Salame, Hillshire Farm, Jimmy Dean, Kahn's, Mr. Turkey, R.B. Rice, Sara Lee, Bistro Collection, Bon Gateaux, Butter-Krust, Chef Pierre, Colonial, Croustipate, Earth Grains, Grandma Sycamore, Heiners, IronKids, Madame Brioche Martinez, Mother's, Ortiz, Rainbo, Rudy's Farm, Rudy's Farm, Sara Lee, Silueta, Sunbeam, West Virginia Brand, Bravo, Butter-Nut Cappuccino, Cafe Continental, Cafitesse Chat Noir, Douwe Egberts Harris, Hornimans, Java Coast, Maryland Club, Moccona Piazza D'Oro, Pickwick, Prima Steamers, State Fair, Bimbo

Dole — 15

Dole

Saputo — 18

Dragone, Frigo, Cheese Heads, Gardenia, Lorraine, Lugano, Saputo, Stella, Treasure Cave, Salemville, Great Midwest, King's Choice,

Campbell's — 19

Campbell's, Pace, Prego, V8, Swanson, Pepperidge Farm

Unilever — 17

Ben & Jerry's, Breyer's, Good Humor, Klondike, Magnum, Popsicle, Skippy, Hellmann's, Wishbone, I Can't Believe It's Not Butter!, Shedd's Spread Country Crock, Promise, Ragu, Knorr, Bertolli, Slimfast, P.F. Chang's Home Menu, Lipton Teas*

HERSHEY'S — 20

Hershey's, Dagoba, Zero, 5th Avenue, Almond Joy, Cadbury*, Heath, Kit Kat, Mauna Loa, Milk Duds, Mounds, Mr. Goodbar, Payday, Reese's, Rolo, Take 5, Whatchamacallit, Whoppers, York, Zagnut, Breathsavers, Ice Breakeres, Jolly Rancher, Twizzlers, Good and Plenty, Bubble Yum

* licensed

Source: Food Processing's Top 100 2011 and company documents and Web sites.

2

THE JUNK FOOD PUSHERS

Corporations, which should be carefully restrained creatures of the law and the servants of the people, are fast becoming the people's masters.
—President Grover Cleveland, Fourth State of the Union address,
December 3, 1888

George Koch, nicknamed "K Street's godfather" by the Washington, D.C., insider rag *Roll Call*, spent twenty-five years—beginning in the mid-1960s—as a power broker for the food and consumer products manufacturing industry. As president of the Grocery Manufacturers of America (GMA) he pioneered the clever tactics used in the game of influence peddling: hardball media strategies, subterfuge, and schmoozing. He and a cadre of young lobbyists shaped the GMA into a force to be reckoned with and feared by elected officials. He was so accomplished, and had such a winning personality, that even his opponents thought he was a nice guy.

Koch trained dozens of lobbyists at the GMA who went on to head major corporations or lobby shops. His philosophy: "Hire 'em young, train 'em well, and move 'em on." In the days of peck-and-tap typewriters he held letters up to the light to see if they had been corrected, in which case the guilty party was asked to type them again. His staff was told that Koch "owned them" from Monday morning until Friday night, and they worked long hours, often joining him at 7:00 A.M. as he ate his morning cheeseburger. He was an enigma: although a conservative Republican, he spent his own money litigating against a local country club for ripping off African American and Latino workers in a cash-skimming scheme, but he also successfully battled the Carter administration's Federal Trade Commission (FTC) on the establishment of a consumer protection agency.[1]

Koch began his career at a time when there was a small brotherhood of industry lobbyists in the nation's capital; they knew one other and worked in a back-scratching environment. He worked in tandem with the legendary lobbyist Bryce Harlow, and went on to hire Harlow's son Larry, illustrating how knowing the right people in Washington is the first rule for going places.

The senior Harlow was the first-ever hired gun in Washington, first working as an in-house lobbyist who represented the White House to Congress for Presidents Eisenhower and Nixon, and then representing Procter & Gamble until he retired in 1978. President Reagan gave him the Presidential Medal of Freedom, the highest civilian honor in the nation.

Larry Harlow managed legislative affairs for Koch at the GMA from 1976 until 1981. According to Harlow, Koch worked his lobbyists hard and found fault if they made mistakes.[2] Then Harlow left to handle legislative affairs at the FTC for the Reagan administration; he worked for three years as part of the team headed by the commission's chairman, James C. Miller, and helped eviscerate antitrust law and advocated allowing food companies the opportunity for unchecked advertising.

Under the Miller team, the Division of Food and Drug Advertising was eliminated and the twenty attorneys working on food and drug issues were reassigned and demoted. Staff resources for policing food ads were drastically cut. An omnibus rule on food advertising was killed and only four complaints were heard, with not one addressing the large national advertising campaigns of the major manufacturers.[3]

Before Miller's arrival the FTC had actively participated in a broad-based group—the Network for Better Nutrition—that was established by President Carter to focus on food and nutrition policies and included government policy-setting agencies. The Reagan administration's failure to support the network led to its collapse. Although Miller refused to challenge deceptive advertising, in 1982 he prepared a comment to the Food and Drug Administration (FDA) arguing that the agency should drop the requirement for sodium information on nutrition labels and leave the issue to market forces.

Harlow's role in legislative affairs at the FTC was to take the Reagan administration's position and lobby Congress. He later ran legislative affairs for President George H.W. Bush. In June 2007, he was named by *Washingtonian* magazine as one of the fifty best lobbyists in the nation's capital.

Some things never change. Lobbyists raise millions of dollars from client firms for the politicians whose favors they seek. Up until recently, some of those millions in fees that lobbyists collected from industry went to the wining, dining, and lining of pockets of legislators and regulators. Today restrictions limit some of these activities, yet influence over elected public officials has never been greater. Trade associations, ever creative, have figured out new ways to establish clout in Washington.

The GMA is the quintessential trade association. During the many decades of the food industry's metamorphosis, it has weighed in on every conceivable issue—from food and agriculture policy to trade, health care, and labor issues. Over the past decades its food-related agenda has included: weakening federal pesticide and toxics laws; stopping the creation of a consumer protection agency; opposing liberal appointees to the FTC; promoting lax global trade and investment rules; allowing dangerous preservatives, additives, and colors to be used in food; weakening antitrust laws; advocating for food irradiation; stopping mandatory food labeling; warping nutrition standards; weakening food safety regulations (under the guise of strengthening them); and promoting genetically engineered food.

In 2007, the GMA merged with the Food Products Association, the organization Cal Dooley directed after his fourteen-year stint in Congress as a free-trade booster; it became the largest trade association for the food, beverage, and consumer products industry in the world. After the merger, the organization kept its well-known acronym, GMA, but changed the "America" to "Association," signaling the global reach of the member corporations.

In 2009, the GMA hired as its CEO Pamela Bailey, who began her career working in various capacities in the Nixon, Ford, and Reagan administrations. More women are being tapped for senior management positions in the food industry, seemingly as part of a new strategy by this cutthroat industry, which caters to women shoppers, to give itself a softer public face. Bailey is a tough advocate for the industry, including opposing country-of-origin labeling, an important and hard-won policy that informs consumers where their food is produced. Proponents of global trade of food, like Bailey, believe this type of labeling discourages consumers from purchasing products grown or produced outside the United States.[4]

Leslie Sarasin was hired by the Food Marketing Institute (FMI)—the powerful trade association for grocery retailers and wholesalers formed by the merger of two smaller trade associations in 1978. Like many lobbyists, she started as a congressional staffer before working for a lineup of trade associations, including the American Frozen Food Institute and the National Food Brokers Association. FMI has been at the forefront of lobbying to: deregulate trucking; pass NAFTA; stop consumer-friendly labeling; weaken antitrust laws; weaken child labor laws; prevent health care reform; weaken organic standards; irradiate foods; and weaken labor standards and worker protections.

Bailey and Sarasin have initiated a GMA-FMI Trading Partner Alliance that will align their agenda. Each year the alliance plans to hold two meetings to target common goals and coordinate cost-cutting practices and sales optimization.[5]

An example of how manufacturers and stores coordinate are the meetings Kraft had over an eighteen-month period with Safeway. One of the problems they tackled was keeping Kraft's sugar-laden Capri Sun lemonade stocked and sales moving, since the product is sold from a floor pallet. The team developed an attractive stacking design that made stocking easy and resulted in a 162 percent increase in sales.[6]

GMA and FMI have also launched a devious campaign for a voluntary front-of-package labeling system that is likely to further confuse and confound consumers by providing incomplete information and nutrient facts that give a false impression of the healthfulness of a food. It is a marketing ploy rather than an honest labeling system. The campaign is designed to derail new federal labeling requirements. Marion Nestle, the well-known nutrition professor and food writer, reports that she listened in on the GMA and FMI teleconference call announcing the label, including the part where the trade association representative said they were "singing 'Kumbaya' " about the label. They called it a "monumental, historic effort" in which food companies "stepped up to the plate in a big way . . . with 100 percent support."[7]

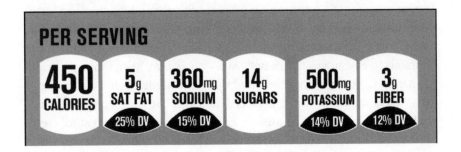

Nestle contends that the goal of the industry effort is "to preempt the FDA's front-of-package food-labeling initiatives that might make food companies reveal more about the 'negatives' in processed foods."[8] The industry is unified in opposing the use of traffic-light symbols—for instance, a red label indicating an unhealthy food.

Unfortunately, the new industry label is merely an abbreviated version of the already required Nutrition Facts label that is on the back of packages; it does not provide additional information or context. For instance, there is no indication that saturated fat can cause heart disease or that sodium is dangerous for people with high blood pressure. The label does not specify if it is desirable to have a high or low number of milligrams or grams of a specified ingredient, and most Americans are not familiar with the metric system used to measure these substances.

Although the industry claims that its Facts Up Front label is easy to use and will address obesity, most Americans have little knowledge about the daily requirements for the different components of food, including the number of calories they need to maintain their weight. It is unlikely that just indicating the number in a serving will help consumers cut back on their total intake.

The International Food Information Council Foundation's *2008 Food & Health Survey: Consumer Attitudes toward Food, Nutrition, & Health* documented that "only 15 percent [of Americans] correctly estimated the recommended number of calories per day for a person their age, height, physical activity, and weight."[9]

Congresswoman Rosa DeLauro, a longtime advocate of informative labeling, rebuked GMA and FMI in a January 24, 2011, press statement about the failures of the label: "Given that negative and positive nutrients will not be differentiated on the package, there is significant risk that these labels will be ignored. An adequate labeling system must clearly alert consumers about potentially unhealthy foods, and should not mislead them into believing that some foods are healthy when they clearly are not."[10]

Facts Up Front is likely to serve more as a marketing tool than as a label. The industry is most interested in listing positive aspects of the food rather than informing consumers about the negatives. The Facts Up Front Web site states: "Manufacturers may also choose to include information on up to two 'nutrients to encourage.' These nutrients—potassium, fiber, protein, vitamin A, vitamin C, vitamin D, calcium and iron—are needed to build a 'nutrient-dense' diet, according to the Dietary Guidelines for Americans." With this guidance, the label could be used to "encourage" consumers to eat ice cream as a source of calcium or to choose sugary cereal with added vitamin C and iron for its nutritional benefits.

GMA's Bailey claims that "this initiative comes on top of the 20,000 healthier product choices we have developed, the responsible marketing practices we have adopted and the tens of millions of dollars we spend annually on healthy-lifestyle promotion."[11] Yet there is little evidence to support the claim that new products, most of which are processed foods, will improve consumer health. A plethora of new offerings at the grocery store is unnecessary, because it is well documented that the key to health is a low-fat diet that is rich in fresh fruits, vegetables, whole grains, and legumes.

In addition to developing deceptive labeling initiatives, the industry trade associations and companies are vehemently opposing even voluntary guidelines about marketing to children that are being written jointly by the FTC, the USDA, and the Centers for Disease Control and Prevention. Children are extremely vulnerable targets of advertising, because they accept ads at face value and do not understand the intent to persuade. According to *Consumer Reports*, "[Y]oung children have difficulty distinguishing between advertising and reality in ads, and ads can distort their view of the world." [12]

The Yale Rudd Center for Food Policy & Obesity documented that aggressive marketing of junk food to children is one of the causes of childhood obesity. The center reported that the fast-food industry spent more than $4.2 billion in 2009 on advertising on TV and radio, in magazines, and with outdoor advertising and other media.

The federal government has proven weak-kneed when it comes to protecting children. In October 2011 David Vladeck, the director of the FTC's Bureau of Consumer Protection, testified about the importance of removing brand mascots, stating, "Those elements of packaging, though appealing to children, are also elements of marketing to a broader audience and are inextricably linked to the food's brand identity." Vladeck proposed a compromise: focus voluntary guidelines on children aged two to eleven instead of up to age seventeen and allow the marketing of unhealthy foods at fund-raisers and sporting events. [13]

Industry representatives, backed by House Republicans, have aggressively killed even voluntary guidelines, because they "are too broad and would limit marketing of almost all of the nation's favorite foods, including some yogurts and many children's cereals," and because the government might "retaliate against them if they don't go along." Not appeased by the Obama administration's surrender on the voluntary guidelines, a lobbyist for the GMA said, "[C]ompanies want the government to prove how these changes will help stem obesity and do a cost analysis looking at the effects through the chain to customers." As a result, the well-known symbols for Kellogg's cereals—Frosted Flakes's Tony the Tiger and Fruit Loops's Toucan Sam—will continue to attract children to the sugar-laden, nutrition-deficient cereals. [14]

Particularly distressing is the GMA study that claims the voluntary standards would cause consumers to eat imported fruits and vegetables, rather than American grains, costing the country $30 billion. General Mills, the sixth-largest food manufacturer in the country, claimed that the "economic consequences [of the guidelines] for American consumers and American agriculture would be devastating," and the company predicted "severe" economic consequences for media outlets.[15] While they were hard-pressed to find an argument against eating fresh produce, it is especially galling that the trade association and industry, the biggest cheerleaders for offshoring food production, are now bemoaning imported fruits and vegetables.

The soda industry has also gone on the offensive against efforts to stop vending-machine sales in schools and has been launching legal attacks against antiobesity advocates. According to Reuters, the international news agency, industry attorneys have filed at least six document requests with public agencies from California to New York. The article quotes an advocate as saying, "It is, in our opinion, an effort to overwhelm or smother government employees, who already have too much to do."[16]

When the prestigious *New England Journal of Medicine* came out with a report concluding that a child's chances of becoming obese increases 60 percent for each can of soda consumed a day, the American Beverage Association went into high gear. The report's authors called for an excise tax of a penny per ounce on soft drinks to create a revenue stream for health programs. In response, the ABA formed the Americans Against Food Taxes coalition. Its dozens of member companies and trade associations—including the GMA, the soft-drink manufacturers, the fast-food chains, convenience stores, the U.S. Chamber of Commerce, the Corn Refiners Association, and Cargill— are spending millions on lobbying and advertising to stop federal and state initiatives, including those in New York, Pennsylvania, Vermont, Mississippi, Kansas, and Alaska.

One of the ways consumers, and especially children, are lured into buying junk food and soft drinks is product placement—the practice of embedding brands in TV shows, movies, videos, and news programs to create in consumers a subconscious emotional connection with the brand. According to the trade journal *Broadcasting & Cable*, two thirds of advertisers employ "branded entertainment"—i.e., product placement—and 80 percent of this is on TV.

Pepsi uses product placement ads to gain brand loyalty in its ongoing war with Coke, by far the highest-selling cola, at 1.6 billion cases in 2010. Pepsi has waged the battle to differentiate itself from its similar tasting and looking competitor by placing products in movies and TV shows such as *Just Go with It* and *The X Factor*.

In October 2011 several consumer advocacy organizations filed a complaint with the FTC against PepsiCo and its Frito-Lay subsidiary for engaging in deceptive digital marketing practices. The advertising campaign on two horror-themed Web sites creates an "illusion that they are entertainment instead of advertising" and that the "sites do not clearly state how and to what extent the data collected on users will be shared." According to Jeffrey Chester, of the Center for Digital Democracy: "Pepsi is really in the forefront of using digital marketing to promote its products, including its snack chip line. They have unabashedly targeted teens." [17]

In 1898, when Caleb Bradham first sold the fountain drink Pepsi-Cola in a Bern, North Carolina, pharmacy, he could never have imagined this type of advertising or that his soft-drink company would morph into the largest food and beverage company in the United States. Bradham was a brilliant marketer and prescient of Pepsi's future reputation for daring advertising that taps into people's deepest emotions; he named the signature drink for the enzyme pepsin and the African kola nut. The name was designed, similar to future marketing ploys, to evoke the exotic. Although it was first advertised as a digestive aid, "an absolutely pure combination" of pepsin and fruit juice, by the early 1920s the original ingredients had been replaced by cheaper substitutes: sugar, flavoring, and hype. [18]

Today PepsiCo is the world's second-largest food and beverage manufacturer, and it is worth $60 billion and operates in two hundred countries. Categorizing Pepsi as a food company is a generous definition, since the majority of its sales are from sugary drinks and salty snack foods. Pepsi, like all of the top food-related corporations, is a marketing machine, with its profits generated through slick advertising and multibillion-dollar budgets designed to convince the targeted consumer that its brand will bring happiness, fun, sex appeal, thinness, health, or all of the above. As a major purveyor of the empty-calorie foods that are becoming ubiquitous around the world, PepsiCo has been scrambling to protect its brand image.

In 2011 the company's increasing profitability was tied to the rising sales volume of snack foods and beverages in the developing world, where PepsiCo hopes to addict people to processed food, high-calorie snacks, and carbonated beverages. Indra Nooyi, the company's CEO, answers critics of their push for junk-food sales: "Right now, good-for-you products are about 22 percent of the [PepsiCo] portfolio. It might rise to 27 or 30 percent of the portfolio ten years from now, but the rest of it is also growing." [19] In 2012 PepsiCo announced it would increase advertising and marketing by $600 million, focusing on twelve beverages and snack foods.

Among the hundreds of products that PepsiCo owns, in addition to its signature drink, are Sierra Mist, Mountain Dew, Mug Root Beer, AMP Energy, numerous fruit punches, Gatorade, SoBe drinks, and, through a partnership with Starbucks, a range of Frappuccino products. Aquafina, its filtered tap water product—falsely marketed as pure—comes in many sweetened and flavored varieties. The Frito-Lay line comprises sixty-five different and mostly salty snack foods, including Lay's potato chips, Doritos, Ruffles, Cracker Jack, Cheetos, and Tostitos. Pepsi subsidiary Quaker Oats' many brands include Aunt Jemima mixes and syrups, Cap'n Crunch, and Rice-A-Roni. Pepsi's Tropicana juices are the largest source of branded juice in the United States.

These thousands of products are produced in facilities throughout the world: 711 in Latin America; 759 in Europe; and 1,465 in Asia, the Middle East, and Africa combined. People concerned about health would probably find the company's portfolio—60 percent junk food sold through glitzy advertising—as nothing to brag about, especially since Nooyi's definition of healthy food is certainly suspect. She recently told Fox Business TV, "Let me first correct you on one thing: Doritos is not bad for you. Pepsi-Cola is not bad for you. Doritos is nothing but corn mashed up, fried a little bit with just very little oil, and flavored in the most delectable way. And Pepsi-Cola was discovered in a pharmacy for a stomach ailment. So these are not bad-for-you products." [20]

Although Nooyi is atypical of multinational CEOs, because she was born in India and later became a U.S. citizen, she is typical in many other regards. She graduated from Yale with an MBA in public and private management, then honed her skills as a corporate strategist at the Boston Consulting Group (BCG), a prominent management-consulting firm that has spawned many CEOs and business leaders who subscribe to the religion of ever-increasing corporate fusion through mergers and acquisitions. She worked at BCG for client companies in textile and consumer goods, including retailers and chemical producers with global operations.

Nooyi, whom *Fortune* magazine has twice named the most powerful businesswoman in the United States, serves on various corporate, foundation, and nonprofit boards. Most notable are the GMA, the U.S.-China Business Council, the foundation board of the World Economic Forum, the U.S.-India Business Council, the Trilateral Commission, the Consumer Goods Forum, and the Peter G. Peterson Institute for International Economics. She was appointed to the U.S.-India CEO Forum by the Obama administration and was formerly a director of the Federal Reserve Bank of New York—the most important bank in the reserve system.

Nooyi is also important in academia in her influential and powerful roles at Yale University; she is on the governing board of the Yale Corporation, its

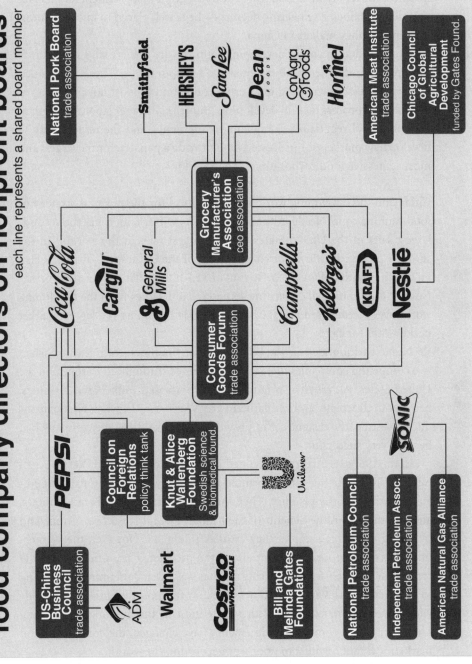

food company directors on nonprofit boards

each line represents a shared board member

policy-making body. As a member of the board of Tsinghua University—the foremost university in China, where both the country's president and its vice president obtained engineering degrees—she is well placed to know the most influential policy makers in China.

China, India, and other developing countries are of increasing importance to PepsiCo and other multinational food corporations. These companies, seeking endlessly increasing quarterly profits, find it most advantageous to produce and process food in developing countries. Cheaper labor, less vigilant environmental regulation, compliant governments, and the opportunity to develop new markets for processed food provide a panacea for the largest and most aggressive food corporations in the world.

China and the developing world are also critical for the business strategy of food multinational Nestlé. Although based in Switzerland, it has the second-largest sales in the United States and the largest sales in the world.[21] In the summer of 2011 Nestlé acquired 60 percent of the Chinese candy maker Hsu Fu Chi International for the hefty sum of $1.7 billion. This acquisition brings Nestlé closer to its goal of earning 45 percent of its sales from the developing world over the next ten years. China is quickly becoming the largest market in the world for candy.[22]

Nestlé was founded in the last quarter of the nineteenth century, when two competing companies merged and expanded throughout Europe and the United States. Although it originally sold condensed milk–based products, baby food, chocolate, and similar items, by World War II, when Nestlé introduced Nescafé, the instant coffee provided to U.S. soldiers, the company had become very successful.

In the 1960s, as birthrates declined in the developing world, Nestlé began rapaciously marketing infant formula in poor developing countries to create a market for the product. They hired women, dressed them as nurses, and had them tout the benefits of their infant formula over breast milk, and by giving away free samples they created a demand. However, the women, often extremely poor, could not afford the product. They would dilute the formula with water from local sources that was often contaminated, leading to diarrheal diseases and sometimes death for thousands of infants. A worldwide boycott initiated in 1977 drew attention to this unethical marketing campaign. While some of Nestlé's tactics have changed, the company is still marketing infant formula to poor women around the world.

Today Nestlé peddles some six thousand brands of foods and beverages in eighty-six countries. The Swiss giant operates 79 manufacturing facilities in the United States and 374 elsewhere under numerous familiar brand names: After Eight, Butterfinger, Carnation, Coffee-Mate, Cookie Crisp, Dreyer's, Fancy Feast, Fitness, Gerber, Häagen-Dazs, Hot Pockets, Jenny Craig, Juicy Juice, Kit Kat, Nescafé, Nespresso, Nesquik, Nestea, PowerBar, a range of Purina pet foods, Skinny Cow Ice Cream, Smarties, and Stouffer's.

Among Nestlé's most aggressive marketing campaigns is the one for bottled water. The former chairman of Perrier, now part of Nestlé's collection of more than seventy global bottled water brands, candidly stated the industry's view on its virtue as a product: "It struck me . . . that all you had to do is take the water out of the ground and then sell it for more than the price of wine, milk, or, for that matter, oil."[23]

The company sells bottled water as the healthy alternative to tap water, even though tap water is much more stringently regulated. The company controls approximately 32 percent of bottled water sold in the United States, and its well-known brands include Perrier, San Pellegrino, Pure Life, Poland Spring, and several others. Nestlé owns seven out of the ten leading brands of bottled water in the United States.

While consumers believe that they are getting high-quality, uncontaminated water, bottled water is so poorly regulated in the United States that consumers have no way of knowing what contamination lurks within. Studies done in independent laboratories have found significant contamination in bottled water. In October 2008, the Environmental Working Group released a report that found mixtures of thirty-eight different pollutants, including bacteria, fertilizer, Tylenol, and industrial chemicals, in ten popular U.S. bottled water brands.[24]

Perpetually underfunded and short-staffed, the FDA, the agency responsible for regulation, has fewer than three full-time employees devoted to bottled-water oversight. The rules apply only to bottled water packaged and sold across state lines, which leaves out the 60 percent to 70 percent of water bottled and sold within a single state.

Fewer than forty states have bottled-water regulations, and in some they are much weaker than the limited federal standards. For the 30 percent to 40 percent of bottled water that the FDA does regulate, it requires that companies test four empty bottles once every three months for bacterial contamination. They must test a sample of water for bacteria after filtration and before bottling once a week. A sample of water must be checked for chemical, physical, and radiological contaminants only once a year. The companies do not have to test the water after bottling or storage.

Nestlé has also been the target of the ire of communities across the nation where the company extracts hundreds of millions of gallons of water from sensitive ecosystems and pays low fees for the privilege. Nestlé's extraction projects have drawn criticism, protests, and litigation in a number of states, including California, Oregon, Wisconsin, Michigan, Florida, and several New England states. Opponents charge that the company is harming the environment by depleting aquifers and drying up wells, lakes, and streams.

In Florida, for instance, Nestlé was charged only $230 for a permit to pump millions of gallons of water from a state park until 2018. The company does not pay taxes or fees to the state or county, even though it makes millions of dollars in profits from the water it bottles from the park and ships throughout the Southeast. The permit was given to the company during a period of drought and against the recommendation of the local water district management staff.[25]

In recent years the company has focused advertising dollars on its Pure Life brand, which is essentially bottled tap water. Between 2004 and 2009, spending on Pure Life advertising increased by more than 3,000 percent; the company's nearly $9.7 million expenditure on Pure Life ads in 2009 was more than any other bottled water company spent on a leading domestic brand, and more than Nestlé's next five spring-water brands combined.

One of the main targets for this advertising is immigrants coming from countries where tap water is unsafe to drink. This population often has low incomes, and in most households the money spent on bottled water could be better used to support other needs. When asked whether Nestlé does market specifically to minority communities, Jane Lazgin, director of corporate communications for Nestlé Waters North America, said, "That's correct. Nestlé Pure Life is a meaningful brand in the Hispanic population." Lazgin also acknowledged that Nestlé Pure Life water "comes from wells or municipal systems."[26]

Pure Life is one of Nestlé's "popularly positioned products" (PPP). A company document prepared by the company's research arm for investors explains that PPP's will be "one of the main growth drivers for Nestlé in the years to come." It goes on: "PPPs target less affluent consumers in emerging markets (UN/World Bank definition—those with an annual purchasing power parity between US$3,000 and 22,000 per-capita) as well as low food spenders in developed economies. Together, they represent some 50% of the world's population. Hence, PPPs target the biggest and fastest growing consumer base in emerging markets as well as important sub-groups in developed markets."[27]

Another product developed by Nestlé for the PPP market is infant formula.

According to the company, "The PPP infant formula is now selling well in the first wave markets: in the Philippines; in Indonesia as LACTOGEN Klasik; in Indochina as LACTOGEN Complete; and in Mexico under the NIDAL brand. Rollout to other markets is planned."[28]

The company was the target of a well-publicized boycott in the late 1970s for violating the World Health Organization guidelines on the advertising of infant formula. As a result of its recent marketing efforts, Nestlé is experiencing a new surge of criticism for its advertising of infant formula in developing countries, where it is often prepared in unhygienic conditions with unsafe water and misunderstood directions. The WHO recommends breast-feeding exclusively for children up to six months of age and supplemental breast-feeding until age two.

Laurence Gray, World Vision's Asia Pacific advocacy director, says of Nestlé's tactics, "Some of the marketing strategy presents formula as better than breast-feeding. It doesn't take into account the circumstances needed to prepare the formula."[29]

Several Laos-based international NGOs, including Save the Children, Oxfam, CARE International, Plan International, and World Vision, announced plans to boycott Nestlé's 2012 competition for a prize of almost half a million U.S. dollars for outstanding innovation in water, nutrition, or rural development projects. They charged that the company visits hospitals and gives doctors and nurses who promote the formula gifts and trips.[30]

Ironically, at the same time Nestlé is targeting the undernourished in developing countries, it is aiming for overfed consumers worldwide. The major marketer of high-calorie candy, ice cream, and chocolate drinks has entered the arena of weight management by purchasing the Jenny Craig line of low-cal foods, which were made famous through TV ads featuring a slimmed-down Kirstie Alley—until she regained the weight and they booted her. Nestlé, which has entered the fitness and beauty market, also owns Lean Cuisine and holds 26.4 percent of L'Oréal, the world's largest cosmetics company.

Nestlé's board chariman, Austrian Peter Brabeck-Letmathe, spent his early career, between 1970 and 1980, working with Nestlé in Latin America, including in Chile, where his portfolio included stopping the nationalization of milk production and dissuading labor leaders from calling strikes in Nestlé facilities during the bloody CIA-led coup d'état that ended the administration of democratically elected reformer Salvador Allende. Brabeck-Letmathe, known either for his forthrightness or for putting his foot in his mouth, told the 2006 Davos Open Forum that the idea of water as a basic human right was "extreme" and that water should have value like any foodstuff.[31]

Brabeck-Letmathe serves on numerous boards, which helps position him to advance Nestlé's agenda. He is vice chairman of Credit Suisse (CS), the giant Swiss financial services company that advises Nestlé on acquisitions. CS and Nestlé have long had an incestuous relationship. The chairman of CS from 1986 to 2000 was Rainer E. Gut, who served as a director of Nestlé from 1981 to 2005 and as the company's vice chairman for his last five years of service.[32] Like Nooyi, Brabeck-Letmathe is on the foundation board of the World Economic Forum. He is also on the boards of ExxonMobil, Roche Holding, Alcon, and the International Association for the Promotion and Protection of Private Foreign Investments, among others.[33]

Brabeck-Letmathe recently told a reporter for *Financial Markets* that genetically modified organisms (GMOs) have never killed anyone, but organic food has—referring to a food-poisoning incident related to organic sprouts. In another interview, he said about organics: "You have to be rational. There's no way you can support life on Earth if you go straight from farm to table."[34]

Not all CEOs are as forthright, even if wrong-minded. Irene Rosenfeld, the CEO of Kraft Foods—the third-largest food company in the United States—is more careful with her words. Perhaps her style is a result of her doctorate in marketing and statistics from Cornell, or because she began her career at a top-tier Madison Avenue advertising agency. She was employed by one of Kraft Foods subsidiaries from 1981 to 2003, including the consumer research division of General Mills, and she left briefly to head PepsiCo's Frito-Lay division from 2004 to 2006. Rosenfeld serves on the boards of the GMA, the Economic Club of Chicago, and Cornell University. Kraft, along with other major food corporations, has been a long-term supporter of Cornell's Department of Food Science.

Kraft is the world's largest seller of packaged foods, and it operates in 170 countries and has 223 manufacturing and processing plants. The company brags that 80 percent of its revenue comes from brands that are the top seller in their categories, and 50 percent of its revenue comes from categories where Kraft's market share is twice its nearest competitor's. Its brands include: Oreo, Nabisco, and LU biscuits; Tang; Milka and Cadbury chocolates; Trident gum; Jacobs and Maxwell House coffees; Philadelphia cream cheeses; Kraft cheeses, dinners, and dressings; and Oscar Mayer meats.[35]

Kraft's origin dates to 1903, when James Kraft started a cheese-delivery business that merged with a larger cheese manufacturer and was eventually acquired by National Dairy Products (NDP) in 1930. Thomas McInnerney, a Chicago pharmacist, had formed NDP by persuading a consortium of investment bankers, including Goldman Sachs and Lehman Brothers, to invest in

a snapshot of corporate influence over university agricultural research in 2012

corporate representative on university board

corporate funding for university departments, schools, and buildings

UNIVERSITY OF CALIFORNIA

Monsanto Chiquita, Dole United Fresh Earthbound Farm Taylor Farms Syngenta, Sysco Produce Marketing Assoc.

Nomacorc, Mars American Vineyard Assoc.

Chevron Technology Ventures

Arcadia Bioscience

Novo Nordisk

IOWA STATE UNIVERSITY

Monsanto Iowa Farm Bureau Pioneer Hi-Bred Summit Group

Monsanto, Dow Deere & Co. Syngenta, Bayer

Iowa Soybean Association

Iowa Cattlemen's Association

National Pork Board

United Soybean Board

TEXAS A&M

Monsanto Pioneer Hi-Bred Cotton Inc.

Chevron Tech National

Cattlemen's Beef Assoc.

National Pork Board

Donald Danforth Plant Science Center

UNIVERSITY OF ILLINOIS

Monsanto

Monsanto, Syngenta, Pfizer Nestle Nutrition Pepsi, Elanco

SmithBucklin & Associates

National Pork Board

UNIVERSITY OF FLORIDA

Pfizer, Intervet Alcon Research Mars, Vistakon

COLORADO STATE UNIVERSITY

Five Rivers Ranch (JBS)

UNIVERSITY OF GEORGIA

Cargill, Conagra General Mills Unilever, Coca-Cola, McDonald's

UNIVERSITY OF MISSOURI

Phillip Morris Monsanto, Dow SmithBucklin & Associates Iams, Pfizer American Veterinary Medical Association

UNIVERSITY OF ARKANSAS

Tyson, Walmart

UNIVERSITY OF MINNESOTA

Cargill

PURDUE UNIVERSITY

Kroger, ConAgra

Dow, Deere & Co. Hinsdale Farms Nestle, BASF

CORNELL UNIVERSITY

Kraft

Food & Water Watch analysis of university grant records obtained through Freedom of Information Act requests of online databases. See Food & Water Watch report "Public Research, Private Gain" at www.foodandwaterwatch.org for more information.

a scheme to consolidate the ice cream industry. By 1930 it was the largest dairy company in the country, demonstrating that the parasitic relationship between the financial services industry and merger mania is not a recent invention. Philip Morris, known best for its tobacco business, purchased Kraft in 1988 and merged it with another of its subsidiaries, General Foods, in 1989. During the last twenty years, Kraft has acquired and sold off various parts of the business. By 2007, Philip Morris—euphemistically renamed Altria in 2003—had completely spun Kraft off as a public company, and today institutional investors own more than 30 percent of Kraft.

In August 2011, eighteen months after buying European candy giant Cadbury, Kraft announced that it would spin off its North American grocery business from its global snacks group. Like PepsiCo and Nestlé, this snack food business would focus on "fast-growing developing markets and instant consumption channels," while the North American business would continue to focus on the grocery sales.[36]

No matter what happens with the Kraft split, the food giant will remain one of the largest and most powerful of the food processors. Along with the other behemoths of food processing, it will play a major role in dictating what Americans eat, unless we are able to make major reforms in the areas of agriculture, food, advertising, and labeling.

Changing the American diet, and thus reducing obesity and improving public health, will require building the political power to challenge the economic forces that have molded the food system and continue to profit from its shape.

food companies in the financial industry

each line represents a food board director on a financial company board

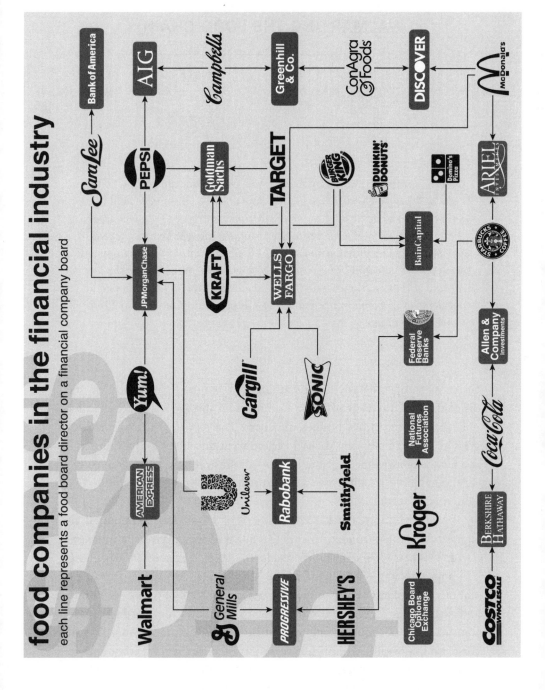

3

WALMARTING THE FOOD CHAIN

Lincoln lamented that his generals were chronically afflicted with the "slows." They rarely advanced. Today Reagan's FTC has a near fatal case of the "slows"—never have so many labored so energetically to produce so little. And if inaction were not consumer injury enough, Reagan's generals on the consumer protection and antitrust fronts display alarming retrograde symptoms—retreating from 60 years of progress in strengthening the law against deception; surrendering whole chapters of the antitrust laws and engaging in economic frolics and detours, corporate over-reaching and consumer exploitation.

So reads the opening salvo in a scathing four-hundred-page report on the dismantling of the Federal Trade Commission, the agency responsible for policing anticompetitive business practices. The report, prepared for Congress by its commissioner, Michael Pertschuk, at the end of his term in 1984, is a blistering assessment of the Reagan administration's destruction of the very foundations of antitrust and consumer protection laws established by Congress.

Reagan's election signaled the new era of laissez-faire economics, an eighteenth-century economic theory that attacked any government intervention in business affairs, including regulation, taxes, tariffs, or antitrust laws. The approach was rehabilitated as the "free market" by Milton Friedman, embraced by Alan Greenspan and the other architects of today's economy, and enshrined in legislative and regulatory policies over the past thirty years. Consolidation of industry, one of its main planks, has the obvious goal of concentrating wealth. Although this was justified for creating economies of scale and economic efficiency, the associated antitrust violations have been documented in recent decades to be major job killers.[1]

Pertschuk, infuriated by the attack on the agency that began in 1981, refused to leave without a fight. He had been the Senate staffer responsible for facilitating legislation banning TV advertising of cigarettes and was appointed by Jimmy Carter to the FTC on the recommendation of influential consumer advocate Ralph Nader. Now retired, Pertschuk says that the tragedy of the assault on the FTC was that the agency had gone through a renaissance in the 1970s.[2]

During the Nixon, Ford, and Carter years, FTC commissioners had built up the agency's antitrust enforcement capabilities, strengthening the regional offices that investigated and developed cases against violators. But under Reagan's appointee James C. Miller (referred to in Chapter 2), the agency waged a "budgetary purge of law enforcement staff" and invited previously shunned practices, including "frequent mergers of competitors, a greater willingness to flout antitrust prohibitions, and a flood of companies petitioning the FTC for elimination or weakening of prior Commission orders."[3]

Chairman Miller was a strident ideologue from the American Enterprise Institute. Pertschuk describes him as coming to the commission with an easily described agenda: "Government bad, private sector good. He was opposed to any intervention in corporate misbehavior or enforcement of antitrust law." Pertschuk goes on to say that Miller "drove us crazy." He was a "very unpleasant individual characterized by the tendency to ingratiate himself with people higher in the hierarchy and bully those that worked for him."[4]

Pertschuk, along with Carter appointee Patricia P. Bailey (a Republican who also was unhappy with Miller's performance), tangled with Miller regularly. Among Miller's faults was his clear disrespect for professional women. Under Miller, the FTC experienced massive budget cuts, staff reductions, and the abolition of whole departments. The agency approved large mergers, failed to enforce antitrust and competition regulations, eliminated antitrust reporting programs, and made radical changes in policy that have facilitated fraudulent activities by corporations.[5]

An example of Miller's philosophy was his response to a Bureau of Competition case regarding faulty life equipment for seamen swept overboard. Miller said that there was no reason for the bureau to bring a complaint against the company selling flawed safety gear, because if men drowned, the company would be sued, and the market would correct itself. Of course, he also opposed class-action lawsuits.[6]

Among the actions affecting food was the abandonment of oversight on advertising to children and the failure to act on predatory pricing, a practice designed to drive competitors out of business. Miller refused to allow the appeal of a decision in a case against the cereal industry that was the most significant monopolization case ever brought by the agency until that time.

The FTC had taken Kellogg, General Mills, and General Foods to court for engaging in practices that excluded smaller companies from competing and for unfairly raising the price of cereal. After a lengthy trial, the FTC lost the case in a decision made by an administrative law judge, but the trial staff wanted to appeal the important and winnable case. They were forbidden to pursue it. As a result, the lack of competition in the grocery industry is many times worse today than it was in the 1980s.[7]

It took just a few years to destroy the advances that had been made in enforcing competition laws during the New Deal era. At that time, Congress and officials appointed by Franklin D. Roosevelt developed policies to encourage competition as a way to improve the economy during the Depression. These populists believed that "natural monopolies" like the provision of water or electricity services were efficient, and that capital- and labor-intensive industries should be larger in size, but that individual companies needed to be forced to compete through strong antitrust laws. The end result was a more diverse and democratic distribution of political and economic power that disappeared as a result of the reactionary policies of the Reagan administration.[8]

Barry Lynn, a journalist and writer, documents in his recent book, *Cornered: The New Monopoly Capitalism and the Economics of Destruction*, the cunning changes made in the criteria used in antitrust investigations. They were neutralized and replaced with ones that supported "economic efficiency." Economic efficiency is a theory developed by the conservative Chicago School of Economics based on the idea that prices are lowered when resources are used to maximize the production of goods and services.[9]

Lynn points out that when federal antitrust regulators look at mergers and acquisitions, they allow concentration under the guise of lowering prices for "consumer welfare." He writes:

> The radical nature of Reagan's attack on antitrust law is, in retrospect, astounding. . . . When the Reagan team published its new Merger Guidelines in 1982, the document formalized two revolutionary changes: it redefined the American marketplace as global in nature, and it severely restricted who could be regarded as a victim of monopoly. From this point on, only one action could be regarded as truly unacceptable—to gouge the consumer. Any firm that avoided such a clumsy act was, for all intents, free to gouge any other class of citizen, not least through predatory pricing and the blatant exercise of power over suppliers and workers.[10]

In a doublespeak that rivals Orwell's Big Brother, competition is touted as the holy grail, but all public and private activities are directed at facilitating consolidation. The evisceration of antitrust law, usually overlooked in critiques about the nation's food system, is at the core of its dysfunction. Shaped like an hourglass, a small cabal of food corporations and retail chains stand between eaters and food producers.

Archer Daniels Midland exemplifies the market power and the political clout to change public policy and the American diet. The corporation paid $279 million in criminal fines, settlements, and litigation expenses stemming from antitrust investigations in 1998, when it was convicted of engaging in a worldwide conspiracy to set prices on lysine, a critical ingredient in factory-farmed animal feed, and citric acid, used as a preservative and an ingredient in soft drinks.

ADM was established in the mid-1880s as a linseed-crushing business and now operates 270 plants worldwide that process grains and oilseeds into ingredients used in food, beverages, nutrition products, industrial processes, and animal feed. Processed foods obtain their taste and texture from the sweeteners, oils, and chemicals derived by the giant multinational from corn, soy, cottonseed, sunflower seed, and other plants.

One of ADM's main businesses is processing ethanol from corn, and the corporation is pursuing biosynthetics for future energy production. The company, often called the ExxonMobil of crop-based fuels, has lobbied voraciously for alternative fuel subsidies. The hiring of Patricia Woertz as CEO in 2006 demonstrates the company's commitment to the fuel side of its business. She is an accountant who previously held executive positions at Gulf Oil and Chevron. As the largest grain processor in the world, no company has benefited more than ADM from finding new uses for corn or from lobbying for commodity crop subsidies.

ADM's other corn business is sweetening junk food. The company pioneered the development and marketing of high-fructose corn syrup (HFCS) in the 1970s because it had the ability to market it. It created a sweetener that enhances processed food by acting as a preservative and increasing the stability of carbonated beverages and condiments, such as ketchup and fruit preserves. HFCS inhibits microbial spoilage, resists crystallization, gives bakery goods a soft texture, and withstands temperature fluctuations.

Food industry consultant Bruce von Stein, formerly employed as a senior marketing executive at Nabisco and General Mills, explains: "If a food company is presented with a cheaper ingredient, they will figure out how to operationalize it. Suppliers of food manufacturing firms have marketing teams literally trying to sell their big customers new product ideas and new ways to use their ingredients in the food they make. They know that food processors are looking for a steady supply of ingredients that are cheap, stable, have a long shelf-life, and process easily." [11]

And that is exactly how HFCS became the sweetener of choice. By 1984, as the price of sugar increased, ADM persuaded PepsiCo and Coca-Cola to announce a switch to HFCS. Of course, the cost of sugar was skyrocketing because ADM had worked with Florida sugar interests to successfully lobby the Reagan administration for a sugar tariff, which has remained in place. Rather hypocritical for the free-trade crowd—but a brilliant move for ADM. Since that time, food manufacturers have embraced the cheap sweetener that is also a key ingredient in the processed foods that make up the bulk of the American diet.

Americans spend 90 percent of their food budgets on processed food—those foods manufactured and sold in a box, bag, can, or carton. The food industry defines processed food broadly and includes fruits and vegetables that have been cut, chopped, cooked, or dried, such as items in a salad bar and canned vegetables, dried fruit, and bags of baby carrots. But a quick perusal of any grocery store reveals that most offerings are ultraprocessed—manufactured through a combination of processes that are prepared with unpronounceable additives that include coloring, preservatives, emulsifiers, binders, flavors, taste enhancers, fillers, and stabilizers. Despite evidence that consumption of these calorie-heavy and nutrient-light foods is associated with high rates of obesity, diabetes, and heart disease, the immense power of the food industry has prevented the changes in public policy necessary to address those connections.

Several recent studies show a relationship between processed foods and cancer. A seven-year study of 200,000 people at the University of Hawaii found a 67 percent higher risk of pancreatic cancer in those who eat processed meat, such as hot dogs or sausages. Studies have linked high-fat diets to breast and colon cancer. Inversely, there is a large body of evidence that conclusively shows that eating low on the food chain—a plant-based diet rich in vegetables, fruits, and grains—maintains health and longevity.

Americans eat half of their meals and snacks away from home, increasingly at fast-food restaurants. In 70 percent of homes with children, all adults work full time. And because Americans work longer hours than the citizens of any other developed country, the heavy advertising of convenience food has found a ready market. A recent survey by Yale University found that 84 percent of parents say that their children have eaten fast food in the previous week.

During the current recession, McDonald's has lured customers with its Dollar Menu. As the largest fast-food chain in the world with 32,000 locations, McDonald's enjoyed rising profits during the past three years. In 2012, the chain plans to invest almost $3 billion in new outlets, and by 2013 it plans to open 2,000 more establishments in China.

McDonald's controls 15 percent of the enormous fast-food market. Ten chains earn 47 percent of all sales in the sector, with Yum Brands (KFC, Taco Bell, Pizza Hut)—the second-largest chain—comprising more than 8 percent. Wendy's, Subway, Burger King, and Starbucks follow in size, respectively.

TOP TEN FAST FOOD COMPANIES

	2010 sales in USD billions	U.S. locations
McDonald's	$32.40	14,027
YUM! Brands Pizza Hut, Taco Bell, KFC, Long John Silver's (sold 2011)	$17.70	19,195
Wendy's Arby's (sold 2011)	$11.35	10,225
Subway	$10.60	23,850
Burger King	$8.60	7,253
Starbucks	$7.56	11,131
Dunkin' Donuts	$6.00	6,772
Sonic	$3.62	3,572
Chick-fil-A	$3.58	1,537
Domino's	$3.31	4,929

VALUE! COMBINE ALL TEN AND CONTROL 47% OF ALL FAST FOOD SALES

Source: QSR Magazine Top 50, USDA ERS, company reports

Since it is composed of some of the largest purchasers of processed food, the fast-food industry greatly influences food manufacturers. The largest chains have played a significant role in the rise of industrialized agriculture and food processing. According to researchers at Duke University, the two industries have co-evolved and can dictate how food is "cultivated, manufactured, packaged, distributed and displayed." McDonald's impact on french fries is a prime example of this codependency.[12]

McDonald's is the largest purchaser of potatoes in the United States, and so it has facilitated consolidation of the french fry supply chain by its need for a tremendous amount of fries with very specific characteristics. Three large companies, ConAgra, McCain Foods, and J.R. Simplot, manufacture the enormous number of fries sold by McDonald's. These companies use the required russet potatoes produced by the giant United Fresh Potato Growers of Idaho. Bayer CropScience and Monsanto provide the specified seeds, herbicides, and pesticides for the contract farmers who produce the potatoes, which then become the uniform size and shape fries consumed at McDonald's and other chains.[13]

When Americans aren't eating out, they are likely to source their food from one of the big grocery giants. No segment of the industry has more influence over the American diet than Walmart and the other three large grocery chains: Kroger, Costco, and Target. In 2011 Target replaced Safeway as the fourth-largest grocery chain, because it has rapidly expanded its grocery selections. Together, these four companies make about 50 percent of all grocery sales in the United States. In the one hundred largest markets in the nation, they dominate more than 70 percent of sales—leaving consumers with few choices.

Changes in the grocery industry began in the 1990s, when large grocery-store chains merged or bought out other regional retailers and large warehouse clubs, and Walmart expanded into grocery products. Grocery-store chains have focused on consolidation, mergers, and takeovers over the past decade, in an effort to compete with the giant food warehouses. The increased efficiency hyped by the industry has not meant lower costs for consumers, but it has meant higher profits for the merged chains. Expenditures on food have risen 12 percent over the past decade, according to the U.S. Bureau of Economic Analysis.

No single company has more impact on what and how food is manufactured than Walmart. Although it opened its first supercenter only in 1988—in which food is sold alongside other retail products—within just twelve years the chain became the largest food retailer in the United States.[14] Today over half of Walmart's business comes from grocery sales.[15] And one out of every three dollars spent on groceries in this country goes to Walmart.[16]

Top 4 U.S. Food Retailers

Walmart 1
stores **4,750** sales **$ 264.2**

Kroger 2
stores **3,624** sales **$ 90.4**

COSTCO WHOLESALE 3
stores **592** sales **$88.9**

TARGET 4
stores **1,767** sales **$ 70.0**

50%
of all grocery sales

net sales in billions of USD

Source: Supermarket News: Top 75 Retailers & Wholesalers 2012;
U.S. Census.

Walmart is so big that it has an unprecedented amount of power in all sectors of the economy. Food is no exception. When there is a single player as large as Walmart connecting food producers and food consumers, individual consumers are no longer the food manufacturing industry's most important customer.

The company continually puts pressure on its suppliers to cut costs. And with Walmart as their biggest customer, food companies have no choice but to comply. When Walmart makes a decision to change the way it does business, the entire industry shifts to keep up. And despite what Walmart would have the public believe, this decision is made with profits in mind. As consumers and policy makers continue to be bombarded with PR messages about Walmart's efforts to help people live better, it is time to look at the actual impact that its rise has had on our food system—and to reconsider whether this model has any place in trying to fix it.

More than just size and market share have enabled Walmart to exercise such considerable control over suppliers. Walmart's success is the result of several very specific ways in which it does business. Walmart's logistics and distribution model is much different from other companies. The primary reason for its incredible growth as a food retailer is because of the way it manages its supply chain.[17] Walmart's model, essentially, is all about sucking money out of the supply chain. Its logistical operations are run primarily through shifting costs and responsibilities to its suppliers. Walmart requires suppliers to adopt supply-chain management, logistics, and data-sharing programs and to manage their own inventory, even on store shelves.[18] It was the first company to bring high-tech information management, and it demanded that its suppliers keep up. They must comply with and use the company's own IT system, which includes automated, scheduled deliveries of products and controls of inventories, which are tracked electronically via a universal bar code.[19] Monitoring this is the responsibility of the supplier, not of Walmart. The company even exercises control over the design of their products by forcing suppliers to meet Walmart specifications in a range of categories, from their ingredients to their packaging.[20]

Contracts with Walmart are nonnegotiable: if a supplier wants to do business with the world's largest retailer, it must accept Walmart's terms without modification.[21] It has shifted the liability for supply disruptions to suppliers. If there are perceived discrepancies with an order, or even if not enough product is sold, Walmart can charge the supplier a fine, known as a "chargeback."[22] These fees, which have become more common in other retail industries as well, can be significant—sometimes in the hundreds of thousands of dollars.[23]

When Walmart began requiring some suppliers to use radiofrequency identification (RFID) tags to keep track of inventory, it made those suppliers pay all of the costs of the technology. RFID tags send out a weak radio signal that allows the item to be scanned and tracked from a distance.[24] The technology to track pallets has been expanded to some clothing and food items as well, including by growers.[25] The estimated cost of adopting it for a grocery manufacturer with $5 billion in sales is about $33 million each year, which saves Walmart billions of dollars.[26] Although it doesn't require the tags on every product, Walmart has forced suppliers who didn't adopt the technology when required to pay fees for untagged goods.[27]

Walmart demands volume. It sells an incredible amount of each food product, much more demand than a small or medium-size producer could ever hope to meet on its own. For instance, Walmart buys one billion pounds of beef each year. For a company obsessed with increasing efficiencies in its supply chain it makes considerably more sense for it to get this meat from a few large meatpackers rather than from numerous small, local suppliers. In addition, these smaller producers are probably less likely to be able to meet and afford Walmart's technological requirements, unlike the bigger players in the industry.

Walmart is the largest purchaser of American agricultural products, and as such has considerable influence over which foods are available to the public, the methods in which they are produced, and the prices paid to food producers. Walmart is now the biggest customer for many of the top food producers and processors in the country, including dairy giant Dean Foods, General Mills, Kraft, and Tyson Foods.[28] Each of these suppliers represents only a very small portion of Walmart's total business, but the relationship is a great deal more important to the supplier. It is such a large customer that they cannot choose to forgo any demands that are made upon them.

The incredibly uneven power dynamic between Walmart and goods suppliers puts it in an excellent position to make demands, and Walmart does. Food processors, meatpackers, and other suppliers cannot sacrifice their sales to major retailers, but the retailers can easily switch to alternative suppliers. The pressure to cut costs has pushed companies like Levi's, Huffy, Rubbermaid, and RCA to close up manufacturing facilities in the United States and move them overseas. It has also pushed food producers such as Vlasic into bankruptcy for failing to make Walmart's price-point demands. This pressure travels all the way down the food chain and has led to increased consolidation in all segments of the food industry.

Lynn gives the example of Walmart's ability to influence Coca-Cola. He says that it "is the quintessential seller of a product based on a 'secret formula.' Recently, though, Wal-Mart decided that it did not approve of the artificial sweetener Coca-Cola planned to use in a new line of diet colas. In a response that would have been unthinkable just a few years ago, Coca-Cola yielded to the will of an outside firm and designed a second product to meet Wal-Mart's decree." [29]

Walmart's negative effects go beyond the food system. Well-known activist and author Anna Lappé, who is involved in stopping Walmart's expansion into New York City, noted in a Civil Eats blog post:

> We also have plenty of evidence now that when Walmart moves into town, the company puts small businesses out of business and sucks capital out of the community. For every dollar spent at a Walmart, only a small fraction stays to benefit the local economy. We've seen enough evidence, too, that the company has a long, dark track record of sex discrimination and workers' rights abuses.
>
> Let's be clear, expanding into so-called food deserts is an expansion strategy for the company. It's not a charitable move. Making a big PR splash about improving the health qualities of its food is a smart tactic to deflect attention from the real impact of Walmart on the quality of life for Americans. (Is it a coincidence that this press conference occurred the same week a new study was gaining attention that tracked health and population data and found links between Walmart expansion from 1996 to 2005 and increased rates of obesity?) [30]

In 2005, during a low point in the company's public image, Walmart suddenly announced it was "going green," and it listed three goals the company would try to achieve over the coming years, including creating zero waste, being supplied 100 percent by renewable energy, and selling more "ecofriendly" products. [31] Since that time Walmart has announced a number of initiatives, in an effort to cut energy costs and waste. The company also obtained the positive image benefits of these decisions and won over many of its critics.

Former CEO Lee Scott noted that the rationale for these initiatives was purely economic, stating, "What Wal-Mart has done is approach this from a business standpoint and not from a point of altruism." [32] Walmart is the largest private consumer of electricity in the United States, so any reductions in electricity usage mean big savings for the company. While Walmart claims it wants to be more reliant on renewable energy, it has also stated that it will not use renewable energy if it is more costly than traditional sources. This is Walmart's line in the sand—once sustainability becomes unprofitable for the company, it will stop pursuing it. In the meantime, the company has not made much progress in meeting its green goals.

The demand from Walmart and other grocery chains for large suppliers has driven the food manufacturing industry to consolidate even more than it had been. Today, the twenty big-brand food manufacturers produce 60 percent of all the food sold in grocery stores. Food processing is one of the biggest single manufacturing sectors in the economy. A merger-and-acquisition mania has created megamultinationals that have offshored food production and led them to access ingredients and labor where it is cheapest. The largest three giant food processors, PepsiCo, Nestlé, and Kraft, manufacture packaged foods that are so ubiquitous in the American diet that most people eat their products every single day.

The food industry has the economic power to dictate not just what we eat but also public policy on a range of issues—from trade and agriculture to nutrition and health. It has extended its political and economic power through a web of business, social, and organizational connections. Corporations hide behind these institutions, sometimes speaking with one voice to cajole, threaten, and unduly influence Congress, the regulatory agencies, and international institutions. The funding they provide to academic institutions guides and defines the scientific research that is the basis for most regulatory decisions made regarding food.

Of these relationships, especially noteworthy are the interlocking directorates of the most powerful companies. The top twenty food companies have 436 shared board members playing decision-making roles in multiple venues. A recent study of global corporate control noted that firms exert control over other companies via a web of direct and indirect ownership relations that extend throughout the world.[33]

Boards bring together individuals with a common economic interest in the corporation, but each individual member has his or her own specific concerns and agendas as well. Boards serve not only as places where the group collectively makes decisions for the one corporation, but also as venues for building the relationships necessary for brokering future mergers and acquisitions, making strategic alliances, finding new employment, and networking for a range of other opportunities. High-level executives from banking and financial services often sit on the boards of food-related companies, and vice

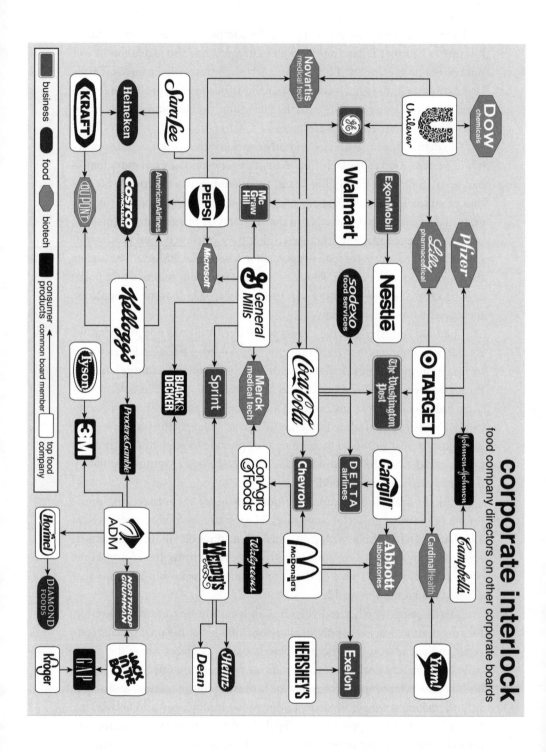

corporate interlock
food company directors on other corporate boards

versa. These relationships are beneficial, because the financial services industry plays a significant role in the food industry.

The agribusiness, food-processing, and grocery industries rely on the services of Wall Street banks, financiers, hedge funds, and private equity to pursue mergers, sell corporate bonds, provide strategic business advice, and even perform ordinary business functions (sort of like the big-business version of the reader's bank accounts). Banks do provide mundane financial services, and many banks are eager to advertise these services to companies that generate billions of dollars in sales. Food companies rely on more extensive banking services than consumers, but the effect is the same. Food and agribusiness companies use commercial banks to perform treasury management functions that include handling both purchases and billing, cash oversight, and investments.

Companies that rely on food ingredients from overseas (think coffee, cocoa, or spices) may need banks to help deal with currency or interest-rate volatility between different markets. Wells Fargo, Harris Bank, Commerce Bank, and others have tailored commercial banking products and services for agribusiness and food companies. The top twenty largest food-processing companies all had more than $5 billion in sales in 2010, so the cash-management fees for the banks handling these accounts can be a significant source of profit.

The past decade's wave of mergers, acquisitions, and corporate takeovers would not have been possible without the Wall Street firms that put the deals together and financed them. Investment banks not only advise the companies involved in structuring the deals, they provide the necessary access to capital. Wall Street banks and investment firms can either lend companies cash to finance a takeover or help issue corporate bonds to fund the deals. During 2010 and 2011 the food, beverage, and tobacco industries issued more than $60 billion in new corporate bonds to raise revenue, either for takeover bids or to expand their businesses.[34] Financial firms also help private companies make initial public offerings (IPO) so they could become publicly traded firms and attempt to raise money on the stock market.

And the more mergers and takeovers there are, the more the banks, private equity funds, and law firms earn in fees. Even during the economic downturn, merger-and-acquisition activity remained brisk. The Food Institute reported nearly one thousand merger deals in the food and beverage sectors between 2009 and 2011, and the annual number of deals reached prerecession levels by 2010.[35]

Big food companies and financial firms are looking at struggling food, beverage, and agribusiness companies as potential targets as the economy starts to improve. According to Neil Masterson, the head of global investment banking at Thomson Reuters, "While we are clearly still in a post–credit crunch environment, M&A [mergers and acquisitions] and private equity firms are returning to robust deal activity, and investment banking fees are on the rise."[36]

These big-dollar mergers involve numerous financial service firms and banks. For instance, in Kraft's $19 billion hostile takeover of the British multinational Cadbury, Kraft was advised by Lazard, the global investment bank, along with Centerview Partners, a new investment bank. Citigroup and Deutsche Bank also served as advisers to Kraft and provided financing: Kraft had to borrow $11.5 billion to finance the deal.[37] Cadbury's financial advisers were Goldman Sachs, UBS, and Morgan Stanley. Morgan Stanley and Citigroup were among the banks that received the largest bailouts during the recent financial crisis.

Another merger that investment banks attempted to broker was one between Diamond Foods, the large snack-nuts company, and Procter & Gamble's Pringles division. Accounting fraud was discovered at Diamond during the negotiations that was designed to make the company's balance sheet look healthier than it was. Since then, Diamond's new management has hired Dean Bradley Osborne, a firm formed by former senior investment bankers from Morgan Stanley, to help facilitate restructuring its debt with lenders. This deal required Diamond to stop paying shareholder dividends, to pay a loan penalty of about $1.5 million, and to pay higher interest rates, which could cost an estimated $3 million more on the remaining unpaid loans. Among its lenders are: Bank of America (the recipient of a large bailout), Barclays, BBVA Compass, BB&T, HSBC, J.P. Morgan, and Siemens Financial Services.

In the wake of the failed merger, Kellogg's announced that it would buy Pringles. The deal was negotiated quickly and was scheduled to close in June 2012. Kellogg's was advised by Barclays Capital and the law firm Wachtell, Lipton, Rosen & Katz. Procter & Gamble was advised by Morgan Stanley and the law firm Jones Day.

Kellogg's chief executive officer, John Bryant, says, "This is a great business that helps us create an even better global snacks business. This is an irresistible asset at a good price. So we moved very quickly."[38]

Like most mergers in the food industry, Kellogg's acquisition has very little to do with the economic efficiency that the Reagan administration championed, or with providing nourishment to the public. It has everything to do with a multinational processed-food company vying to increase the size of its vast junk food monopoly. It is about the avaricious financial service companies earning millions of dollars in fees and interest on the loans. It's about high finance and profit, not feeding people.

Changing the dysfunctional food system and reclaiming our political system means that we must also fight for the reinvigoration of antitrust law. Lynn makes the case that we have no choice but to fight.

> [I]f we choose to ensure the health and flexibility of our economy and our industrial systems and our society; if we choose to protect our republican way of government . . . [w]e must restore antitrust law to its central role in protecting the economic rights, properties, and liberties of the American citizen, and first of all use that power to break Wal-Mart into pieces. We can devise no magic formula or scientific plan for doing so—all antitrust decisions are inherently subjective in nature. But when we do so, we should be confident that we act squarely in the American tradition. . . . We should act knowing that the ultimate fault lies not with Wal-Mart but with our last generation of representatives, who have abjectly failed to enforce laws refined over the course of two centuries. We should act knowing that much similar work lies ahead, against many other giant oligopolies, in many other sectors. We should act knowing that to falter is to guarantee political and perhaps economic disaster.[39]

PART III

The Produce and Organics Industries: Putting Profits Before People

The production and sale of healthy food, from fruits and vegetables to organic products, suffer from the same consolidation and corporate control that plagues conventional foods. A growing share of food, including fruits, vegetables, and organic products, is sourced from developing nations such as China and Mexico as a consequence of the trade agreements negotiated for the benefit of large economic interests. Retail chains such as Walmart are so large that they demand scale all along the distribution chain, resulting in the creation of large produce companies that source from industrial farms in California as well as around the world. Many of the same large companies that produce and sell processed food have bought small companies in the organics industry and merged them into their empires. Not only is this increasing the cost of organic food, but these multinationals are lobbying to weaken organic standards, so that cheaper synthetic ingredients can be used. Creating a regionalized food system necessitates understanding all of the issues involved in growing and marketing fruits and vegetables as well as in producing organic foods.

4

THE GREEN GIANT DOESN'T LIVE
IN CALIFORNIA ANYMORE

And it came about that the owners no longer worked on their farms. They farmed on paper; and they forgot the land, the smell, the feel of it, and remembered only that they owned it, remembered only what they gained and lost by it.

—John Steinbeck, *The Grapes of Wrath*

We're no longer in the farming business.... [W]e sell a branded line of products.
—Tom Loveless, senior vice president of Fresh Express's parent company, Performance Foods, in the *Los Angeles Times*, August 19, 2002

Like real estate, growing vegetables and fruit is all about location, location, location. Ever been to Iowa in the summer, when the hot, dry winds sweep across the almost treeless landscape? Several weather-producing systems collide here in the center of the corn belt, often producing late spring freezes and scorching summers, with a dash of drought, flood, hail, and tornadoes, and early, frigidly cold winters. The Iowa climate is favorable for the tall prairie grass habitat that dominated the land before the pioneers came.

"Until the rise of factory pig farms in the state, diversified farms dotted the Iowa landscape," according to Brother Dave Andrews, a senior representative for Food & Water Watch and the former executive director of the National Catholic Rural Life Conference, which is located in Des Moines. Andrews says that Iowa's weather and countryside are especially suitable for small livestock farms. Corn, a warm-season grass, grows well in the state's hot summers, and it was sustainable when livestock farms were small enough for farmers to use the animal manure for growing their own grain to feed their livestock.

"Sixty years ago, Iowa was almost self-sufficient, with most farms growing livestock, vegetables, and fruit. But by the 1970s, as the grocery chains moved in, farmers couldn't compete with produce grown in California," says Andrews.

The same was true in other states across the country before interstate highways, diesel-fueled trucks, and chain grocery stores changed both the economics of the food system and expectations at the dinner table. Prior to World War II even Montana had the infrastructure to be self-sufficient, growing 70 percent of the food consumed in the state—a range of fruits, vegetables, and livestock. Two hundred canneries, mills, dairies, and slaughterhouses employed four hundred people across the state in 1947, before refrigeration and freezing made it possible to haul fresh and frozen produce from California.

In contrast to Iowa and Montana, California's fertile soil, warm climate, and river water made California a produce superstar. Historian Steven Stoll says that these environmental attributes gave the arid valleys and isolated coastal plains of the state a "natural advantage" for producing fruits and vegetables. In 1880, after global competition caused wheat prices to collapse on the world market, resourceful "orchard capitalists" turned to fruit production in the state's long, fertile Central Valley.[1]

In the 1880s adherents of Chicago School free-market economist Edwin Nourse seized the opportunity to make farming a business by specializing, mechanizing, and marketing. This new type of grower surveyed the soil, studied climate patterns, and manipulated the natural environment to grow cash crops for the eastern market—creating the proving ground for intensive agriculture. The growers formed voluntary associations to lobby the state legislature to invest in agricultural research and to develop strategies for marketing the fruits of their labor. Monoculture—growing one crop on a large scale—made crops susceptible to disease and pests and drove the development of the agrochemical industry in the state. By harnessing chemicals, migrant labor, and technology, California came to dominate the market in perishable products. But none of this would have been possible without the diversion of river water.[2]

Although they are unarguably a calamity for the environment, California's fourteen hundred dams are monuments to engineering. Throughout the twentieth century, public investment in dams, dikes, and diversion made subsidized water available and California's agricultural bounty possible. In contrast to other states, California has a distinctive water distribution system, with three thousand public and private water suppliers distributing subsidized water through federal, state, and local water projects. Low-cost water allows the state to be the largest fruit and vegetable producer, with four hundred agricultural commodities to market. California supplies almost half of U.S.-grown fruits, nuts, and vegetables and over 90 percent of the nation's almonds, artichokes, avocados, broccoli, and processing tomatoes. Nine of the USDA's top ten agricultural counties in the United States, in terms of the market value, are located in California—five in the Central Valley. The top produce cash crops are grapes, almonds, berries, and lettuce.

The Central Valley stands alone among the agricultural regions in California, producing 8 percent of the nation's crops on 1 percent of the country's farmland. Its northern region, the Sacramento Valley, receives twenty inches of water a year, and the flow of water is regulated to prevent flooding. The larger southern part, the San Joaquin Valley, is semiarid and extremely reliant on irrigation and groundwater pumping; one sixth of the irrigated land in the United States is in the Central Valley. Farmers are charged between $2 and $20 per acre-foot for irrigation water—about 10 percent of the cost of providing water, and taxpayers make up the difference.

The availability of water has made this long stretch of flat land one of the most productive places on Earth. After years of squabbling about which government agency would have the pleasure of reworking nature, the Central Valley Project was created in 1933 to dam northern rivers, create reservoirs, and regulate the flow of the water to the entire valley through a series of canals, aqueducts, and pumps—preventing flooding in the north and providing irrigation to the dry south. The project—completed only in the early 1970s—irrigates 3 million acres of farmland and provides drinking water to 2 million households.

The southern half of California has always jealously sought to rob the north of its precious water resources. In the 1950s, after decades of lobbying by the state's biggest agricultural interests, the largest state-built water project in the United States was constructed to provide water for drinking and irrigation. In Southern California, a million acres of agriculture get 30 percent—much of it for a handful of billionaires growing almonds and other nuts for export. The California State Water Project, which shares some infrastructure with the Central Valley Project, transports river water from the northern part of the valley to the San Joaquin Valley, and farther south to the Imperial Valley.

Well-known author and water expert Maude Barlow says that water projects have had a profound impact on the environment, destroying rivers, habitat, and wildlife, including the once plentiful chinook salmon. She notes, "Dams hold back the natural flow of sediment in rivers, eventually burying riverbeds and blocking water channels. Regulating the flow of water to prevent floods negatively impacts ecosystems that have adapted to this weather pattern. Wetlands are dependent on the nutrients that floods deposit, and many animal species require floods for various lifecycle stages, like reproduction and hatching."

Lloyd Carter, a former UPI and *Fresno Bee* reporter, has been writing about California water issues for more than thirty-five years. Reporting on water in California means stepping on many powerful toes when you refuse to compromise the truth, and Carter has made his share of enemies. His reporting on the devastating effects of selenium-contaminated irrigation water on wildlife garnered several journalism awards. Carter charges: "The decision to irrigate the selenium-laced, alkali desert of the western San Joaquin Valley is one of the great mistakes of America's environmental history. Despite more than sixty years of searching for a salt disposal solution for this godforsaken desert, and the expenditure of hundreds of millions of dollars of taxpayer money, no economical, safe disposal solution is in sight, and the western valley continues to slowly march toward the same fate as Mesopotamia: a salted-up, lifeless desert."

Referring to a University of California study, Carter notes that the San Joaquin Valley may be home to some of the nation's richest agricultural resources, but half of the people who live and work there face elevated levels of air and water pollution, coupled with poverty, limited education, language barriers, and racial and ethnic segregation.

The Central Valley is ranked as the most polluted air basin in the country; children residing there have a 35 percent higher chance of suffering from asthma. The area's geography—a valley with tall mountains on three sides—creates a pool for the pollution created by agriculture, including emissions from the barrage of large diesel trucks hauling produce. Allowing the agricultural industry to operate in a way that pollutes the air reduces its costs and is an indirect subsidy.

Carter adds that this is also true of the low-wage work, which is done under harsh conditions by undocumented workers. He explains, "A few dozen farming dynasties benefit the most from the huge numbers of low-paid, seasonal harvest workers that live in grinding poverty."

Kathy Ozer, executive director of the National Family Farm Coalition, agrees. She notes, "Large industrial vegetable and fruit growers also benefit from the use of low-wage labor, often undocumented. Some would label illegal labor a 'subsidy' as well."

While the issues around farm labor are complicated and beyond the scope of this book, it is fair to say that without migrant labor the produce industry would not exist. Unquestionably, a handful of giant produce companies are growing very rich on the structure of the industry in California; however, small and midsize growers are being squeezed out. In spite of low-wage labor, the profitability of produce production has diminished for all but the very largest growers, driving many smaller and midsize producers out of business.

Growing produce, an extremely perishable commodity, is a risky business; achieving a fair price is even more daunting. Over the past fifteen years, the economic clout of Walmart and other large grocery retailers has restructured production and distribution, as suppliers have had to consolidate to match the scale of their largest buyers. In this environment, even with all of California's advantages, it is becoming increasingly difficult for small and midsize produce growers to be profitable.

Roberta Cook, an agricultural economist at the University of California–Davis, says: "Growers and shippers are price takers; they typically are not large enough to set prices. Produce is generally harvested and shipped daily, and changes in weather can affect both supply and demand, making markets very volatile. This, combined with high production costs, makes produce a risky business, and prices may not always cover total costs. It requires substantial capitalization to withstand low prices."

She goes on: "Growers receive the market price minus charges for harvesting, packaging, and marketing and in some cases cooling, along with other handling charges and mandated-marketing or other institutional fees."[3]

Expenses for farming are increasing—fuel, seeds, fertilizers, and other inputs are becoming more expensive as these industries also become more consolidated. Transportation for crops has become a major problem for growers; few independent truckers are still in business, and the demise of the railroads means there is no other option for long-distance transportation.

Cooling crops immediately after harvest exemplifies the infrastructure problems that growers face. Most growers cannot afford the specialized equipment for cooling or otherwise readying crops for market, which can include washing, sorting, grading, and packaging of produce before it is loaded into trucks and transported. Crops such as oranges, apples, tomatoes, and onions are often kept in cold storage before they are sold, in order to wait until the best price can be achieved. Some fruits and vegetables require immediate refrigeration at an optimum temperature and humidity levels that differ for each crop but are necessary to maintain quality. Other crops, such as celery, lettuce, and grapes, can be packaged for sale in the field before they are cooled, to lower growers' costs.

Growers of most crops have affiliations with a packer/shipper—a company that arranges harvesting, prepares the crops to sell, and markets it. These companies usually grow large acreages of crops themselves and have affiliations with other growers. Each service provided to the grower diminishes the final price they receive. Depending on the contractual arrangement, packer/shippers sometimes hold crops in cold storage longer than is necessary, in order to collect higher fees.

Already large packer/shippers are increasing in size and scope in order to supply retail chains. Larger firms have access to the capital necessary for investing in the facilities, equipment, and staffing required by large chains, including massive warehouses located at airports or transportation hubs. They are able to provide high volumes of produce by growing their own crops and selling produce grown by affiliated farms. These suppliers procure year-round produce by having relationships with foreign companies and can ship directly to the retail chains' distribution warehouses. These can be joint ventures, exclusive relationships, or contractual arrangements. Many packer/shippers grow crops in Mexico to lengthen their growing season and have year-round produce.

Cook comments: "Scale is increasingly important for grower/shippers. California firms dominate, followed by Florida and Washington. Large retail and food-service buyers demand year-round supply, increasingly provided by U.S. grower/shippers, which import during the off season." She goes on to say that fresh produce trade patterns are largely determined by identifying the ideal growing locations for each product during the different seasons.

Of the more than three thousand grower/shippers that operate in the United States, most are privately owned. These companies are diverse, specializing in different crops, marketing strategies, and foreign affiliations. Many companies are focused on specific retailers, helping them develop special produce products not offered elsewhere to attract customers. It is noteworthy that Monsanto is a member of United Fresh, the lobby arm of the produce industry.

According to Cook, "seed companies are striving to develop more output-specific consumer traits; in some cases in conjunction with growers and shippers in order to capture more of the value chain; in some cases

developing exclusive marketing relationships with retailers, supporting retailer differentiation."

Sun World, a large packer, shipper, and marketer, also has a significant
licensing and breeding operation to develop new varieties. The company has
a large international network, with offices in Italy, Australia, Chile, Mexico,
and South Africa. It farms on 23,000 acres and has relationships with 950
domestic and international growers. The company breeds and grows a variety
of fruits and vegetables. Pandol Brothers, another packer/shipper, supplies
grapes and tree fruits, and has operations in California, Chile, Peru, Brazil,
and Mexico. California Giant Berry Farms grows and ships over 20 million
trays of berries annually from farms in California, Florida, the Pacific Northwest, Mexico, Chile, Argentina, and Uruguay.

"Fresh cut" is a category of shipper geared to selling fruits and vegetables to
food service and retail stores. Taylor Fresh Foods, a privately owned company,
is among the largest suppliers of prepackaged salads and fresh-cut fruits and
vegetables, and it grew large because of numerous acquisitions. The company
has plant locations in Arizona, Colorado, Florida, Maryland, Mexico, Tennessee, and Texas. Although it was originally focused on food service, it has
moved into retail, grown through numerous acquisitions, and boasts more
than $1.3 billion in annual sales.

A handful of large publicly traded multinationals that initially sold tropical fruit grown on plantations now also dominates fresh and packaged produce. Dole has its headquarters in California, is the largest produce company
in the world, and employs sixty thousand full-time and seasonal workers to
produce three hundred products in ninety countries. Although the company's
highest volume of sales comes from bananas, it markets a large number of
frozen, fresh, and packaged fruits and vegetables.

Chiquita Brands, previously known as United Fruit, not only is well known
for its bananas but is also the largest producer of greens for private-label
brands, for which it competes with Dole. As one of the largest processors of
private-label canned vegetables, Chiquita sources internationally. One of its
subsidiaries, Fresh Express Inc., leads the market in the consumer packaged-
salad business, producing 40 million pounds each month.

Nonetheless, even the largest suppliers of produce are affected by the
market power of the retail giants. Walmart, Kroger, Costco, and Safeway
control more than 50 percent of grocery sales, giving them the market power
to dictate prices to suppliers. The largest retailers have streamlined and
centralized purchasing systems that require suppliers to provide high

volumes of year-round produce. Facing added pressure during the economic downturn, they are seeking to squeeze additional costs out of their supply chains.

Produce suppliers must pay a variety of fees, promotional allowances, and other charges beyond the cost of the actual product to do business with the largest chains. Retailers require complicated sales arrangements that include high-volume discounts and automatic inventory replenishment. They want services that lower their costs, such as having bar-coded price lookup (PLU) stickers applied to the individual pieces of produce delivered to their distribution centers. Each retailer has complex and different food safety requirements, including the use of a range of third-party food safety certifiers. These companies specialize in assessing a supplier's compliance with the grocery chain's standards.

This account management approach to sales includes chain-specific research on target customers, the design of displays, the positioning and mix of products on the produce aisle, and the development of branded value-added products, often targeting children. Produce suppliers and their trade associations are even supplying research on demographics and the consumer psychology of buying different varieties of produce.

Produce is important to these retailers, contributing 10 percent to 12 percent of their total store sales and almost 17 percent of net store profits, according to Cook's research. They demand guaranteed high volumes of uniform produce of a consistent quality—no blemishes allowed. The pressure that retailers put on suppliers means that large and small growers are told what to grow to achieve a long shelf life and be shippable. This compromises diversity, taste, and even nutrition, since vitamins decrease each day after harvest. The grocery store tomato is a good example. It has been bred to be shipped, stored, and looked at—not to be delicious or even worth eating for anyone who has tasted a tomato grown locally.

Retail chains have demanding requirements for specific packaging, transportation, cooling, and specialized containers. They often refuse shipments because they fail to meet a specific requirement. They are particular about the appearance of produce: consumer research shows that freshness is an important factor for consumers in selecting the place they shop. Produce departments are becoming larger and more diverse, with many value-added products. Specialized packaged and marketed products, such as prewashed lettuce or carrots and cabbage grated for coleslaw, increase profitability. Many retail chains use produce to attract consumers and to differentiate their stores from others. Some even require that the quality exceed USDA grading standards.

All of the economic pressures put on packer/shippers are passed on to growers, resulting in less profitability for them. Fees for cold storage, packing, and wholesale marketing services reduce revenue. The demand for standardization and low prices is making it increasingly difficult for small and midsize growers to make a living. More diverse marketing opportunities existed in the past, such as large wholesale terminal markets.

Historically located at city railway terminals, a range of produce wholesalers and brokers bought and sold crops that were ready for delivery immediately or in the near future. Large numbers of independent sales took place, and the prices paid were transparent. But, much to the detriment of smaller growers, few wholesale terminals are still in business. Only thirteen still operate today, selling to small grocery stores, restaurants, and institutions.

Growers have also used cooperatives as a strategy to enable them to control price and marketing. Cooperatives are geared to a specific type of produce and provide a variety of services, including cold storage, packing, transportation, technical assistance, and marketing. Although the co-op is run collectively, many small and midsize growers are dissatisfied with large ones, such as Sun-Maid (raisins), Blue Diamond (nuts), and Sunkist (oranges). Large co-ops are dominated by and run for the benefit of the very largest growers and operate more like large corporations than member-owned and -operated entities.

The many risks and difficulties of selling produce has led to "contract agriculture." More than 20 percent of vegetable and fruit production in the United States is done through some kind of contract arrangement between the grower and a "first handler" of the crop—usually either a packer or a food-processing company. Growers are at a disadvantage when negotiating a contract with a large firm that has legal representation and often uses a standard contract.

The types of arrangements used differ, depending on the segment of the produce industry. Agreements between growers and packer/shippers usually do not include a set price; they use the market price at the time of the sale. Contracting for processed fruits and vegetables usually includes a set price, which is lower than a grower might receive on the open market. Contracts have been used by large international companies like Chiquita for decades. Global vegetable and fruit sourcing has led to an increase in contract agriculture in developing countries for crops beyond bananas, pineapples, and other tropical fruits.

Along with the consolidation of the retail industry, global trade in produce is an important factor in the reshaping of the produce industry. Today the United States imports more fruits, nuts, and fresh and processed vegetables than are produced domestically, both in tonnage and dollar value. Between 2001 and 2010, the country had a net trade deficit of $2.6 billion in fruits and nuts, $19.5 billion in processed fruits and vegetables, and $18.5 billion in fresh vegetables.

Retailers' demands for year-round produce that has the same characteristics and quality in January as in July has not only changed American eating habits and expectations, it has completely changed the produce business. These large retailers and their trade associations, such as the Grocery Manufacturers Association (GMA), which represent the interests of retailers, were among the most vocal supporters of global food trade during the policy debates around the United States' membership in the World Trade Organization (WTO), both for lower costs and procurement opportunities. Trade agreements not only facilitate exportation by foreign companies, they enable American corporations to establish low-cost arrangements for shipping fruit and vegetable products back to the U.S. market.

Global trade and the rise of the retail grocery chains has doubled the amount of fresh and processed fruit and vegetable imports since the 1990s. Although imported produce once consisted primarily of tropical fruits and fresh vegetables during the winter months, now Americans are eating more of these imported fruits and vegetables year-round. Crops such as tomatoes, cucumbers, potatoes, and melons that can be grown in the United States are being replaced on store shelves by imports—during the U.S. growing season. Retailers are always looking for new ways to lower costs, including sourcing produce directly from the developing world.

Produce trade involves shipping by freighter from halfway across the world to U.S. ports. Some high-value items, such as asparagus, raspberries, and cherries, are even shipped by air. International trade deals such as the North American Free Trade Agreement (NAFTA) and a raft of regional and bilateral trade pacts have facilitated these surging imports. Trade in produce has grown faster than the trade of other agricultural commodities because of the series of new trade deals.

The creation of the WTO in 1995 made it easier for countries worldwide to export fruit and vegetable products to the United States. After the agreements enforced by the WTO went into effect, the number of countries exporting the top fifty types eaten in the United States increased by a third, to 110 countries that were exporting fruits and vegetables to the United States. In 2000, after the trade deal with China was approved, its share rapidly expanded. China is now one of the five largest exporters of these products to the United States.

The GMA pushed for the Central American Free Trade Agreement in 2004 to gain "new avenues for imports of key ingredients for food processors" and promoted the Peru trade deal to get "access to duty-free imports of seasonal vegetables."[4] In 2004, the Chile free trade agreement went into effect, allowing Chile's fruit sector to ship more grapes, cherries, peaches, and berries to the United States. The trade pacts with Caribbean countries and the four Andean nations—Bolivia, Colombia, Ecuador, and Peru—also lowered U.S. agricultural tariffs and increased imports.

American fruit and vegetable farmers are threatened by the rising volume of imported produce. This is because the U.S. consumer market is the primary destination for American fruit and vegetable production. Imports have displaced domestic production in the produce aisle and contributed to a decline in the number of farms. Imported produce is more likely to have food safety issues than the domestic equivalent.

Rising imports are displacing domestic farm products from supermarket shelves. And as imports have skyrocketed during the past twenty years, U.S. exports have seen minimal growth. Free-trade proponents at the USDA promised U.S. fruit and vegetable growers that export opportunities would expand significantly with new trade agreements. The reality did not live up to this prediction. American producers face growing low-cost import competition, but any growth in exports never made up for the lost domestic market.

Instead, the United States has become a net importer of many commonly farmed—and consumed—fruit and vegetable products. The USDA's promised balance between exports and imports was a mirage for farmers. Food & Water Watch studied fifty products that grow in temperate or subtropical crops (such as citrus fruits) that are cultivated commercially in the United States. Rising imports of these crops has contributed to the decline in cultivated acreage for these crops in this country. The United States lost more than a quarter million cultivated fruit and vegetable acres between 1993 and 2007, as imports of these products nearly tripled.

Many free-trade proponents have contended that produce imports complement domestic production by providing fresh items during the winter months, after the domestic growing season ends. This proposition suggests that American producers are not displaced by imports because the competitive crops enter the U.S. marketplace when farms are dormant. Of course, winter imports do compete head-to-head with the winter fruit and vegetable production in Florida. After NAFTA and the agreements enforced by the

WTO went into effect in the mid-1990s, the majority of produce imports did enter during the winter. But now imports that compete with domestic crops enter the U.S. market year-round.

Asparagus provides a telling example for other fruit and vegetable farmers. Proponents of asparagus trade under NAFTA and the U.S.-Peru Free Trade Agreement initially claimed that fresh asparagus imports would complement the domestic production by providing it to consumers year-round. Instead, Peru grew to become a year-round producer and exporter of fresh asparagus, even shipping it during the part of the growing season when California growers have historically produced it for the American market.

These trade agreements were also promoted as a strategy for addressing poverty in the developing world. Pro–free trade advocates cite the benefits for farmers in these countries. Unfortunately, nothing could be further from the truth. Growing crops for exports not only results in less food production for domestic consumption; it also often results in the removal of peasant farmers from their small landholdings by large agribusiness interests, so that large acreages of land can be cultivated for export crops.

Many proponents of global food trade, who assert that it helps lift peasant farmers out of poverty, have ignored this trend. When peasant farmers in places such as Mexico are driven off their plots of land into urban slums, they often find no employment or are paid extremely low wages.

Anuradha Mittal, director of the Oakland Institute, who has written extensively on this issue, says that growing for export has displaced small farmers and facilitated the concentration of landholdings by rich landlords. "Displaced from their lands, farmers have been forced to eke a miserable livelihood in cities where they form the core of cheap labor for the sweatshops."

Mittal notes that it has also resulted in land grabs—the purchase or lease of vast tracts of land in poor, developing countries by private investors to produce crops for export. She says, "Rapid acquisitions of crucial food-producing lands by foreign private entities pose a threat to rural economies and livelihoods, land reform agendas, and other efforts aimed at making access to food more equitable and ensuring the human right to food for all."

Despite the promise of low-cost produce from global trade, imports overall have not improved the affordability of produce for Americans either. Even adjusted for inflation, most retail prices increased between 2001 and 2010 for produce such as navel oranges, strawberries, Red Delicious apples, tomatoes, potatoes, and broccoli. Only banana and iceberg lettuce prices fell over this period.

Free trade has also resulted in a loss of American jobs. Lower U.S.

tariffs combined with loosened investment rules have encouraged U.S. food-processing companies to invest in factories overseas and to shutter plants in the United States. This means that the produce grown for processing moves to the country where the new plants are located. Foreign plants in the developing world generally operate under weaker environmental and workplace safety regulations, which reduces production costs for American-owned factories. Lower labor costs have been a key factor in the foreign investments and plant relocations of U.S. food-processing companies.

Several large American food-processing companies invested in Mexico, as NAFTA went into effect, to take advantage of lower wages and weaker environmental rules. Green Giant began to shift its production there from California, eventually closing a Watsonville, California, frozen-food factory. Green Giant's Mexican workers earned about $4.30 each day, compared to the $7.60 an hour that workers earned at the Watsonville plant. As these U.S. investments increased, the share of imports that are essentially shipments between food company affiliates or subsidiaries increased, with different corporate divisions shipping ingredients or products to one another across national borders.

The share of processed fruit and vegetable imports that comes from foreign operations of U.S. companies and transnational corporate affiliates has been rising: between 2000 and 2007 it grew to nearly 35 percent. About half of the processed fruit and vegetable imports from NAFTA partners Mexico and Canada between 2000 and 2007 were from corporate affiliates. This means that every other can of imported tomato paste or imported package of frozen sweet corn was manufactured at a U.S.-owned factory in Mexico or Canada and shipped to the United States. These same export arrangements have emerged under trade deals with China and Chile. Imports of processed produce from corporate affiliates in China quadrupled, to more than 20 percent, and from Chile they rose 74 percent between 2000 and 2007.

Another consequence of produce trade is a proliferation of high-profile outbreaks of food-borne illnesses that highlight the potential hazards. Fresh produce presents unique safety considerations for growers, shippers, and regulators. Polluted irrigation water, contamination from livestock operations, inadequate sanitation conditions for farmworkers, cross-contamination in packing and processing plants, and breakdowns in cold storage during long-distance shipping can expose fresh produce to pathogens and allow them to multiply to dangerous levels. And because fresh produce is often eaten raw, the opportunity to kill bacteria through cooking is lost.

Fresh produce is a significant source of *Salmonella* infections, which cause an estimated 36,000 cases of food-borne illnesses each year. Regulators have had enormous difficulties responding to such outbreaks. The sources of half of all food-borne illness outbreaks from produce are never traced back to their packinghouse, supermarket, or farm source, because perishable fresh fruits and vegetables are often eaten or thrown away before they can be tested and positively linked to a source.

Potentially, imported fruits and vegetables have higher risks than domestic produce. Imports from some developing countries may be grown under less sanitary conditions and face weaker environmental rules and indifferent regulatory oversight than in the United States. For example, when investigating the conditions at Mexican farms that were the source of several *Salmonella* outbreaks, the FDA found that the "Mexican cantaloupe are indeed manufactured, processed, or packed under gross insanitary conditions," with inadequate environmental safeguards on the farms or regulatory oversight by the Mexican government. Similarly, the USDA found that China's farmland in many rural areas is "dangerously polluted" but nonetheless operates under weak environmental rules that are barely enforced.

Over the past decade, China has become a significant supplier of imported produce to the United States. Increasing exports have meant that the value of imported Chinese fruit and nuts quadrupled between 2000 and 2005, and the value of vegetable imports nearly tripled. China is likely to be a major player in global vegetable markets for years to come: between 2000 and 2004, China added 5.7 million acres of vegetable production, more than the United States' entire vegetable acreage (4.7 million acres) in 2007.

China's farm and food-processing sectors are plagued with problems that contribute to safety concerns for consumers. Far fewer pesticides are banned in China than in the United States or Europe, meaning that pesticides banned in America may be immigrating to the United States on Chinese crops. The USDA reported that produce from China presents significant risks, noting: "Chinese fruits and vegetables often have high levels of pesticide residues, heavy metals and other contaminants. Water, soil, and air are dangerously polluted in many rural areas as a result of heavy industrialization and lax environmental regulation."

Considering the problems with food safety issues, it is especially troubling that increasing amounts of organic fresh and processed fruits and vegetables are coming from China and other countries in the developing world. Private companies paid by the grower verify that organic practices are being used in growing the produce. This third-party certification system is already problematic in the United States, because of the variability in how organic standards are interpreted and production is verified. Issues around certification are magnified in countries with weaker regulatory systems. The duplicitous sales of faked organic produce, grown more cheaply than the genuine item, create more unfair competition for organic growers in the United States struggling to compete.

The ability of multinationals to procure either organic or conventional produce at a lower price in the developing world is increasing competitive pressures on produce growers in the United States. This increasing control of production by a few large players hinders the nation's ability to shift to the more regional and sustainable food advocated by the local-food movement. Independent grocery stores and smaller chains that have been more willing to buy and distribute locally grown produce have been driven out of business, and those that remain are competing with "everyday lower prices."

Unfortunately, an even more bloodthirsty environment in the grocery industry is on the horizon. Walmart and other large and highly capitalized firms are continuing to expand their reach, capturing more market share. Participants at *Supermarket News*'s 16th Annual Roundtable said that "the industry could be in for a new round of consolidation activity next year [2012] in which some major national chains go after some of the larger regional operators."[5]

Fulfilling the prediction, in March 2012, BI-LO and Winn-Dixie merged to become the ninth-largest supermarket chain in the country, with 688 grocery stores throughout the Southeast. Gary Giblen, managing director of Aegis Capital in New York, summed up the situation in the grocery industry: "Like sharks that die if they stop moving forward, these chains are going to grab growth one way or another."[6]

Walmart continues to drive consolidation through its aggressive moves into urban areas, utilizing a "smaller format" called Walmart Neighborhood Markets. The company's president and chief executive officer, Mike Duke, wrote in his April 16, 2012, letter to shareholders, "There is no doubt Walmart is the best-positioned global retailer." He noted that net sales in 2011 increased by 5.9 percent, to $443.9 billion.[7]

In 2011, First Lady Michelle Obama and several good-food advocates embraced the announcement that Walmart would make produce more affordable as part of its "Initiative to Make Food Healthier and Healthier Food More Affordable." A key element of the program, according to the company's press release, is "Making healthier choices more affordable, saving customers approximately $1 billion per year on fresh fruits and vegetables through a variety of sourcing, pricing, and transportation and logistics initiatives that will drive unnecessary costs out of the supply chain."[8]

This is the very strategy that already has driven produce growers out of business and changed the structure of the produce industry. Walmart shifts the cost of production to the producer to increase its profitability. A story in the company's 2012 *Global Sustainability Report* discusses a project it engaged in with Tuskegee University and the USDA to identify local growers, which resulted in an agreement to supply shelled peas to the megaretailer. While Walmart is proud of this project and its work to overcome the challenges, it was Tuskeegee that picked up the cost for the specialized shipping containers used to transport the shelled peas, not Walmart. The report notes: "With no previous access to a large retailer, these farmers did not have the financial resources necessary to pay for the containers, labels and more they would need to make this partnership a reality. Determined to open the door to this potentially long-lasting economic opportunity, Tuskegee University covered these initial expenses."

The healthier food initiative is just one more example of Walmart's attempt to ride the wave of interest and support for local agriculture. The superstore defines local as having been produced in the same state it is sold in. Although most in the local-food movement would argue with the adequacy of this definition, even if this goal is taken at face value, Walmart must work with large shipper/packers to accomplish it. Furthermore, the chain's goal for local production was averaged for the entire United States, so it is likely that their sales in California, Florida, Texas, and Washington enabled them to achieve it.

Walmart's relationship with Frey Farms provides a good illustration of the company's loose definition of local. In a 2006 press release Walmart touted its "commitment to purchase from local growers for distribution to stores in their areas in support of locally grown agricultural products," and it mentioned Frey Farms in particular, "to provide visibility to [Walmart's] commitment to Illinois agriculture and growers around the country."[9]

Frey Farms' Web site, however, notes that it is "a year-round supplier of fresh fruits and vegetables," and that "[t]he Frey Farms Produce headquarters is located in rural Wayne County northeast of Mt. Vernon, IL and is supported by strategic shipping locations in Illinois, Indiana, Missouri, Georgia, Florida and throughout the Midwest. . . . We service our customers by way of direct store delivery routes and shipments to major distribution centers."[10]

This is not intended as a criticism of Frey Farms, which is no doubt a well-run, privately held produce company specializing in the growing, packing, shipping, and marketing of watermelons, pumpkins, cantaloupes, and ornamentals. Frey Farms' organization is similar to many of the larger produce operators in the Central Valley of California. But the produce from this centrally managed company, with farms located in different regions of the country, does not meet most people's definition of local agriculture.

In a short YouTube video titled *Growing Business Together, Family Farm Supplies Wal-Mart*, the viewer is told that "Frey Farms has grown along with Walmart and now supplies millions of pumpkins, watermelons and canta- loupes to Walmart stores across the country."

One of Frey Farms' owners, Sarah Talley, advised businesses in a Harvard Business School newsletter article about how to negotiate with Walmart. Among the tips she gave: "Do not let Wal-Mart become more than 20 percent of your company's business. It's hard to negotiate with a company that con- trols yours." Which gives a pretty good indication of the size of Frey Farms.[11]

It is time for the mavens of the sustainable-food movement who have praised Walmart to take another look at the long-term effects of consolida- tion. Large, centrally managed, industrial produce operations do not qualify as local and sustainable agriculture. And more consolidation in the grocery industry from Walmart's pressure to provide "everyday low prices" can only have a chilling effect on local vegetable and fruit production.

The mounting pressure on the produce industry in California, which is causing the loss of small, midsize, and even some larger produce farms, bears witness to this fact. Even in a state with near-perfect conditions for growing fruits and vegetables—subsidized water, good land, and a mild climate— farmers are struggling to make a living. Only large-scale, well-capitalized packer/shippers with the financial and technological capacity to meet the growing demands of the grocery industry can survive, much less thrive.

Without significant changes to antitrust law and trade policy, the drive to cut costs and use economies of scale will trump efforts to grow a substantial percentage of vegetables or fruit at a more local level. It is high time that a strong antitrust platform be added to the good-food agenda. The longer that monopolization is allowed continue and advance, the more difficult it will be to reclaim the food system.

5

ORGANIC FOOD: THE PARADOX

You can make a small fortune in farming, provided you start with a large one.

—Anonymous

In the early 1970s, when I was in college, I lived in a commune, where we grew organic vegetables and cooked huge, stupendous meals for ourselves and large numbers of visitors. It wasn't the sort of thing you might see on TV or in the movies—we were a wholesome lot: enthusiastic, idealistic, earnest, and, of course, naive. But more than anything, we were optimistic about the future. And a big part of that positive outlook had to do with the future of organic agriculture and the potential for living in harmony with nature and creating an alternative economy. But little did we know at the time that organic food would stray far from its principled origins to become a highly consolidated, globally traded business—with all the pitfalls associated with highly industrialized food products.

We never could have envisioned the trajectory of organic food from a small homegrown affair to an industry with almost $30 billion of sales in 2011 that is dominated by the largest food companies in the country. Our aspiration for organic food was part of a broader vision for a society that is socially and economically just and ecologically sound. Organic food embodied integrity: more than eschewing the use of agrichemicals, it signified a set of principles by which to live.

The roots of this movement can be traced back to the founding of Rodale Press in 1930 by J.I. Rodale, a visionary who promoted organic food and healthy living. He started publishing *Organic Farming and Gardening* magazine in 1942—a magazine still published today as *Organic Farming*. In 1962, Rachel Carson's *Silent Spring* changed the consciousness of the growing youth movement about the toxicity of agricultural chemicals. By the mid-1960s and early 1970s, the counterculture had spurred the growth of natural food stores and food co-ops.

Rodale Press established voluntary organic guidelines in 1972, and organic growers associations were formed in some states, each with its own set of guidelines. In 1989, a scandal over Alar, a dangerous chemical used on apples, helped stimulate a broader market for organics. As the demand for organic food skyrocketed, organic producers saw the need for replacing the patchwork of standards with a uniform federal one. A broad coalition successfully lobbied for passage of the Organic Foods Production Act, which was adopted as part of the 1990 Farm Bill.

Authority for the new National Organic Program was given to the USDA, an agency historically opposed to organic agriculture. It took the agency two years to establish the mandated National Organic Standards Board (NOSB), an advisory committee appointed by the secretary of agriculture to develop standards that are given final approval by the USDA.

In 1997, under the Clinton administration, the USDA unveiled a set of standards developed at the behest of the food industry, rather than using the ones painstakingly negotiated by the NOSB through a series of public meetings and consultations with the public. Monsanto, the Grocery Manufacturers Association, the Biotechnology Industry Organization, and other agribusiness interests used their influence with the administration to have the use of genetic engineering, food irradiation, and sewage sludge included in the organic standards. The public outcry could be heard from coast to coast. More than 275,000 Americans sent comments to the USDA—a record number for the agency—almost all saying no to the inclusion of the "big three" in organics.

Responding to the public pressure, Clinton's secretary of agriculture, Dan Glickman, announced on December 20, 2000, that the new standards "specifically prohibited the use of genetic engineering methods, ionizing radiation, and sewage sludge for fertilization." Even so, many were disappointed, because the standard did not address a wide range of areas that the organics community felt should have been included to protect the integrity of organic products.

Dr. Phil Howard, a professor at the University of Michigan who has written extensively on the organics industry, says the USDA standard was "fairly strict with respect to prohibiting other unacceptable inputs (such as antibiotics and synthetic fertilizers and pesticides), but also removed references to the higher ideals of organic found in some regional certification systems."

Howard charges that "the national standard creates a 'ceiling' by prohibiting organic certifiers from enforcing stricter standards than those required by the USDA, even those they previously maintained."[1]

Small organic producers around the country argued that the standards were designed to exclude small farmers and favor industrial farms, because organic certification required too much recordkeeping and paperwork. Farmers must track every time a new crop is planted, which for a large industrial farm is not difficult. For a community-supported agriculture (CSA) venture that might grow forty different crops and rotate them in different areas in a season, the recordkeeping is onerous. Certification can also be expensive, so many small farmers who use organic practices choose not to go through the process. These small producers have counted on local customers who can visit the farm and see for themselves how food is being produced without relying on a USDA organic label.

Adoption of the federal standards changed the organic industry dramatically. The largest food manufacturers saw an opening for a profitable niche market without having to adopt the overall ethic of organic agriculture. An opportunity was lost for creating a new type of food system, because the failure to address scale issues in the standard has allowed the big food processing companies to gobble up organic brands. The economic pressure for low-priced organic food has created the same inequities for independent processors, small retailers, farmers, and farmworkers that exist in conventional agriculture.

Moreover, government involvement lent the necessary credibility for the big corporations to develop an interest in organics, while the guidelines allowing industrial production (minus the use of agrochemicals) made consolidated processing and distribution feasible. In a sad twist of fate, the type of agriculture that our great-grandparents practiced prior to 1950 is more expensive today than the chemical-intensive agriculture used by conventional growers.

The new guidelines allow organic products and ingredients to be sourced from around the world, wherever they are cheapest, giving large companies like Dean Foods and Hain Celestial an advantage. The Stonyfield Farm chairman and CEO, Gary Hirshberg, admits that to produce organic dairy at a competitive price his company has to buy milk powder from New Zealand, nine thousand miles away. "It would be great to get all of our food within a ten-mile radius of our house," he says. "But once you're in organic, you have to source globally."[2] This cost-cutting practice is not sustainable, and it is contrary to what most organic consumers expect.

The failure to include livable-wage language in the USDA standard meant that organic agriculture would mirror the same injustices faced by workers in conventional agriculture. By failing to include many of the principles that were originally part of the organic philosophy, the organic label had lost its soul.

Further, while the USDA sets the standards, third-party "certifiers" actually inspect and approve farmers and processors as organic. Many of the integrity questions about the organic standards boil down to whether different certifiers allow different practices to be used. Some nonprofit certifiers do an excellent job, while others that cater to the larger operations are less vigilant. The USDA's failure to properly manage certifiers, especially under the Bush administration, has exacerbated some of the standards' weaknesses.

Nevertheless, despite the flaws, the development of a government-ordained organic standard was critical for providing consumers with a clear, labeled alternative to foods produced using pesticides, herbicides, and dangerous technologies. Ironically, the millions of dollars spent by the conventional food producers in marketing their organic lines has awakened a larger number of Americans to the danger of toxic agrochemicals. Consumers of organic products report overwhelmingly that they purchase them because these items are healthier—a situation that agribusiness interests view as dangerous. Although organic food makes up only 4 percent of food sales in the United States, in a Hartman Group survey in 2010, 75 percent of consumers reported buying organic food in the preceding year, and 22 percent bought it monthly.

Yet the weakness of a standard that is the lowest common denominator, and that eliminated the strictest levels of certification, has created a paradox. Organic food, catapulted into popularity as an alternative to a corporate-controlled food system, is now largely controlled by the largest food companies in the world. Today, fourteen of the twenty largest processors of food have acquired organic brands or introduced organic versions of their products. Whole Foods Market dominates the U.S. natural food retail sector, and one company, United Natural Foods, Inc. (UNF), controls distribution.

A fever of mergers, acquisitions, and strategic alliances swept through the organics industry as the companies that dominate the food-processing industry began enlarging their operations. Consolidation of organic brands snowballed between 1997 and 2002 when some of the largest and most successful companies were sold to conventional food processors. General Mills acquired Cascadian Farm and Muir Glen; Kellogg's acquired Morningstar Farms/Natural Touch; Heinz invested $100,000 in Hain Foods, which merged with Celestial Seasonings; and Unilever bought Ben & Jerry's. As the organics industry became more concentrated between 1990 and 2005, it grew from $1 billion a year to almost $15 billion.[3]

Top U.S. Organic Food Processing Companies

1 *Dean* — Horizon Organic, Silk, The Organic Cow, Alta Dena **$877**

2 THE HAIN CELESTIAL GROUP

Almond Dream, Rice Dream, Soy Dream, Arrowhead Mills, Boston's*, Breadshop*, Casbah, DeBoles, Celestial Seasonings*, Earth's Best, FreeBird Chicken, Garden of Eatin', The Greek Gods*, Hain Pure Foods*, Health Valley, Hollywood*, Imagine, Little Bear Foods*, Ethnic Gourmet*, Maranatha, Mountain Sun*, Nile Spice*, Rosetto*, Sensible Portions*, Spectrum Naturals, SunSpire, Terra*, Walnut Acres Organic, Westbrae Natural*, WestSoy, Yves Veggie Cuisine
natural foods brand **$524**

3 DANONE — Stonyfield, Brown Cow, Organic Oikos, Organic Activia **$304**

4 Earthbound Farm — Earthbound Farm **$299**

5 Amy's — Amy's Kitchen **$266**

6 ORGANIC VALLEY — Organic Valley, Organic Prairie **$244**

7 CLIF BAR — Clif Bar, Luna **$168**

8 General Mills — Cascadian Farm, Muir Glen, Larabar **$111**

9 NATURE'S PATH — Nature's Path, EnviroKidz, Optimum **$NA***

10 Pacific — Pacific Natural Foods **$NA***

$ 2008 total **organic food** sales in millions of USD (Mintel—2008 Organic Food)

* exact sales data unavailable—these private companies are in the $50–$200 million bracket (*Nutrition Business Journal* 2007)

The USDA's "streamlined" certification process for international and domestic trade in organics means that organic food and ingredients can be sourced from anywhere in the world. Imports of organic foods and ingredients have increased dramatically. The USDA reports that half of the foreign-sourced organic products come from China, Turkey, Mexico, Italy, and Canada. The agency has certified 27,000 producers and handlers worldwide—approximately 16,000 in the United States and 11,000 in other countries.[4]

According to the 2009 Cornucopia Institute report *Behind the Bean*, when the USDA first traveled to China to audit organic certifiers in 2007, the federal inspectors audited only four and visited just two farms. The limited audit found multiple problems with compliance with U.S. organic standards, raising questions that have never been answered about China's ability to maintain the integrity of organic products.

Unsurprisingly, the outsourcing of organic production and the transformation of organics into big business has been a stealth operation: few consumers know that the hundreds of brands they see at Whole Foods are controlled by a small group of companies, often owned by conventional food processors. The parent companies of organic brands rarely use their name in advertising their organic subsidiary. As expert marketers, the giant food corporations understand that organic consumers do not trust them. And for good cause.

As the industry became more consolidated, the companies' ability to weaken the organic standards increased. The Organic Trade Association, founded in 1985 to lobby on behalf of the industry, increasingly came to be controlled by the largest players in the industry—companies that had no commitment to the spirit of organic agriculture. In October 2005, at the behest of its large corporate members, the OTA successfully lobbied Congress to weaken the standard by allowing the use of nonorganic and synthetic additives in foods labeled organic.

Eden Foods, the oldest organic food company in the nation, decried the violation of organic principles. Their Web site explains: "In a back room deal, the Organic Trade Association lobbied Congress to legalize the adulteration of organic food with basically any toxic additive a manufacturer may want to use, including substances that do not need to appear on ingredient panels. . . . [A]gribusiness influences prevailed."[5]

Since 2005, controversy over putting synthetic additives into organic food has continued to mount. The NOSB, increasingly controlled by large economic interests, has weakened the standard by designating synthetics as non-synthetic and approving materials that are clearly problematic, like the use of the antibiotics tetracycline and streptomycin in fruit production.

Sodium nitrate exemplifies the type of chemical that should never have been approved for use in organics. It is a quick-fix fertilizer that boosts plant growth, but in the long term it has a detrimental effect on soil fertility, and therefore on soil quality—a violation of a central principle of organic agriculture. The fertilizer is mined in Chile, and so not only has to be transported a long distance, but also leaches easily out of soil, creating the potential for water pollution.

In another recent dispute at the NOSB, the OTA successfully lobbied to continue designating corn-steeped liquor (CSL) as a non-synthetic fertilizer. CSL is a synthetic chemical by-product of producing high-fructose corn syrup, corn starch, and ethanol, and it is created through the use of sulfur dioxide during the wet milling process of corn. Several NOSB members issued a minority position statement: "The issue of determining whether CSL is synthetic or non-synthetic may appear to be a technical issue for experts to decide, of no interest to the organic consumer. But the synthetic/non-synthetic determination really is a foundational issue in the determination of allowable inputs in organic production."

Even more controversial has been the inclusion of synthetic additives in organic infant food, especially since more than one hundred infants have had adverse reactions to these products. Additives provide a marketing advantage for differentiating products and for justifying higher prices. According to Marion Nestle, professor of nutrition at New York University, this inclusion is all about competition for market share. She notes, "Even if the health benefits are minimal or questionable, they can be used in advertising."

The Cornucopia Institute filed legal complaints against infant formula companies and Horizon milk, owned by Dean Foods, for including unapproved additives in dairy products. Among the products that contain the synthetic fatty acids omega-3 (DHA) and omega-6 (ARA) are Earth's Best (Hain Celestial) and Similac Advance Organic (Abbott Laboratories). The synthetic fatty acids are manufactured by Martek Biosciences and created using hexane during the fermentation of algae and fungus. Hexane is also widely used in the production of other soy-based organic foods, including veggie burgers and protein bars. Hexane, listed as an air pollutant by the EPA, is a highly explosive by-product of gasoline production, as well as a neurotoxin. Grain giants Cargill, ADM, and Bunge, as producers of soy products, are major emitters of the chemical.

In November 2011, the NOSB voted to allow the use of DHA and ARA, as long as hexane is not used in manufacturing them. Although hexane use had never been approved in organics, manufacturers have used it for several years, because the USDA has never taken any enforcement action. The compromise vote by the board, allowing the use of DHA and ARA if hexane is not present, could create more problems, because it opens the door for even more dangerous solvents to be used in manufacturing.

Cargill also produces "enhanced" ingredients for processed organic food. Using its deep pockets and fourteen hundred food scientists, the corporation has been able to develop a wide range of additives that provide companies the opportunity to market their products as a healthier choice. Cargill uses hexane and acetone in producing soy lecithin—an ingredient included in a range of organic foods. The company also produces xanthan gum, an approved ingredient for organic food that acts as a stabilizer that is created from corn syrup through a fermentation process. Cargill also produces citric acid from the by-products of processing corn, using sulfuric acid in the final step of production. Citric acid is widely used in processed foods, including in organic foods and beverages, as a preservative and flavor enhancer.

It is not surprising that Cargill has teamed up with the large organic food company Hain Celestial, which uses Cargill's "enhanced" ingredients in processed foods. In 2003 the companies announced a strategic alliance. Irwin Simon, chairman, president, and chief executive officer of the Hain Celestial Group, said in the joint press release, "By combining our expertise, we bring two industry leaders together to focus on the development of the next generation of functional beverages, one of the hottest categories serving today's health-conscious consumers." Ted Ziemann, president of Cargill Health & Food Technologies, added, "We are excited about the opportunities to explore new formulations with our family of ingredient brands." Among the Hain Celestial products that use Cargill ingredients are Rice Dream, Soy Dream, and WestSoy nondairy beverages.[6]

Hain Celestial, the second-largest natural and organic food company in the country, was originally founded in 1926. It was reorganized in 1993 by Simon—who had honed his management skills at Slim-Fast and Häagen-Dazs—and has been expanded through more than a dozen mergers and acquisitions. Today Hain earns half a billion dollars in revenue annually and has fifty food and personal care brands. While Hain's top customer is still Whole Foods, 70 percent of its products are mass-marketed at grocery chains and other outlets.

In 2010 mass-market retailers sold 54 percent of organic food. This figure does not include organic private-label sales—products sold under a brand name owned exclusively by a retail store—a growing segment of organics. Products sold under store brands are not unique to the store, because they

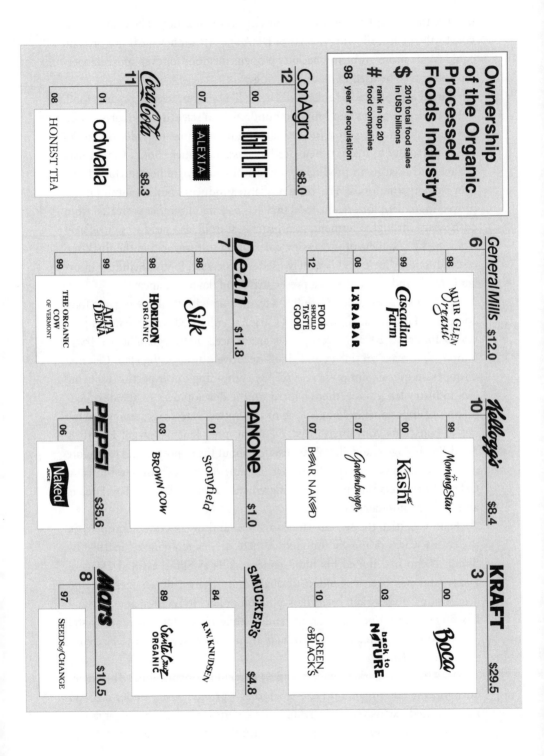

Ownership
of the Organic
Processed
Foods Industry

$ 2010 total food sales in USD billions

rank in top 20 food companies

98 year of acquisition

ConAgra $8.0
00 LIGHTLIFE
07 ALEXIA

12

Coca Cola $8.3
01 odwalla
08 HONEST TEA

11

General Mills $12.0
98 MUIR GLEN Organic
99 Cascadian Farm
08 LARABAR
12 FOOD SHOULD TASTE GOOD

6

Dean $11.8
98 Silk
98 HORIZON ORGANIC
99 ALTA DENA
99 THE ORGANIC COW OF VERMONT

7

Kellogg's $8.4
99 MorningStar
00 Kashi
07 Gardenburger
07 BEAR NAKED

10

DANONE $1.0
01 Stonyfield
03 BROWN COW

PEPSI $35.6
06 Naked JUICE

1

KRAFT $29.5
00 Boca
03 back to NATURE
10 GREEN & BLACK'S

3

SMUCKER'S $4.8
84 R.W KNUDSEN
89 Santa Cruz ORGANIC

Mars $10.5
97 SEEDS of CHANGE

8

are manufactured by companies that supply their product to multiple out-lets. Information about private-label products is usually not disclosed by retail chains, so sales figures and other data are not available. Requests about private-label manufacturing made to the manufacturers of organic food dur-ing the course of writing this book netted little information. General Mills did respond that its organic subsidiaries do provide private-label products.

Dairy is the organic sector most strongly affected by private-label sales. Nearly 75 percent of the conventional milk market is from private-label sales, and organic milk is following the same path. Dean Foods, the largest organic company in the country, owns Horizon Organic milk and Silk, the largest producer of soy milk. As the largest conventional milk producer, Dean, which provides conventional and organic private-label milk, controls almost 40 per-cent of the conventional milk market. Aurora Organic Dairy, widely criticized for its large factory farms, produces private-label milk for Safeway and Costco.

Earthbound Farm, the fourth-largest organic company in the country, produces one hundred varieties of organic fruits and vegetables—some for private-label brands. According to the company's president, Charles Sweat, Earthbound is increasingly providing fresh-cut private-label produce. Besides packaging salads, he notes that other "private-label commodities" include romaine hearts, celery, broccoli, and cauliflower. Sweat said that he expects a rapid growth of private labels in "commodity vegetables" over the next four years, with private labels becoming a larger part of Earthbound's business.[7]

Trader Joe's, a secretive retailer with $8.5 billion in revenues and privately owned by a German family, pioneered the private-label phenomenon, but the store releases virtually no information about its operations or products. According to an article in *Fortune* in August 2010, the company, with 344 stores in 25 states, requires suppliers to sign a nondisclosure agreement. The chain typically carries four thousand products, 80 percent of which are store brands. This results in the average store selling $1,750 in merchandise per square foot, almost double the ratio of Whole Foods, allowing the company to have no debt and enabling it to fund growth from its own coffers. Among the companies producing products for Trader Joe's are PepsiCo and Danone's Stonyfield Farm.[8]

Natural food stores like Trader Joe's account for 39 percent of organic food sales. But within this category, Whole Foods Market dominates the market. Since acquiring its competition over the past twenty years—including Well-spring Grocery, Bread of Life, Bread & Circus, Food for Thought, Fresh Fields, Mrs. Gooch's, and Wild Oats Markets, among others—the chain has had no national competition. It began inauspiciously in 1978 as a small natural food store in Austin, Texas, that college dropout John Mackey and his girlfriend Renee Lawson Hardy called Safer Way Natural Foods. Two years later they formed a partnership with another natural food store, creating Whole Foods Market. In 1992, after aggressively acquiring other stores and chains, the company went public.

Whole Foods' most contentious acquisition was of its closest competitor, Wild Oats Markets. It was challenged by the Federal Trade Commission for violating federal antitrust laws, but after two years of legal wrangling the deal was blessed. But Mackey's reputation was damaged by the FTC discovery that over a number of years he had been using an alias on the Yahoo! investment message board in an attempt to boost Whole Foods Market's reputation and to trash its competitors, especially Wild Oats.

Today Whole Foods Market, with 311 stores and $9 billion in revenues, caters to a high-income clientele willing to pay a premium for food perceived to be of a higher quality and for the stores' ambience. Mackey, a self-proclaimed libertarian, has pursued a fiercely antiunion strategy. His philosophy is summed up by his August 2009 editorial in the *Wall Street Journal*, damning health care reform. He wrote that "the last thing our country needs is a massive new health-care entitlement. . . . A careful reading of both the Declaration of Independence and the Constitution will not reveal any intrinsic right to health care, food or shelter. That's because there isn't any. This 'right' has never existed in America."[9]

Andrew Wolf, managing director for BB&T Capital Markets, says, "Whole Foods has been an acquisition story more than a new-store growth story." He notes that Trader Joe's is no longer a threat, because Whole Foods has chosen to challenge Trader Joe's selectively on prices in categories where the latter has maybe two items and Whole Foods has ten, by matching Trader Joe's on the two shared items. By maintaining a higher market price on the other eight items, it has "taken away some of Trader Joe's value advantage through category management." Wolf says that Mackey is very competitive and adaptive—for instance, he doubled sales per square foot at Whole Foods when competing with Wild Oats.[10]

Advocates criticize Whole Foods for one of the company's most profitable practices: it primarily sells conventional foods under the false illusion that they are better than products sold at conventional grocery stores. When shopping in the rarefied atmosphere of a Whole Foods Market, consumers falsely perceive that they are buying products screened to be healthier or even organic, and will pay a higher price.

Ronnie Cummins, national director of the Organic Consumers Association, is one of the company's biggest critics. He charges that the store has "row after row of attractively displayed but mostly nonorganic 'natural' foods" that are marketed by a "sleight of hand" to seem "almost organic." Truth be told, most of the products sold at Whole Foods Market, even its best-selling private label, the 365 house brand, are not organic.[11]

Consumers who buy meat at Whole Foods are paying a premium for a product they definitely believe is better. In February 2011, the retail chain announced a five-step animal-welfare rating system for meat products that would be managed by an independent third-party certifier: the Global Animal Partnership (GAP), a nonprofit organization actually founded and primarily funded by Whole Foods. In this less than honest five-tier rating system, Whole Foods Market sells the bulk of its meat from the lowest tiers, giving a false illusion about its level of commitment to animal welfare—a top priority of many of its customers.

Andrew Gunther, formerly employed by Whole Foods Market in the UK as the senior global specialist for "animal compassionate product procurement and development," confirms: "GAP has a very low threshold for entry into their program, allows feedlots, and does not audit slaughter plants. It has transportation standards that allow cattle to be hauled for up to twenty-five hours without rest, feed, or water." Currently the program director at Animal Welfare Approved—the nonprofit certification program started by the respected advocacy organization Animal Welfare Institute—Gunther notes that the GAP program is confusing to consumers and gives credibility to a production system known to be welfare-negative.

Gunther says that one of the most disturbing aspects of the program is the marketing of the five-tier rating system. The publicity pictures show bucolic photographs of cattle and chickens ranging and foraging in farmyards and on pasture. He says, "This is deceitful, because the vast majority of the animals sold under the GAP program are grown conventionally: caged hens with no access to pasture or to the outdoors, and cattle raised in vast feedlots." Gunther argues that this is why we need "truly transparent and independent standard setting, rather than programs that just give industry a pat on the back."

One of the least transparent aspects of the organics industry is distribution, the supply chain that enables manufacturers, especially smaller companies, to sell their products to stores across the nation. Whole Foods' primary supplier since 1998 has been United Natural Foods, Inc. (UNFI), the largest publicly traded company involved in purveying organic and natural food in the nation. The two companies recently extended the contract making the UNFI the distribution company for Whole Foods through 2020.

Although Whole Foods accounts for 35 percent of UNFI's sales, the distributor also services independent natural food stores and co-ops that have few other choices for obtaining organic or natural products, because of the consolidation of distributors. UNFI maintains warehouses that stock products that are then distributed to stores. Small manufacturers are dependent on contracts with the distributor to get their products to market. Conversely, small retailers often have to pay a premium price for products because of their dependence on this distributor. The tiny number of companies in distribution—and their ability to charge high prices to retailers—is one of the chief reasons that many food co-ops and small natural food stores have been driven out of business. They cannot compete.

During the 1970s and 1980s, a variety of distribution companies served a diverse number of mostly small natural food stores, buying clubs, and food co-ops. In 1983, there were twenty-nine cooperative-owned distribution companies, twenty-three regional distributors, and one national distributor. Now the little-known UNFI dominates the natural and organic distribution business. Its smaller competitor, Dutch-owned Tree of Life, which has offered a more narrow range of products, was purchased in 2010 by the specialty food distributor KeHE Distributors. According to *Nutrition Business Journal*, "The acquisition upsets the balance of power in retail. In a smaller universe, manufacturers have fewer options for taking their products to market."[12]

Since its founding in 1977, UNFI has grown from a single natural food store to a giant supplier of organic and natural products, supplying more than 6,500 stores in forty-six states. Norman A. Cloutier quickly refocused his business on distribution for the quickly growing organic industry, pursuing acquisitions and mergers of regional distributors, including Stow Mills, Albert's Organics, Blooming Prairie, Northeast Cooperatives, Select Nutrition, Roots 'N Fruits, Millbrook Distribution Services, and others. The most significant merger was in 1996, with the large western distribution company Mountain People's Warehouse. UNFI, which went public in 1998, has been able to underprice the remaining local or regional distributors.[13]

While UNFI is primarily a distributor of wholesale products, it owns twelve retail stores through a subsidiary, Earth Origins Market. Located in Florida, Maryland, and Massachusetts, UNFI uses these outlets to try out new products and marketing and promotional programs before selling these services to its customer base.

Nature's Best, the largest privately owned natural and organic food distribution company, competes head-on with UNFI in a dozen states west of the Rockies. It serves approximately two thousand stores as a wholesaler/distributor of health and natural food products, including grocery, refrigerated, frozen, and bulk foods; supplements; and personal care, herbal, medicinal, and pet products.

Russell Parker, a senior vice president of procurement and marketing at Nature's Best, says, "Before Tree of Life left the West, we had to compete with UNFI and Tree of Life every minute. It kept us sharp. Now there's only one evil empire we have to keep our eye on." Parker adds: "We are very sober about UNFI. But because we are so small, we can change instantly." [14]

To be competitive, the company recently consolidated its operations into one large building, cutting labor costs by 40 percent. According to Brian McCarthy, senior vice president of operations: "In order to maintain our service, we were deploying more and more people. . . . We were making price concessions in order to survive." [15] Parker has said that Nature's Best is "open to the possibility" of acquiring another distributor, although "there aren't that many opportunities out there." But he says the privately owned company would not consider a buyout offer. [16]

One of Nature's Best's biggest clients is Phoenix-based Sprouts Farmers Markets, a chain that operates in more than a dozen states in the Southwest, going up against Whole Foods in some locations with smaller stores. Over the past two years, since it was acquired by an equity firm, Apollo Global Management, the chain has grown significantly. Sprouts was merged in 2011 with Henry's Farmers Market, a California natural food store owned by Apollo, creating one of the largest natural food grocers in the western United States, with ninety-eight stores and annual revenues in excess of $1 billion. In March 2012, Apollo merged Sprouts with Sunflower Farmers Market—boosting its operation to 139 natural food stores. [17]

The company is majority-owned by funds affiliated with Apollo and operates under the banner of Sprouts in Nevada, Utah, New Mexico, Oklahoma, California, Arizona, Colorado, and Texas. The merged company employs ten thousand people and plans to open thirteen new stores during 2012. A senior partner at Apollo, Andrew S. Jhawar, commented on the merger: "We feel incredibly fortunate to be able to bring together the management and operations of these two growth-oriented grocery. . . . This is a combination that makes great sense given the rapid growth in demand for natural and organic products and the complementary nature of the geography of the two companies." [18]

Apollo is a publicly owned equity firm that owns entertainment companies,

real estate operations, retail stores, and fast-food restaurants, and it has historically specialized in leveraged buyouts and purchases of corporations that are
being restructured after a default or bankruptcy. No one can doubt that selling
organic and natural food has become extremely profitable with the entry of
private equity firms into the business. The deals are made with expansion and
further consolidation in mind, facilitated by the infusions of large amounts of
capital to open new stores and maintain and upgrade old ones.

For instance, in April 2012 the private equity firm Oak Hill Capital Partners acquired an 80 percent interest in Earth Fare, an organic and natural food
store based in North Carolina with twenty-five stores in the Southeast and
Midwest. The deal, which uses Nature's Best as a distributor, was worth about
$300 million. Another equity firm, Monitor Clipper Partners, had acquired
the chain in 2006; it maintains a significant minority interest in Earth Fare.[19]

Tyler Wolfram, a partner at Oak Hill, said: "Earth Fare is well positioned for
expansion given robust consumer demand for natural and organic food. . . .
We look forward to working closely with Jack Murphy [Earth Fare's CEO]
and the rest of the management team to support Earth Fare's next phase of
growth."[20]

Unsurprisingly, as organic and natural foods trend upward in profitability, Walmart has gotten into the action. The behemoth, known for its
extremely efficient distribution network, announced in 2006 that it would
expand the number of organic products it sells and price them at 10 percent
above conventional products, virtually guaranteeing a further cheapening of
the organic market by increasing its dependence on industrial agriculture,
processing, and distribution.

Walmart, as usual, had done the numbers. The move into organics was
part of a marketing strategy to attract wealthier consumers and to improve its
image. While some food advocates cheered Walmart's conversion to organics, others could see the writing on the wall—one more nail in the coffin of
organic food's integrity, as the supergiant pushed down its cost and quality.
The processed-food industry rejoiced. Companies began introducing organic
versions of the foods they already produced—organic Keebler crackers, Ragu
pasta sauce, and Nabisco's Oreos. Mars introduced Dove Organic Chocolate,
exclusively for sale at Walmart.

When Walmart talks about organic, it is different from what many consumers expect from organic products. It means partnering with big food
companies making organic versions of the processed foods that are already
on Walmart's shelves—like Rice Krispies and Kraft Macaroni & Cheese—processed foods made "organic" by replacing high-fructose corn syrup with cane

sugar and removing preservatives. The executive in charge of perishable food at Walmart admitted that the move into organics is simply a merchandizing scheme, and that "organic agriculture is just another method of agriculture— not better, not worse." [21] Walmart is not concerned about the principles behind organic agriculture, and will accept the bare-minimum requirements for what it takes to make a product fit the USDA's organic labeling requirements. Milk is a good example.

Organic Valley, the largest organic dairy co-op in the country, stopped selling its products to Walmart because of the pressure to cut prices. The farmer-owned coop supplied Walmart for approximately four years, selling 1.3 million gallons of fluid milk to the company each year. Yet this represented only 3.6 percent of Walmart's total milk sales for that year. Organic Valley decided that it no longer wanted to be a supplier to Walmart after a shortage of organic milk meant that it would have to short other customers in favor of Walmart. Organic Valley executives feared that they would become so reliant on Walmart that the box store could force them to lower prices, and thus pay their dairy farmers less.

When Organic Valley was considering ending its supplier agreement with Walmart, Dairy giant Dean Foods was waiting to get in line. Dean's Horizon organic milk brand, produced on factory farms, was willing to offer lowball prices to Walmart—prices so low that Organic Valley could not begin to compete. The company knew that Walmart was interested only in price, not in whether the dairy farmers could have smaller herds, use better practices, and still make a living.

Walmart's own private-label organic milk brand, provided by Aurora Organic Dairy, is produced by cows living on factory farms—thousands of cows raised in tight conditions without adequate access to pasture. The cows eat predominately grain during all but two months out of the year, violating organic principles and the actual standard that requires the animals to have access to pasture.

Aurora Organic Dairy was formed by the founders of Horizon, Mark Retzloff and Marc Peperzak, specifically to supply private-label organic dairy products. Aurora is vertically integrated from the production of milk to the bottling stage, and it uses scale to keep prices low. In 2007, after the Cornucopia Institute filed several legal complaints against the company, the USDA investigated its operations. Aurora was sanctioned for fourteen "willful" violations of federal organic law, but the Bush administration stepped in to keep the company from being decertified.

CEO Retzloff says, "We are firm believers that there is a place for larger scale operators in the organic sector." He goes on to say, "So for those who criticize and attack Aurora Organic Dairy for being larger scale, we can only point to the many benefits scale brings to animals, people, and the planet, and say that organic is simply a better way than how the other 98 percent of dairy products are being produced in this country."[22]

Retzloff misses the point: organic agriculture has always been about appropriate scale—and principles. Dairies with four thousand to five thousand cows like Aurora defy the basic tenets of land stewardship, ecological balance, and animal welfare. The monopoly control of dairy and soy milk by Dean Foods, a company that also controls a large percentage of conventional milk production, is contrary to the original vision of organic dairies that sustain the farm family and strengthen the local economy. The globalized corporate model of organic yogurt production that requires the use of powdered milk produced on the other side of the world epitomizes the weaknesses of the globalized corporate model of organic production.

The same is true of the vast acreage of organic fruits and vegetables produced in a few counties in California by large, heavily capitalized corporations that can outcompete local organic growers around the country. Investment by equity funds in corporations like Earthbound provide the capital for expansion and an expectation for high returns on investment. The availability of subsidized water makes it possible for Earthbound's industrial-size operation to grow lettuce, fruit, and vegetables on thirty thousand acres to be shipped across the nation.

Corporations that source globally can provide the large quantities of organic fruits and vegetables required by Walmart and other larger players in the grocery industry. Lower labor costs and weaker environmental laws make organic farming operations in the developing world profitable even with high transportation costs. Ingredients for organic processed foods can often be procured at a better price outside of the United States. And since large multinational corporations are focused on profit, the riskier process for organic certification overseas is not a major concern.

The largest multinational food corporations entered the organic market not because of commitments to its principles but because they saw a profitable niche market for products that can be sold at a premium. The purchase of organic companies by General Mills, Kraft, and Kellogg's creates the pressure on them to continually increase quarterly profits and to develop new products. Organic Oreos and other processed organic food—high in calories and low in nutrition—do not meet the spirit of organic principles. These multinational firms lobby for the use of synthetic additives and nonorganic ingredients. This might be expedient for the industry, but it signifies a loss of integrity for the definition of organic. It is unfair to those who spent several decades building the organics industry and to the many family farmers who adhere to the best organic practices.

Consumers of organic food pay a premium price because they believe in the original underlying values: local, ecological, chemical-free, healthy, and produced ethically. People are willing to pay more for organic, because they believe that environmentally friendly practices and the additional labor required to replace chemicals for weed control cost more. Once they catch on, consumers will not be willing to pay a higher price so that profits can be siphoned off to increase the earnings of multinational companies that have no commitment to the spirit of organics. It is time to correct the dishonesty and lack of integrity that kills the golden goose.

This will require food advocates to move beyond the hackneyed explanations for the cost and availability of organic food. It means addressing the control of the industry by Walmart, the large processed-food companies, and equity firms. Reforming organics is part of the larger battle to address the monopolization and control of our food system by multinational corporate interests. We must reengage people in building the political power to strengthen the organic standards so that they encompass the original vision for a system of high ethics and sustainability.

PART IV

Deregulating Food Safety

The multinational meat industry is continuing to use its power and influence to deregulate food safety and lobbying for the use of unsafe chemicals and drugs in producing animals. Food poisoning from tainted meat has increased greatly since the industry dramatically increased slaughter line speeds and lobbied successfully to reduce the role of federal meat inspectors. Rather than using safe processing techniques and ensuring cleanliness, the industry is substituting chemicals and technologies such as irradiation to remove deadly bacteria. Further increasing the negative impacts of industrialized animal production, the largest companies are continuing to use antibiotics to increase growth on their factory farms, leading to the creation of antibiotic-resistant bacteria. Public health communities are continuing to sound the alarm about the consequences for human health if the efficacy of antibiotics is compromised. Industrialized animal production must end as part of creating a safe and healthy food system.

6

POISONING PEOPLE

A government, for protecting business only, is but a carcass, and soon falls by its own corruption and decay.

—Amos Bronson Alcott (1799–1888),
Transcendentalist writer and philosopher

It was exciting. Bill Clinton had been elected president and the future held promise for tackling the regulatory rollbacks that had changed the face of America over twelve years of Republican rule. But despite the positive rhetoric, it quickly became clear that Clinton had no intention of changing the privatization and deregulation agenda that had swept the country. And no place was this clearer than at the U.S. Department of Agriculture (USDA).

In January 1993 Clinton appointed the first African American secretary of agriculture to head an agency notorious for racial discrimination. Advocates wished they could celebrate the momentous event, but Mike Espy was no progressive. As a fellow member of the conservative Democratic Leadership Council, Espy was well acquainted with Clinton, and he was one of the first congressmen to endorse him. But Espy's track record was not comforting. He had appeared in ads for the National Rifle Association, spoken in favor of the death penalty, supported Reagan's policies on funding the Nicaraguan Contras, voted for Republican-sponsored budget cuts, and voted against environmental regulations.

Late in January 1993, just as Clinton was entering the White House, America's biggest food safety scandal ever was making big headlines. Jack in the Box, the fifth-largest hamburger chain in the country at the time, had served contaminated meat, sickening six hundred people and killing three toddlers. *E. coli* 0157:H7 was the culprit.

From the food industry's point of view, Espy was an ideal candidate for the job of deregulating food-safety regulations under the guise of reform. The meat industry had long wanted to get USDA inspectors out of the way, because if they saw contamination, they could stop the line and thereby cut industry profits. If global food trade was to proceed, a new system was necessary to reduce government inspections. The United States had the most vigilant system of inspecting meat in the world, and it was slowing down "progress." While deregulation had long been in the works, the hamburger that poisoned consumers in the Pacific Northwest was both a public relations challenge and an opening for changing the inspection system.

The meat-and-food-processing industry was among the biggest proponents of "free trade," and it needed to "harmonize" food standards with other countries to prepare for the globalization it was lobbying for through trade agreements. The vertically integrated meat companies saw both the potential for new markets in the developing world and locations for factory farms where environmental regulations were scant. Clinton and Espy, both fans of deregulation and liberalizing trade, were easily persuaded that the harmonization the meat industry wanted was necessary. The stage was set to make big changes, and together Clinton and Espy represented a dangerous combination.

Of course, it is improbable that either man really meant to create a system that would lead to food poisoning for thousands of Americans. It was a case of blind faith in technology and a belief that industry would put public health above private profit. However, neither man knew much about food safety. In fact, Espy had not set out to head USDA—he had hoped to be secretary of health and human services. And while he didn't know much about food safety issues, he knew a lot about deal making.

Clinton, as governor of Arkansas, the state where Tyson Foods is headquartered, had firsthand experience with the political power of big poultry. CEO Don Tyson had supported him in his first run for governor in 1978. However, when Clinton tried to make some progressive reforms as governor, including raising fees on big truck rigs, Tyson wasn't happy, and Clinton lost his support for the 1980 reelection campaign. But Clinton was a quick study. He learned his lesson about trying to bring moderate reform to Arkansas, and especially with anything affecting poultry. Billionaire Don Tyson helped reelect him in 1982, and supported him during his next ten years as governor and in his presidential campaigns.

Hillary Clinton had her Tyson ties as well. James B. Blair, a family friend and an attorney for Tyson Foods, placed many of the cattle futures trades that netted Hillary both hundreds of thousands of dollars in profit and a scandal when she became first lady.

It would be naive to think that Bill Clinton—the advocate for globalization—had not been lobbied by the poultry industry about replacing USDA meat inspectors with another system that relied on industry self-inspection. Clinton came into office with a free trade agenda, and under his watch, besides the passage of landmark trade legislation, a new food safety system was to become the law of the land. That new system, with an obscure and hard to remember acronym, is called the Hazard Analysis & Critical Control Points system (HACCP).

Yet the story of how the meat-inspection system was weakened did not begin with Clinton or Espy. It began in 1959, when NASA contracted with Pillsbury to feed U.S. astronauts in a way that would not endanger their health through food poisoning. Howard Bauman, chief food scientist at Pillsbury, teamed up with the army's Natick Laboratory to create a program that eventually morphed into the deregulated meat inspection program now known as HACCP.

Bauman and his colleagues originally designed a commonsense approach to eliminating food safety problems methodically and developed a system that, if used properly, would identify the critical processing points at which food safety problems could be monitored, verified through microbial testing, and remedied. "Treatments," such as irradiation or chlorine, would be used to eliminate the bacteria.

Irradiation had emerged as a solution to bacterial contamination because the army's Natick Labs were simultaneously being funded to promote the technique—blasting food with radioactive elements like cobalt-60 and cesium—as an easy way to preserve food for soldiers. It was at the behest of the army that, in 1963, the Food and Drug Administration (FDA) approved the irradiated bacon that became a staple of the military diet. But it was revealed that lab animals fed irradiated food suffered numerous health problems including premature death, a rare form of cancer, tumors, reproductive problems, and insufficient weight gain. In 1968 the FDA rescinded the army's permission to serve irradiated bacon to military personnel.

Yet by the mid-1980s the technology had become largely accepted, after almost thirty years of funding and promotion by the federal government. Today, new technologies that do not use radioactive isotopes are also promoted for irradiating large volumes of food, and the globalized fruit and vegetable industry has jumped on the irradiation bandwagon because the technique increases shelf life. The USDA's Animal and Plant Health Inspection Agency promotes irradiating imported produce to keep invasive insects, like fruitflies, from reproducing.

It is doubtful that even Pillsbury's Bauman, a proud technocrat, could have foreseen the future applications of irradiation or the quality control system that he developed. He surely believed that the system, not yet named HACCP, would be used to improve food handling, and that astronauts would be safe from food poisoning as they ate irradiated strawberry and peanut cubes, non-crumb cake, and rehydrated spaghetti.

In the spring of 1971, Bauman's expertise in solving food safety problems was called upon by his boss to help solve a crisis. Pillsbury's farina cereal for babies had been found to contain shards of glass, creating a public relations nightmare. In response to the ensuing scandal and recall, Pillsbury announced a new food safety system based on Bauman's work at Natick. Soon afterward, a scourge of botulism poisoning from low-acid canned food began plaguing the canning industry.

The FDA needed to take action. The well-publicized food safety program initiated by Bauman at Pillsbury attracted its attention, because the agency was looking for ways to deal with large food-manufacturing facilities— something Congress had not envisioned when giving the agency its mandate.

Bauman was asked by the FDA to hold a training course for the canning industry in 1973, which was called "Food Safety through the Hazard Analysis and Critical Control Point System," the first use of the HACCP terminology. That same year, the FDA, which regulates all foods except meat, poultry, and processed eggs, wrote regulations requiring that the canning industry use a HACCP-type approach for low-acid foods.

While the botulism poisonings of the early 1970s were tragic, the incidents were easy to document and the solution was easy: low-acid vegetables must be pressure-canned to kill the disease-causing bacteria. But the health crisis that unfolded twenty years later, in the winter of 1993, created out of Jack in the Box's poisoned hamburger crisis, was a whole new kind of problem. It struck at the heart of American culture, putting a food most people thought of as an American icon in a whole new light.

A scary new type of bacterium had emerged called *Escherichia coli* or *E. coli* 0157:H7. The bacterium was first documented in 1982 creating a toxin that causes 5 percent to 10 percent of those exposed to it to become seriously ill with kidney failure—most often affecting the young and old—and in some cases to die. Cattle feces is the most common source of the bacteria.

While we do not know the origin of this strain of E. coli, we do know that the industrialized food system is responsible for its proliferation. Cattle spend their last three to four months crowded together in megasize feedlots, where they wallow in their own waste. Because they arrive at the slaughter facility covered in fecal matter, from the first step of killing the animal, throughout processing, fecal bacteria are dispersed in the meat product.

Also, the digestive tracks of bovines are suited for foraging on pasture, and for eating hay in the months that grasses do not grow. But, in contrast to their natural diet, while at the feedlot beef cows are fattened on a calorie-high diet of corn, soy, cotton meal, ethanol waste, and other ingredients that create fat-marbled meat. Compromising the meat supply even further are extremely fast slaughter lines. Large slaughterhouses can kill and butcher four hundred cows an hour, using extremely high-speed slaughtering methods, and as a matter of course some fecal material remains on the carcasses. Hamburger is especially vulnerable to contamination, because it is ground in enormous batches that contain parts from thousands of cows that originated in feedlots in multiple countries.

No wonder that the emergence of this deadly new pathogen sent the Clinton administration into a tailspin. While the Jack in the Box scandal was about beef, it didn't take too much analysis to realize that any changes to meat inspection rules would also affect poultry. But Clinton was walking a tightrope between the meat industry and public opinion. Overnight, people had become afraid of meat.

Three days into his new office, Mike Espy was hit by the crisis. He reported directly to Clinton on the outbreak. On February 6, 1993, he said that the USDA's Food Safety and Inspection Service (FSIS) "will prepare a 'revolutionary' strategy to create a meat inspection program more capable of combatting threats from a host of harmful bacteria." Among his prescriptions: "accelerate federal approval of irradiation for use on beef." [1]

Espy went on to announce that the USDA had asked the FDA to legalize irradiation of beef, veal, pork, and lamb. Poultry already had been approved for irradiation in 1990. As part of the American Meat Institute's push for irradiation, the industry's trade association had submitted a petition to the FDA asking that the agency legalize the technique's use for meat. Irradiation was the silver bullet they had been looking for since it would help erase the sins of dirty meat and fast slaughter lines. Irradiation was viewed by many in the industry as the ticket to reducing their liability under the self-regulatory regime of HACCP.

It is clear that the Clinton administration had HACCP in mind from the beginning. In the spring of 1994, Mike Espy interviewed Mike Taylor—master of the government-industry revolving door. Taylor had been deputy commissioner for policy at the FDA since 1991; he had written not only the deceptive-labeling guidelines that prohibit dairies from labeling milk rBGH-free, but also the HACCP guidelines for seafood. Seafood safety had become a major issue in the previous two decades, since the Centers for Disease Control (CDC) was reporting that almost 3 percent of food-borne illnesses were from seafood. HACCP was promoted to the seafood industry as a way to put the industry in charge of maintaining safety through improving its procedures for processing.

The seafood regulations mandated under the Clinton administration in 1997 were based on the HACCP guidelines that were first developed by Taylor in 1991. HACCP had been described by an official from the FDA Office of Seafood as "something that the industry would do, while FDA would examine how well these establishments were doing it."[2]

And Taylor—with his long history of bouncing between the industry lobby shop King and Spalding, agribusiness giant Monsanto, and the FDA—was a perfect choice to finesse morphing USDA's inspection into a self-regulating HACCP system. He was an experienced lobbyist who could talk smoothly to the industry and the consumer advocacy community with comforting words. As Espy began suffering a free fall from a corruption scandal, Taylor stepped in to do cleanup. During the next two years as administrator of USDA's FSIS and acting undersecretary for food safety, he became HACCP's biggest cheerleader.

However, Taylor had to convince consumer organizations that HACCP was the right program. He also had to create an environment for convincing the media (and through them the public) that the new system would prevent more Jack in the Box–like incidents. He handled this task skillfully by talking about the creation of a science-based food safety system, and in the fall of 1994 he told the American Meat Institute that *E. coli* 0157:H7 would be regulated as an adulterant in ground beef. This single act, along with the access Taylor granted consumer groups to high-level agency meetings, convinced many inside the Beltway that HACCP was the answer to food safety.

While Taylor strategized on how to pacify all sides of the debate, Espy was dealing with the corruption allegations. He had made some dangerous enemies in the beef industry, including a large feedlot operation that filed a lawsuit against USDA for unfair practices that were seen as favoring the poultry industry at the expense of beef. After it was revealed that Espy had solicited and received gifts from Tyson Foods and other companies that USDA regulated, he announced his resignation in fall 1994. An investigation culminated in a thirty-nine-count indictment against Espy for accepting $36,000 in gifts, for which a jury later acquitted him.[3]

Dan Glickman, a ten-term congressman from Wichita, replaced Espy as secretary of agriculture early in 1995. He was considered a shoo-in: he had spent eighteen years serving on the House Agriculture Committee, and he knew how to work an old boys' network. But Glickman's confirmation became heated, as investigators spent three months reviewing his parking tickets, credit card expenses, and bounced checks at the disgraced House Bank.

As Glickman began to take stock of his new job, HACCP promoter Taylor was already working behind the scenes to undo more than a century of meat inspection regulation. Taylor was either naive or cynical—he was about to trade inspection for treatments. The meat industry wanted inspectors out of the slaughterhouses. USDA inspectors had the power to stop slaughter lines when they saw contamination, and that cost money.

Meat production had changed dramatically because of technological advances allowing animals to be killed and processed at lightning speed. Just as electricity, expanded railways, and refrigeration revolutionized meat production in the mid-1800s, new technologies were dramatically changing the industry in the late twentieth century. The meat industry disparaged USDA inspectors for using an outdated "poke and sniff" method of inspection when they examined each and every animal carcass, slowing down the line. One small problem for the meatpackers was that federal law requires each animal carcass to be inspected by a USDA employee. But the industry was lobbying for a new system that put industry regulators in charge of safety. History was repeating itself.

The first meat inspection law that was passed, at the urging of meatpackers in 1884, was designed to counteract the bad press about filthy and diseased meat exported from the United States to Europe. Much like today, a handful of companies, called the Chicago Beef Trust, dominated the industry. Not only were they selling filthy and adulterated meat, but the trust used monopoly practices to keep farmers from selling cattle at a competitive price. Then, out of the blue, Upton Sinclair's novel *The Jungle* sent shock waves reverberating from coast to coast, as the muckraking press of the day wrote about the disgusting conditions in which meat was produced. Theodore Roosevelt sent investigators to Chicago. Once Sinclair's descriptions were confirmed, Roosevelt supported legislation for federal inspection. The Meat Inspection Act of 1906 mandated inspection of each animal carcass and is the basis of regulation and safety standards for red meats and pork today.

In 1957, as a result of explosive growth in the poultry industry, Congress passed the Poultry Products Inspection Act. In 1968, the Wholesome Poultry Products Act amended and strengthened the earlier law, mandating that poultry be inspected continuously, from slaughter through processing.

In contrast, the FDA's regulatory functions were established with the passage of the Pure Food and Drugs Act in 1906, which passed at the same time as the law mandating meat inspection. Originally part of USDA, the FDA in 1940 became part of the agency that is the Department of Health and Human Services today. While the USDA regulates meat and poultry, the FDA is responsible for shell eggs and all other foods except meat and poultry. However, the FDA has always been poorly staffed and had many fewer inspectors— approximately twelve hundred today, compared to the USDA's six thousand line inspectors and twelve hundred veterinarian inspectors.

From the 1970s onward, the meat industry's resistance to USDA inspection increased as technology made large, industrialized slaughterhouses possible. Lobbying during the Ford administration resulted in the commissioning of a study by the D.C.-based consulting firm Booz Allen Hamilton to make recommendations for changing the meat inspection system. The study was released during the Carter administration, and its focus was on improving cost-effectiveness and eliminating unnecessary interference with commerce. The study recommended cutting back on the role of government meat inspectors and encouraging "corporate quality control."[4] It also proposed creating a monitoring system that "places the burden of proof of compliance with Federal laws and regulations on the industry. It is their responsibility to provide acceptable evidence of this compliance."[5]

Stan Painter knows firsthand how the meat industry pressured elected officials and the agency to deregulate USDA meat inspection. Painter, born and raised in Alabama, speaks with a pleasing Southern inflection that belies his passion for protecting Americans from the ravages of food poisoning. As chairman of the National Joint Council of Food Inspection Locals (NJC), an affiliate of the American Federation of Government Employees (AFGE, the union that represents USDA meat inspectors), he has been fighting to protect the integrity of the inspection system for decades.

Painter has spent his career in slaughterhouses—mostly poultry. He began as a line worker and on the sanitation cleanup crew in a poultry processing plant that supplied meat to fast-food giant Kentucky Fried Chicken. After being promoted to quality control officer and eventually to supervisor of the plant, in the mid-1980s Painter was encouraged by a USDA veterinarian to apply for a position as an inspector with USDA's Food Safety and Inspection Service. He has worked for the agency since 1985.

Painter says that the pressure to deregulate meat inspection began early in his career as a USDA inspector. In 1985, during the Reagan era, the Gramm-Rudman-Hollings Act passed, mandating across-the-board budget cuts. FSIS personnel began to feel the pain of reduced funding in 1986, because 90 percent of the agency's budget was in salaries. According to Painter, the first thing the Reagan appointees to FSIS ordered in response to the cuts was to do away with inspectors at the beginning of the slaughter line. Those inspectors checked poultry for things that are gross and disgusting but that do not typically kill poultry or people.

Consumers would no doubt be shocked to know that, as a result, today they are eating chicken with external blemishes, tumors, cancers, and gaping wounds oozing pus. Since 1986, there is no USDA inspector present to check if a bird arrived at the slaughter line with a disease or was already dead from injury or disease on arrival.

Meanwhile, during Reagan's tenure, as pressure mounted around seafood safety, Congress gave the Department of Commerce and the National Marine Fisheries Service (NMFS) the task of developing a new model of safety inspection for seafood. They recommended that HACCP be used as a voluntary inspection program. In 1988, as the deregulation frenzy continued, the FDA, FSIS, NMFS, and the U.S. Army's Office of the Surgeon General joined together to create and fund the National Advisory Committee on Microbiological Criteria for Foods (NACMCF), which endorsed HACCP in 1989.[6] Serving on the NACMCF committee was Howard Bauman, originator of the HACCP concept.

During 1990, under President George H.W. Bush's administration, FSIS held consultations with industry, the USDA inspectors, and the public regarding HACCP. Five formal meetings, called "workshops," were held in 1991 and 1992 to develop the deregulated inspection program. The meat industry was given a prominent role in developing the proposed regulatory changes that were debated. The USDA meat inspectors and consumer groups were solemnly promised that HACCP would not replace inspection but would augment and modernize it, through further safety procedures and the addition of microbial testing.

Painter explains that they were told their role as inspectors wouldn't change with the adoption of the new program. But, he says, "the ink wasn't dry on the paper before the agency started making it a replacement."

By the time Clinton came into office, deregulation was well under way, and it had his blessing. In May 1993, Espy directed FSIS to initiate a rulemaking to establish HACCP in all meat and poultry plants. By the time of Glickman's arrival, the timetable for rewriting the regulations had been established.

During the summer of 1996, during barbecuing season, Glickman gleefully released a statement to the press: "President Bill Clinton announced a new food safety rule that will revolutionize meat and poultry inspections. . . . It is a complicated name for a simple idea. . . . HACCP puts safety first by putting prevention first."[7]

In October 1996, Glickman appointed Tom Billy to be the administrator of FSIS; he had been associate administrator since 1994. Before that time, he was at the FDA directing the Office of Seafood—where he worked with who else but Michael Taylor in implementing HACCP for seafood. By 1996, having performed his magic, Taylor was leaving USDA to go back to the lobby-shop law firm King & Spalding, where he had represented Monsanto before joining the FDA for the second time. He said in an agency press release that the implementation of HACCP "could not be in better hands."[8] Painter calls Taylor and Billy "partners in crime." They had worked together at the FDA on HACCP, and they were committed to deregulation. In his new role, Billy moved ahead to implement HACCP in meat processing.

USDA considers slaughtering and processing two different and separate operations. The 1996 regulations mandated HACCP for meat processing and recommended it for slaughter plants. Processing is the step after a bird is killed and gutted, which involves butchering, preparing, and packaging different cuts of meat, and it is carried out either in a different facility or in another part of a very large plant. Tom Billy was committed to seeing HACCP in all phases of meat production to line up with the trade rules the Clinton administration was pushing.

Billy served an important role in globalizing food trade through his positions at Codex Alimentarius, the agency that creates the internationally adopted food standards used by the World Trade Organization. Concurrent with his position at FSIS, Billy in the mid-1990s served as vice chairman of Codex, during the time that the standards for irradiation and HACCP were adopted. In 1999, he became chairman of Codex for four years, eventually leaving FSIS to work full time at Codex.

In 1997, as part of the international trade agenda, Codex released the "International Code of Practice: General Principles of Food Hygiene." As a result of the World Trade Organization phytosanitary agreement, which limits protective measures by individual member countries, HACCP became the safety system of the world's globalized food system. It replaced protective regulations with weaker ones that were based on the lowest common denominator. For instance, the rules were designed to allow a chicken processed in China or Mexico to be sold in the United States or in any other country in the world. The purpose of instituting HACCP became immediately clear: it put the industry in the driver's seat on food safety and meant that giant food companies could move processing to the developed world.

In the United States, the USDA inspectors very quickly renamed the acronym HACCP "have a cup of coffee and pray." Just as the meatpackers had hoped, it created a company "honor system" in which inspectors monitored plant records rather than inspected meat. If HACCP had indeed been adopted in addition to inspection and had required companies to identify critical points where contamination could occur, it would have been a positive move. If it had required the industry to develop sanitation plans that remedied the situation and allowed inspectors to use real-time microbial testing, the new system would have improved meat safety. But this was never the intention.

HACCP reduced the role of USDA inspectors and created a new paperwork function. Inspectors in processing were told not to stop the line for contamination, but to wait until the meat product reached the end of the line, where a "treatment" would take care of the problem. Treatments such as ammonia, chlorine, and trisodium phosphate were encouraged during processing, and irradiation was promoted as an "end-of-line treatment." And there was no way that inspectors could chase contaminated meat to the end of the line to see if it had been treated successfully.

Painter explains that in 1970 line speeds for poultry slaughter were forty-six birds a minute, while today they have advanced to 140 birds per minute and can be as high as 210 birds per minute in some large plants. Rather than act when witnessing a potentially hazardous situation, which slows down the line, inspectors are told "to let the system work," meaning maybe a later step will catch the contamination.

Not only does this create food safety issues as birds whiz by, it also creates a very dangerous work environment. Painter recounts several macabre safety incidents: a woman whose thumb was caught in the equipment and pulled off; a worker cut in half while cleaning the "chicken chiller"; several people killed from exposure to the carbon dioxide "treatment"; and a man ground alive when he fell into a grinding machine.

In June 1999, the Government Accountability Project (GAP), the nonprofit whistle-blower organization, designed a survey for federal meat and poultry inspectors who worked in HACCP plants. The survey results were incorporated into a report written in 2000 by GAP and Public Citizen called "The Jungle: Is America's Meat Fit to Eat," of which I was one of the authors.

The survey offered proof that HACCP had indeed become a system that took the inspectors away from the front lines of inspection. The inspectors documented that they could no longer take direct action against contamination, including preventing feces, vomit, and metal shards from entering the food system. Inspectors reported that they have been instructed not to document violations they have observed of company employees performing slaughter duties.

Among the quotes from the inspectors:

> "Instead of taking action immediately, we are instructed to 'let the system work.'"
>
> "It's a big paper chase . . . dot the 'i,' cross the 't.' That is all that counts."
>
> "Plant managers say the rule is—there are no rules! We [plant managers] write our own regulations."
>
> "Two sets of records are being kept by [the meat plant]; one set to show USDA inspectors (looks real good); and one set for their own use."
>
> "Many things go on—especially things on the floor; they just pick it [contaminated meat] up and put it in for human consumption."

While HACCP had been mandated only for processing, Painter says that USDA had always planned to introduce it into slaughter. By 1998, they had established a pilot program for slaughter called HACCP-based Inspection Models Project (HIMP). Its goal was to take inspectors off the line in slaughter facilities and have them review company records of inspection done by company employees. AFGE national president Bobby Harnage said, "This is a back-door attempt to change administratively what Congress would never consider changing legislatively—significantly weakening the entire meat and poultry inspection process. We're not going to sit back and let the USDA abdicate its responsibility to American consumers."[9] The AFGE filed suit against the USDA for breaking the law that stated that an inspector must examine each animal carcass.

Despite the lawsuit, in 1999, thirty plants began participating in the program to reduce the number of inspectors in slaughter plants. In 2000, the District of Columbia Circuit Court of Appeals ruled against the USDA, saying that HIMP violated federal law. The judges wrote in their decision: "The government believes that federal employees fulfill their statutory duty to inspect by watching others perform the task. One might as well say that umpires are pitchers because they carefully watch other throw baseballs." [10] In response, the agency redesigned the HIMP project to position an FSIS carcass inspector at the end of each slaughter line. The union appealed this action, but in 2001 the court found that the redesigned program met the statutory requirements.

Felicia Nestor has worked directly with USDA meat inspector whistle-blowers for the past fifteen years. It was never a career she set out to have. Nestor put herself through City University in New York as an art and music major by waitressing, and eventually worked as a photographer. She says that while taking professional photographs at the United Nations, she heard incredible people speaking about civil rights—an experience that inspired her to go to law school. While in Washington, D.C., finishing her degree at Georgetown Law School, she met someone from GAP, where she began as a volunteer photographer before joining the staff.

Nestor says she had no background in food safety, but when she first started working for GAP in the mid-1990s she had a completely open mind about the situation. She could see that the inspectors were raising red flags about many issues but that they had no support from consumer groups who went right along with the USDA administration.

Now, in retrospect, Nestor can see that everything the inspectors were concerned about has come to pass, and that the consumer groups that signed off on HACCP were completely wrong. She says in large part it was preju-dice against blue-collar workers that caused these groups to be persuaded by smooth-talking officials that HACCP would be better than having inspectors on the front lines.

Nestor states that nothing illustrates how the "consumer groups were snowed than when they signed on to HIMP." GAP and Public Citizen were the only groups to speak out against privatized inspection during slaughter—the period during which most contamination occurs. FSIS administrator Tom Billy engaged in a propaganda campaign focused on convincing advocacy groups like the Consumer Federation of America that the new system was based on new scientific methods, and that the inspectors were just afraid of losing their jobs.

Nestor laments the refusal to listen to the USDA inspectors: "Over the years, I've spoken personally with hundreds of concerned inspectors. None of them have been in danger of losing their jobs because of an agency pro-gram, but they do worry about the food that their parents, their children or grandchildren, and their neighbors eat. USDA depends on the uninformed, knee-jerk, antiunion, or anti–blue collar prejudice to push these deregulatory programs through."

She explains that USDA managers have never worked inside a meat or poultry plant and have had no experience with industry attempts to cut corners without being caught. The only experience they have inside the plants is an occasional dog-and-pony show that a company will put on, before which plant employees scour the plant from top to bottom, and during which managers run the production lines at about half the normal speed.

Nestor recounts her experience with a Gold Kist poultry plant that began operating under the HIMP pilot program. Everything blew up when she was able to convince the *Austin Statesmen*, and eventually Cox Newspapers, to cover the scandal, because the company had the contract for providing chicken nuggets to the school lunch program. Inspectors reported that birds were being slaughtered with a line speed of two hundred birds per minute, and that diseased birds with tumors, oozing wounds, and other health problems were being processed for schools around the country. When the inspectors informed the USDA chain of command of the problem, they were admonished, and nothing was done.

Under HIMP, contamination is removed through the use of disinfectants such as ammonia, chlorine, and trisodium phosphate. As Stan Painter says, the inspectors in those plants are "window dressing." Today, there are approximately thirty HIMP plants processing chicken, swine, and turkey.

Nestor also has written extensively on the problems with the microbial testing that the USDA promoted in order to gain support for the new system. She says the agency has consistently misled the public about its pathogen-sampling programs. She uncovered the fact that the USDA was not testing product daily for salmonella, as consumers had been misled to believe during the campaign to get public buy-in for the new program. Instead, most plants receive fewer than sixty tests per year. Nestor says, "The agency has been heavy on rhetoric about its 'science-based' programs, yet light on effective scientific methods, and obtuse, even deceptive about its practices."

GAP and Public Citizen's exhaustive, five-month review of USDA's own records, obtained under the Freedom of Information Act and published as *Hamburger Hell*, concluded that there was no factual evidence, based on the testing program, for USDA's reassurances that the food supply has become safer for consumers of ground beef. Using the agency's own test results, Nestor and co-author Patty Lovera found that the agency was taking less than 1 percent of all of its ground beef samples from the large plants that produced 85 percent of the raw ground beef supplies. It was taking 60 percent of the samples at the smallest plants, which produced less than 1 percent of all ground beef.

USDA gave the large plants a pass and blamed the smallest processors for contamination problems. As a result, in just the first few years of the new policy, USDA forced over 40 percent of the smallest grinders out of the market. To this day both the FDA and the USDA discriminate against small food processors. For instance, the FDA wastes resources patrolling for sales of raw milk (or cheese produced from the milk) that consumers buy directly from the producer, instead of using resources to deal with the major food safety issues that exist at large, industrial food-processing plants.

The USDA has never made it possible for agency inspectors to use the many new tools that are available for microbial sampling. Nestor, who continues to work with whistle-blowers, says that inspectors could easily detect sources of contamination prior to the food entering the market, but "their bosses just won't authorize them to do so."

Under George W. Bush's administration, Elsa Murano became undersecretary for food safety and deregulation took a new form: risk-based inspection (RBI). This program also was geared toward removing inspectors from the front lines of inspection, based on flawed microbial testing. Murano had run the Center for Food Safety at Texas A&M University and had begun her career working in the food irradiation center at Iowa State, along with her husband, Dr. Peter Murano. Elsa Murano, with cultlike trust in the technology, attempted to change the already weak irradiation label to read "cold pasteurization" to remove irradiation on the label.

During her tenure as undersecretary, Murano's husband was appointed deputy administrator for special nutrition programs at the USDA's Food and Nutrition Service. In that capacity, he was responsible for the National School Lunch Program, where he promoted the use of irradiation for the meat used in the program. Public Citizen launched a successful nationwide campaign that resulted in no school district ever purchasing irradiated meat, and in maintaining the word "irradiation" on the labels of irradiated food.

Irradiation was viewed by the Bush administration's USDA as the silver bullet for preventing food poisoning under the RBI system. In Bush's second term, the undersecretary for food safety, Richard Raymond, a medical doctor who had formerly been Nebraska's chief medical officer, began a series of daylong meetings with stakeholders to promote the newest deregulation scheme. In 2006, he called RBI the "natural evolution" of FSIS procedures. In essence, the proposed program would have removed meat inspectors from plants with good testing scores and focused inspections on plants with poor records. The program would be based on ranking meat products by inherent risk and using chemical washes and irradiation to destroy bacteria.

It was an outrageous proposal, because not only were the salmonella and *E. coli* testing records they planned to use flawed, but the testing had been done in small plants, many of which had since been shut down, rather than the large ones that produced most of the meat consumed by Americans. A September 2006 USDA inspector general report identified as many as 865 establishments nationwide that had no testing data for salmonella.

There is no doubt that part of the USDA's enthusiasm for RBI was resource-related. The agency told the media that it was considering allowing "virtual inspection" of plants—companies would e-mail records so that agency personnel could examine them without ever coming to the plant.

Fortunately, Congresswoman Rosa DeLauro, chairwoman of the House Agriculture FDA appropriations subcommittee, brought the scheme to a screeching halt. She was able to add an amendment to an Iraqi war supplemental budget bill that FSIS could not spend even another dollar of taxpayer money until the Office of the Inspector General audited the agency's inspection system, including the microbial testing. In 2010 the OIG spent six months auditing the program and issued part one of the report in March 2011. The report was very critical of the agency's plan and concluded that FSIS must thoroughly reevaluate its testing. A second phase of the investigation is in progress.

But perhaps nothing demonstrates the Bush administration's failure to put public health first more than mad cow disease, the common name for bovine spongiform encephalopathy (BSE). First identified in the UK in 1986, when ranchers noticed their cows getting sick and being unable to walk, the USDA has never fully dealt with the safety issues related to this frightening disease. Now present in Europe, Asia, and North America, BSE has killed more than one hundred people, forced farmers to preemptively kill millions of cattle, and devastated the beef industry in some countries. Scientists believe the disease is spread when cattle eat nervous system tissues, such as the brain and spinal cord, from other infected animals.

In 2008 President Barack Obama came into office facing a food safety scandal that was a result of Bush-era policies. School lunches had fed children meat from sick and abused cows that were at a higher risk of having mad cow disease. A meat plant in Chino, California, the second largest supplier of beef to the National School Lunch Program, was found to be serving meat from tortured "downer" cows, which are so sick or crippled that they cannot get up. Making the scandal even more sensational, under the Bush administration the company had been named the USDA "supplier of the year" for 2004–5 and had delivered beef to schools in thirty-six states.[11]

A Humane Society of the United States undercover investigator filmed workers at the midsize plant shoving cows violently with forklifts, using electric prods in sensitive areas, and employing other repulsive methods to make the diseased and sick dairy cows that are used for cheap meat walk through inspection. Wayne Pacelle, president of the organization, said that their investigator "found cows—in all stages of the handling and pre-slaughter process—being tormented to get them to stand and then walk toward the kill box."[12]

Foreshadowing the future lack of timely action, it took the new president a year to respond. Obama eventually banned the use of downer cows for meat, closing the loophole that the Chino plant was exploiting. While a step in the right direction, the move did not stop other practices that can spread mad cow disease, such as allowing cows to eat waste from the floors of poultry houses, cattle blood, and processed leftovers from restaurants. The administration is also not doing adequate testing for BSE in the United States and is allowing cattle in from countries such as Canada that have had reoccurring cases of the disease.

The failure to act quickly and decisively on important issues is an ongoing characteristic of the administration. Obama also waited until January 25, 2010, a full year into his presidency, to announce the nomination of Elisabeth Hagan, who had been chief medical officer at the USDA, as the permanent undersecretary of FSIS—a length of time much criticized by his opponents. Congress finally made her appointment permanent in September 2010. Hagan had been trained at Harvard and taught and practiced medicine before joining the senior staff of FSIS in 2006. Although the agency under Hagan has been somewhat more receptive to concerns of the advocacy community, Tony Corbo, lobbyist for Food & Water Watch's food program, which has been fighting for more stringent meat inspection regulation, assesses FSIS as follows: "The Obama administration has made some long overdue updates to the rules for meat inspection, like expanding the list of pathogens that are considered adulterants in ground beef. But they have not stood up to the meat industry strongly enough to slow the momentum toward deregulation that has prevailed for decades."

In the meantime, the FDA was in deep trouble dealing with a series of massive food recalls. And Michael Taylor would be coming back for another turn of the revolving door.

7

ANIMALS ON DRUGS

Corruption is like a ball of snow, once it's set a rolling it must increase.
—Charles Caleb Colton (1780–1832), British writer

It was déjà vu. Michael Taylor, the ever-ready maestro of the revolving door, was back at the FDA—this time appointed by President Barack Obama to be deputy commissioner of foods. The *New York Times* politely observed that Taylor "migrated among government, industry and academia." A former strategist for Monsanto, Taylor started in July 2009 as a senior adviser to Obama's new FDA commissioner, Margaret Hamburg, and in January 2010 he was appointed deputy commissioner of food.[1] This newly created position gave Taylor authority over all food-related work at the FDA, including oversight of the Center for Food Safety and Applied Nutrition. The FDA had faced one food scandal after another—massive recalls had become the new normal.

The FDA was established with the passage of the 1906 Pure Food and Drugs Act, a mandate that has grown exponentially as the food system has become more consolidated and globalized. The advent of processed food, produced in huge volumes, and the abundance of produce sourced from just a few locations have made food safety a whole new ball game. The FDA has come to regulate all foods except meat, poultry, and processed eggs.

In 1906, the Pure Food and Drugs Act was focused on preventing unscrupulous companies from selling mislabeled products—for instance, maple syrup that was 90 percent glucose with a coal tar–based maple flavoring or jelly advertised as quince made mostly of glucose and coal tar essence of quince. Like today, these were contentious issues, but unlike now, most food in the early twentieth century was prepared "from scratch" at home, and produce during most of the year came from local or regional sources.

The FDA today is understaffed and underresourced in today's globalized world, where most food is processed. Most food-processing facilities are not inspected, and the FDA has relied on issuing guidances to the food industry, which are not backed up with enforcement. More often than not, the FDA reacts to problems instead of trying to prevent them. Hundreds of recalls have taken place over the past decade, but some are particularly memorable in their scope.

- On September 14, 2006, the FDA told Americans to stop eating bagged spinach because of contamination by a virulent strain of *E. coli* that killed at least five people after a painful, bloody illness sickened more than 205 people in twenty-six states, leaving them vulnerable to future health problems.[2]
- In March 2007, due to adulterated pet food, thousands of pets died from kidney failure. More than 5,300 brands of pet food had been contaminated with Chinese-produced wheat gluten that had been tainted with melamine to give the false appearance of a higher level of protein.
- An enormous national salmonella outbreak in peanut products began in 2008 and sickened more than 630 people in forty-three states, killing nine. The incident was linked to a sole Georgia processing plant owned by the Peanut Corporation of America that had opened in 2005—a facility that had never been inspected until after the outbreak. The illness affected 275 companies, and almost 3,500 products were recalled. One of the companies involved declined to recall its products, highlighting the lack of food safety authority at the FDA. Later, in 2009, a similar salmonella outbreak took place in pistachios.
- During the summer of 2009, a refrigerated-cookie-dough shortage took place after Nestlé's Toll House dough sickened at least sixty-six people in twenty-eight states from *E. coli* 0157:H7. The massive amount of dough was prepared in the company's Danville, Virginia, plant, but the source of the *E. coli* was never found.
- In March 2010 the FDA announced a nationwide recall of black pepper. Over the previous several years, pepper-related recalls had been initiated because of salmonella contamination. The largest incident related to black pepper occurred when 1.24 million pounds of pepper-coated salami was recalled because of salmonella poisoning, which affected 238 people in forty-four states and the District of Columbia.
- A massive recall announced on March 4, 2010, was remarkable for the number of foods it involved and for the FDA's lack of spine in dealing

with the company responsible—a producer of the flavor enhancer hydro-lyzed vegetable protein (HVP). According to nutritionist and food writer Dr. Marion Nestle, not only did the company not take immediate action, but the FDA failed as well: "[F]rom January 21 until at least February 20, the company continued to ship HVP potentially contaminated with Salmonella. Then, over the next *six* days, the FDA had to beg Basic Food Flavors to issue a recall. The company may have started notifying custom-ers on February 26 but the FDA did not announce the recall until March 4, weeks after the first findings of Salmonella."[3] (Emphasis in original.)

• The largest egg recall in history took place in 2010. Half a billion eggs, produced in just two facilities, were recalled that August. Two rodent-infested Iowa egg farms caused almost two thousand traceable illnesses from salmonella and sickened nearly sixty thousand people nationwide.

This rash of high-profile, large-volume food recalls brought the issue of reforming the FDA to a head during the 111th Congress, when, after a strange and convoluted path, the FDA Food Safety Modernization Act finally passed in December 2010 and was signed into law by Obama on January 4, 2011. The rancorous debate over the bill began during 2009, with the initial passage of a controversial House version of the bill. The debate moved to the Senate, where after a vicious and prolonged process companion legislation passed on November 30, 2010, and moved on for reconciliation with the House bill.

At this point, the legislation hit a snag, because Senate sponsors had added tax provision, which according to the U.S. Constitution can originate only in the House. The bill was then inserted by the House into a budget bill and sent back to the Senate, by then deep into a tax debate, where one of its harshest critics, Republican senator Tom Coburn from Oklahoma, threatened to fili-buster it if it was not removed. At the last moment, leaders of both parties in the Senate, fearing they would be blamed for food-poisoning deaths, agreed to pass a revised version, and the House passed the final version shortly before Christmas 2010. It was a theatrical finish to a bill crafted with drama.

Industries had lined up against consumer organizations and outspoken advocates such as Connecticut congresswoman Rosa Delauro. As chairwoman of the House Agriculture FDA appropriations subcommittee, Delauro had been leading the charge for adequate funding of food safety at the FDA and USDA. Her demands for regular inspections of food facilities, adequate trace-ability for contaminated food, and standard imported-food regulations were viewed as radical, job-killing government interference.

Senator Coburn, a physician who often takes anti–public health positions, had editorialized in *USA Today* that the bill would "impose new and invasive regulations" and expand "duplicative" bureaucracy. He went on to declare: "For the past 100 years, the free market, not the government, has been the primary driver of innovation and improved safety. Consumer choice is a far more effective accountability mechanism than government bureaucracies."[4]

The David-and-Goliath battle was vicious and marked by an ongoing misinformation campaign by the industries. The food-processing industry— represented by the Grocery Manufacturers Association, which is a well-funded organization with a staff of one hundred—had no desire to change the status quo. It was dead set against mandatory recalls and frequent inspections.

United Fresh, the lobbying organization for the produce giants, argued that scale is unimportant and that large produce growers should have the same regulations as the small farmer who sells directly to consumers at a farmers' market. Other debates raged over registration fees, and again industry demanded that Kraft should pay the same as the small independent cheesemaker.

Considering the gravity of the situation at the FDA, and the contentious battle taking place over the legislation, it is no surprise that Michael Taylor was back. In rotations between government and industry he had become the consummate damage-control expert for Democratic administrations. He was back to help shepherd the bill as it moved through the Senate and to manage the contentious process of writing the rules for implementation after passage. No one is more accomplished at schmoozing industry or mainstream consumer advocates than Taylor. FDA staff say—off the record—that Taylor must be involved in every decision and is a continual bottleneck to making progress on implementing the bill.

While the law certainly doesn't represent an overhaul of the food safety system, it has provided the FDA with mandatory recall authority. The other significant part of the bill is the establishment of a schedule for FDA inspections of food-processing facilities—a measure that is long overdue, since many manufacturing plants have never been inspected, but that is completely inadequate. There are 190,000 registered food facilities in the United States and 230,000 foreign ones. The new law mandates inspection once every three years for "high-risk" facilities and once every five years for "low-risk" ones.

But the devil is in the details. The legislation directs the FDA to double its inspections of foreign food facilities that export products to the United States every year for five years, beginning with six hundred foreign facilities in fiscal year 2011, bringing the number to nineteen thousand in fiscal year

2016. Industry has vehemently opposed inspections, and with this schedule it has been granted its wish. The other problem is the definition of risk. Taylor's staff is still in the process of determining how the risk categories will be defined.

The other big win in the legislation was the Tester amendment, offered by Senator Jon Tester, a rancher from Montana. An exemption from inspections is given to farms or small food businesses grossing less than $500,000 per year that sell a majority of their food products directly to consumers, restaurants, or grocery stores within a 275-mile radius from their place of business or within the same state. The produce industry tried vainly to strip this provision, and it is a testament to grassroots activism that it remained in the bill. United Fresh and nineteen other produce organizations sent a letter denouncing the amendment and calling Tester names.

Unfortunately, the new legislation does not deal with many of the important issues that are challenging public health because of the globalized and industrialized food system. While the FDA spends resources and staff time on busting small cheese producers for using raw milk, we are facing a crisis of antibiotic resistance that is caused first and foremost by the industrialized livestock industry. The FDA has the power to stop factory farms from using low-dose antibiotics to promote growth, yet it has refused to take sufficient action, even though, at some time in the near future, antibiotics may be rendered useless against infection.

In August 2011 the biggest contaminated meat recall to date was ordered by the third-largest turkey-producing corporation, Cargill. In twenty-six states, 36 million pounds of ground turkey contaminated with a strain of salmonella resistant to multiple antibiotics sickened dozens of people. The poisoned meat was produced at a single plant in Arkansas, demonstrating the flaw with a food system that consolidates production and deregulates safety. *Salmonella* Heidelberg, the offending bacteria, is a "superbug" that has mutated to become resistant to antibiotics. For most people, diarrhea, vomiting, and nausea mark salmonella poisoning, but it can cause a serious infection of the blood that can be fatal.

And worse is coming, according to a multinational team of scientists who documented illnesses caused by *Salmonella* Kentucky in Europe, the Middle East, and the United States. This new strain of salmonella is resistant to Cipro, the powerful antibiotic that is usually used to treat the illness. The primary carrier: poultry.[5]

Dr. Robert Lawrence is not surprised that poultry is contaminated with antibiotic-resistant bacteria. The organization of which he is the founding director—the Johns Hopkins Center for a Livable Future (CLF)—has a research staff that has investigated and written extensively on the growing threat of antibiotic resistance and its relationship to industrialized animal production. CLF is an interdisciplinary group of faculty and staff that focuses attention on equity, health, and the Earth's resources.[6]

Lawrence's stamina and energy are more reminiscent of someone in the early years of their career than a person who began his career in the late 1950s. Not only a medical doctor and an expert on antibiotic resistance, he is now an activist academic willing to publicly challenge agribusiness and factory farms. In addition to filling his position at CLF, he is a professor at Johns Hopkins Bloomberg School of Public Health and professor of medicine at the Johns Hopkins School of Medicine. His staff says he wears many hats: one day he might be in Oklahoma testifying as an expert witness against Tyson for polluting a million-acre area along the Illinois River, and the next day he could be in South Africa for a meeting on HIV.

Lawrence is the son of a minister, and he recollects his father telling him that he "could do whatever he wanted as long as it was socially useful." He says that in the narrow confines of his life at that time, "socially useful" meant being a minister or a doctor. He chose doctor, although he thought he might practice as part of a ministry in Africa.

Lawrence has had a long and distinguished career; after graduating from Harvard Medical School, he practiced tropical medicine in Latin America and ran the first multiracial primary care facility in North Carolina. In 1974 he was appointed as the first director of the Division of Primary Care at Harvard Medical School, and then he was recruited to run the Rockefeller Foundation's Public Health Program for Africa, Asia, and Latin America. It was during this time that he became interested in agriculture, because he worked in an atmosphere where "the silos were coming down" in the grant making for health, agriculture, the environment, and population.

By the time Lawrence was recruited by Johns Hopkins he was a convert to sustainable agriculture and convinced that industrialized animal production was a major cause of the health problems plaguing Americans. One of CLF's most important missions is combating antibiotic resistance. According to the Centers for Disease Control and Prevention (CDC), 2 million people in the United States contract resistant infections each year, and ninety thousand of them die. Almost all bacterial infections are now resistant to the specific antibiotic that was initially the most effective treatment for it.

The livestock industry is engaged in a shameful misuse of antibiotics. CLF's Dr. David Love examined FDA data and calculated that 29 million pounds are used each year. His analysis showed that animal agriculture was responsible annually for almost 80 percent of the antibiotics used.[7]

The threat of antibiotic resistance emerging for the "wonder drugs" of the twentieth century was identified early on. Lawrence cites the 1945 Nobel Prize lecture by Alexander Fleming, who discovered penicillin and warned of antibiotic resistance: "It is not difficult to make microbes resistant to penicillin in the laboratory by exposing them to concentrations not sufficient to kill them."[8] Lawrence goes on to say that the habitual use of low doses of antibiotics in animal feed is the precise formula for developing antibiotic resistance. Low doses of antibiotics are able to eliminate only the most susceptible bacteria in a survival of the fittest contest that promotes the reproduction of antibiotic-resistant strains. Humans come into contact with these strains through their food, the air, the water, and the soil.

Feed companies and factory farms, Lawrence says, have unrestricted access to these drugs, with no government oversight. In 2008 CLF partnered with the Pew Commission on Industrial Farm Animal Production to produce policy recommendations in the report "Putting Meat on the Table: Industrial Farm Animal Production in America." The commission recommended phasing out the use of antimicrobials currently added to feed in food animal production, in order to preserve antibiotics for treatment of infectious diseases in people.

Cases of the bacterial infection known as MRSA (Methicillin-resistant Staphylococcus aureus) are now killing between seventeen thousand and eighteen thousand Americans a year, and this is likely related to the use of low-dose antibiotics in swine and other animal production.[9] MRSA is a type of staph infection that does not respond to the antibiotics commonly used to treat the disease. In the past it was a hospital-acquired infection, but increasingly it is acquired outside of the medical setting. The medical establishment is aware of what is going on, according to Lawrence, but it's just not doing everything necessary to stand up to the drug and livestock industries.

Even the conservative American Medical Association has passed a resolution against the use of nontherapeutic antibiotics. Lawrence wrote to the directors of the National Institute of Allergy and Infectious Diseases at the NIH and of the CDC about his concerns and received written confirmation from both that the misuse of antibiotics in industrial food animal production is directly linked to antibacterial resistance in human pathogens.[10]

Another activist physician, Dr. David Wallinga, the senior adviser in science, food, and health at the Institute for Agriculture and Trade Policy (IATP), says that MRSA can be found in some farm operations and retail meats, as well as in previously well people. Wallinga says that the MRSA bacteria often lives in the nose and on skin: "People can carry the bacteria unknowingly and without getting sick, but it also can cause serious human infections of the bloodstream, skin, lungs (pneumonia), and other organs. . . . Rising numbers of people are falling ill with a kind of staph untreatable with these drugs." [11] He adds that a 2009 study found MRSA highly prevalent in 49 percent of swine and 45 percent of swine workers for a large-scale commercial confinement company with farms in Iowa and Illinois. Wallinga also notes that a Canadian study found pigs carrying MRSA on almost half of Canadian pig farms tested.[12]

While the European Union and the most respected health agencies in the world, including the World Health Organization, agree that the use of antibiotics as a component of animal feed to promote growth should be banned, the United States has failed to take strong action. It is not just the livestock industry that lobbies to prevent legislation or regulation from hampering its use of low-dose antibiotics, the powerful pharmaceutical industry finds the misuse of antibiotics highly profitable. This combined political influence has stymied legislative and regulatory action on antibiotics.

Wallinga and Lawrence are both advocates of legislation that would phase out the nontherapeutic use of medically important antibiotics in livestock. The Preservation of Antibiotics for Medical Treatment Act (PAMTA), most recently introduced by Congresswoman Louise Slaughter (D-NY), has been introduced in thirteen different iterations over seven sessions of Congress— every Congress from the 106th to the 112th. It was named the Preservation of Essential Antibiotics for Human Diseases Act of 1999 in the 106th and subsequently referred to as the Preservation of Antibiotics for Medical Treatment Act. The bill has had only one hearing during this entire period of time.

The regulatory authorities—especially the USDA—also have been unwilling to take sufficient action to protect the effectiveness of antibiotics. President Obama's secretary of agriculture, Tom Vilsack, told the National Cattlemen's Beef Association, "USDA's public position is, and always has been, that antibiotics need to be used judiciously, and we believe they already are."

In a September 7, 2011, Government Accountability Office (GAO) report, "Antibiotic Resistance: Agencies Have Made Limited Progress Addressing Antibiotic Use in Animals," the USDA is quoted as saying, "Currently, there is insufficient scientific information available to make important policy decisions regarding use of antibiotics for growth promotion purposes." [13] Twenty-four public interest organizations wrote in response to Secretary Vilsack about their "grave concerns" concerning the agency's position, stating that "USDA has been inconsistent at best in recognizing and accepting the significant scientific evidence supporting the existence of an overuse of antibiotics in animal agriculture." [14]

The group's response went on to note several instances that showed the USDA's refusal to take the life-threatening loss of antibiotics seriously, including the removal from the USDA's Web site of a July 2011 technical report summarizing the robust literature on antibiotic resistance. At the May 2011 Future of Food conference at Georgetown University, Vilsack responded to a questioner about what action the agency would take on antibiotic resistance, stating: "I'm not quite sure. How do you basically legislate that?" He added, "It's not as easy as it appears." [15]

The real power to limit use of antibiotics, however, resides with the FDA—the agency that approves the use of the drugs. Among the recent issues the 2011 GAO report addressed was what actions the FDA has taken to mitigate the risk of antibiotic resistance. The report criticized the FDA for the "lack of crucial details necessary to examine trends and understand the relationship between use and resistance." It chastised the FDA for collecting data from drug companies on antibiotics sold for use in food animals without showing which antibiotics are used or for what purpose. The report summarizes the problem:

> FDA . . . faces challenges mitigating risk from antibiotics approved before FDA issued guidance in 2003. FDA officials told GAO that conducting post approval risk assessments for each of the antibiotics approved prior to 2003 would be prohibitively resource intensive, and that pursuing this approach could further delay progress. Instead, FDA proposed a voluntary strategy in 2010 that involves FDA working with drug companies to limit approved uses of antibiotics and increasing veterinary supervision of use. However, FDA does not collect the antibiotic use data, including the purpose of use, needed to measure the strategy's effectiveness. [16]

Pressure mounted on the FDA when, on March 23, 2012, it lost a lawsuit. A federal judge ruled that the agency must act on a proposal it made in 1977 to prevent two antibiotics important to human medicine—tetracyclines and penicillins—from being given routinely to healthy livestock. After citizen petitions in 1999 and 2005 and a lawsuit filed in 2011, the FDA took action, quietly withdrawing the proposal just before Christmas. But the judge ruled that the agency actually had to address the concerns it had identified over thirty years earlier. Drug manufacturers will have a chance to make their case that the antibiotics are safe to feed routinely to livestock. But if they aren't able to (and science indicates they won't), the FDA must withdraw its approval of subtherapeutic uses of the drugs.

Not long after, on April 6, 2012, the FDA banned most subtherapeutic uses of one class of antibiotics, cephalosporins. These are used to treat food-borne illnesses in humans, especially children, as well as pneumonia and skin and soft tissue infections. Salmonella resistance to cephalosporin drugs is on the rise, putting the public at risk.

Five days later, the FDA announced another voluntary initiative to promote the "judicious use" of antibiotics in livestock. The agency released the final "Guidance 209: The Judicious Use of Medically Important Antimicrobial Drugs in Food-Producing Animals," along with more clarification about how to make it work. The FDA provided direction for transitioning from the use of over-the-counter antibiotics in animal feed to a new system requiring oversight by veterinarians. But without regulations in place, we have no guarantee that the pharmaceutical industry and livestock producers will voluntarily stop the use of the drugs in animals.

The FDA continues to be conflicted. It is unwilling to completely ban the dangerous and inappropriate use of antibiotics. And it currently insists that industry voluntary efforts will address this public health issue. Wallinga says, "Politics is holding public health hostage. The FDA has effectively done nothing over thirty years after first labeling this a problem. We need federal legislation—PAMTA—despite the fact that it makes the big pork producers unhappy. Unfortunately, they scream louder than the AMA and the doctors calling for PAMTA to pass."

Failure to pass the Preservation of Antibiotics for Medical Treatment Act is not surprising. The drug-health industry spent $2.3 billion lobbying from 1998 to 2011, and of this amount, $1.5 billion was from the pharmaceutical manufacturing group. Total campaign contributions to federal elected officials between 1990 and 2011 amounted to $131 million, with 66 percent going to Republicans and 34 percent to Democrats.

The Animal Health Institute (AHI) is one of the trade associations that lobby for the veterinary medicine side of the drug industry. Among the gimmicks used by the AHI is Celebrity Pet Night, where for the past fifteen years members of Congress and their staff are invited to a reception to mix and mingle with celebrities. The event features a Cutest Pets on Capitol Hill photo contest. This is just one example of how the animal drug industry hides behind the cuddly image of medicine for household pets.

AHI also represents companies that have promoted the use of another dangerous feed additive: arsenic. The FDA approved the use of the arsenic-based drug roxarsone as a feed additive in 1944, when Franklin D. Roosevelt was president. Industry researchers had discovered that roxarsone promoted growth, increased feed efficiency, and gives the appearance of health by brightening the color of flesh. Between 1995 and 2000, 70 percent of broiler chicken producers used roxarsone feed additives. Its use is prohibited only in organic chicken production.

Arsenic is an element that does not break down in the environment; instead, it combines with other elements to form compounds. Nearly 90 percent of the arsenic fed to chickens is excreted through urine and feces. An estimated 2 million pounds of roxarsone are fed to chickens each year, contaminating much of the estimated 26 billion to 51 billion pounds of waste that broiler chickens produce each year. Most of that waste is applied to fields as fertilizer, causing arsenic to leach into soil, water, and crops.

Arsenic-based feed additives are also used in the turkey and hog industries to prevent disease and promote growth, but there is far less research on the public health and environmental impacts from its use in these industries. One study has found evidence of inorganic arsenic in waste lagoons on large hog operations where arsenic feed additives are used.[17]

Poultry farmers also use arsenic to control a common poultry disease known as coccidiosis that is caused by the coccidian parasite. Affected birds experience a variety of symptoms, including diarrhea, impaired food absorption and growth, immune suppression, and even death. While not all chickens infected with coccidia die, their meat and egg production is impaired, leading to significant economic losses.

Arsenic poses problems both in the chicken meat itself and in chicken waste. U.S. chicken consumption has increased significantly over the last several decades, and new studies demonstrate that arsenic residues may be higher in chicken meat than has been previously known. More research is necessary to understand just how much arsenic Americans consume in chicken. Arsenic is also present in chicken waste, where it converts to more dangerous forms than those originally used in the feed.

Chronic exposure to arsenic is associated with increased risk for several kinds of cancer, including bladder, kidney, lung, liver, and prostate. It is also associated with increased risk of cardiovascular disease and diabetes, as well as neurological problems in children. Each exposure contributes to a person's total arsenic exposure, and sources such as the American Cancer Society urge the importance of reducing arsenic exposure from any venue as much as possible.

The FDA set allowed levels for arsenic residues in poultry in 1951, and these rules are long overdue for reconsideration, particularly because Americans' consumption of chicken has increased substantially since that time. In the 1940s Americans ate less than twenty pounds of poultry per person per year on average; by 2008 that had tripled to nearly sixty pounds per person. African Americans and Latinos generally eat more chicken than Caucasians and Asians, and are thus at greater risk of arsenic exposure. According to epidemiologist Dr. Keeve Nachman, science director at the Center for a Livable Future at Johns Hopkins University, the tolerance levels "predate our current understanding of the human health effects of exposure to arsenic."

In 2006 a study by IATP tested arsenic levels in the chicken meat sold at grocery stores and fast-food outlets. Of the 151 retail packages tested, 55 percent had detectable levels of arsenic. The range of brands sampled included some certified organic and others from companies that do not use arsenical feed additives. Of the non-premium and nonorganic brands, 74 percent of the retail chicken tested had detectable levels of arsenic. Of the ninety orders of fast-food chicken tested, arsenic was detectable in all samples.

The USDA Food Safety and Inspection Service (FSIS) is responsible for monitoring various residues in meat and poultry, but the agency has failed to take appropriate action to determine arsenic residues in chicken meat. In total, FSIS tested 5,786 of the approximately 72 billion broiler chickens produced between 2000 and 2008—that amounts to only one in every 12 million chickens being tested.[18]

In 2010 the USDA inspector general released an evaluation of the FSIS National Residue Program and reported that "it is not accomplishing its mission of monitoring the food supply for harmful residues."[19] Two criticisms stand out. The first is that the FSIS fails to recall meat even when it finds evidence of veterinary drug residues. The second is that the FSIS, the EPA, and the FDA fail to coordinate effectively to prevent the public from harm by establishing relevant standards.[20] The demonstrated existence of arsenic residues in chicken meat is a case example of oversight failure and insufficient monitoring to protect consumers.

Tyson Foods and Perdue, two of the largest U.S. poultry companies, claim to have stopped using arsenic compounds in 2004 and 2007, respectively.[21] However, they continue to lobby for the right to use it. In testimony before an agricultural subcommittee of the U.S. House of Representatives, Steve Schwalb, Perdue's vice president of environmental sustainability, stated, "Perdue agrees to make every effort not to use arsenic compounds in its feed, but may use it where the health of the flock is a concern and other non-arsenic techniques fail to restore the flock to health in a timely manner."[22]

"The science doesn't support a ban right now," said Schwalb. "If people believe it's a safety issue, then they can take it up with the FDA." [23]

Perdue vehemently opposed a state ban on arsenic feed additives in Maryland, where, after a three-year campaign, Food & Water Watch helped pass legislation in 2012 to ban the use of roxarsone in chicken feed. Maryland, the eighth-largest producer of chicken in the United States, was the first state in the country to take steps to restrict the use of arsenic in animal feed, although the poultry industry did get several loopholes included in the bill that it will try to use to reintroduce arsenic-based drugs in the future.

Multiple studies and industry estimates suggest that between 70 percent and 88 percent of broiler chickens receive arsenic additives in their feed. [24] Even the industry estimated in 2011 that nine out of ten chickens consumed had been fed arsenic. [25]

The EPA addresses maximum levels of contaminants in the environment as well as specific instances of severe, localized contamination. In 2001, the EPA reduced the maximum contaminant levels for arsenic in drinking water from fifty parts per billion (ppb) to ten ppb, with compliance required by January 2006. [26] While the action to reduce arsenic exposure is laudable, the risk of cancer from arsenic levels at the new standard is still fifty times higher than the risk allowed for many other carcinogens. [27]

In response to publicity regarding new studies on arsenic in 2007, an FDA spokesperson stated that the agency "has no data to suggest that there have been any adverse health effects in humans" because of roxarsone in chicken feed. [28] The lack of evidence seems to have more to do with a failure to look for it than a lack of adverse effects. While the drinking water standard for arsenic has been strengthened, the standards for arsenic residues in poultry have remained unchanged by the FDA for nearly sixty years. [29]

Concerns about arsenic exposure prompted Representative Steve Israel (D-NY) to introduce the Poison-Free Poultry Act in Congress in 2009. To date this legislation has not moved forward. The combined political power of the drug and livestock industries is formidable.

Feed-additive production has become extremely concentrated, like all aspects of agribusiness. As of 2000, the pharmaceutical company Alpharma Animal Health (Alpharma) was the top producer of antibiotic feed additives and the second-largest producer of anticoccidial drugs. In 2008, King Pharmaceuticals acquired Alpharma. Alpharma is the producer of more than half of roxarsone products, and just six companies produce more than 90 percent of them.

Two years later, Pfizer, the largest drug company in the world, bought King

Pharmaceuticals in a $3.6 billion deal. Pfizer, the maker of drugs like Viagra and Celebrex, is facing the loss of patent protection for the cholesterol drug Lipitor and is seeking new revenue sources. It was most interested in gaining access to King's pain treatment division and other drugs in its pipeline. Rumors abound that the industry giant will sell King's animal health division.

The eventual sale of this division may be why on June 8, 2011, just thirty days after announcing the acquisition, Pfizer stated that it would suspend the sale of roxarsone, based on new FDA data that found arsenic in the livers of chickens fed the drug. Pfizer voluntarily suspended sales, giving the FDA cover for not banning use of the drug. Advocates fear that it will be brought back on the market after stockpiles of it are used or in the event that it is sold to another company. During the thirty-day lead-up to the suspension of sales, large quantities were available to the poultry industry.

A possible sale after the voluntary suspension of 3-Nitro, roxarsone's trade name, has been mentioned in relation to a lawsuit brought against Pfizer by one of its subcontractors. In October 2011 the Chinese chemical company Rong-Yao, which manufactures roxarsone for Pfizer, filed a breach of contract lawsuit against the company for $20 million in a federal district court. Dr. Rener Chen, Rong-Yao's general manager, noted in a release, "The timing of Pfizer's decision to voluntarily suspend its sales of 3-Nitro, under the veil of the FDA study and the associated FDA pressure, is suspect as it comes at the same time as reports in the industry that Pfizer is looking to completely sell off all of its Animal Health division and remove itself from this market altogether."[30]

Consumer advocates will be watching to see if the drug is brought back on the market if Pfizer does indeed sell the Animal Health division to another company. Unfortunately, if this is the case, it is unlikely that the FDA will ban its use. FDA lacks the fortitude to stand up to the excessive political power of the economic interests benefiting from weak regulation. When the agency does act, it is to beg the food, meat, and drug industries to cooperate through voluntary programs. The FDA is unwilling to take the actions necessary to ensure that dangerous residues like arsenic are removed from the American diet.

The FDA and its political masters are even willing to sacrifice the efficacy of antibiotics, the miracle drugs that have saved millions of lives, for the sake of the meat industry. The agency's reckless refusal to take decisive action banning the nontherapeutic use of antibiotics on factory farms not only is causing a health crisis today, from antibiotic-resistant superbugs, it is risking the health of future generations.

The FDA is cowed by industry pressure, but it is also underresourced to deal with the deadly threats that are the result of a food system out of control. With only twelve hundred inspectors to oversee all food except meat and poultry, the agency does not have the staff to inspect it vigilantly. It does not have sufficient funding to do laboratory tests for residues or to make sure that foods imported from the developing world are free of dangerous toxics or agrochemicals. Each year, sufficient funding for the agency is threatened during the increasingly antagonistic and partisan budget debates. Until the FDA is adequately funded, it will be unable to make sure that the health and safety of Americans are protected.

No one can fully escape from the impact of the FDA's failures or from the ill effects of the dysfunctional food system. In a large and industrialized nation, everyone is dependent upon the federal regulatory system to some degree. No one can grow all of the food that they eat, unless they live an entirely subsistence lifestyle. While shopping at the local farmers' market and knowing where your food comes from is part of the solution, it does not protect you from all the dangers lurking at the grocery store, at the restaurants you patronize, or when you go to Grandma's for a holiday dinner. Most of us purchase produce from the grocery store when it is out of season locally; at the very least, we depend on stores for staples. We eat at many places where we are not in control of the shopping list.

Even individuals who only purchase organic produce, avoid consuming industrialized meat and processed food, and shop at a natural food store for most products depend upon a protective and alert regulatory system to ensure that the products are free of deadly bacteria, chemicals, and residues. As organic products increasingly come from China, extremely health-conscious consumers still rely on the FDA's vigilance as we face more risks from chemical contamination. And anyone who shops for food and looks at the ingredients list depends upon the FDA for the creation and enforcement of transparent and effective labeling.

In the long term, while we should avoid processed food, shop locally, and get to know our local farmers if possible, the best solution is to build the political power to reform not only the food system but the regulatory system that governs it. In a large and complex society that has more than 300 million people, it is crucial to have a protective and fair regulatory system overseen by a federal agency willing to guard the health and safety of consumers in the face of political pressure.

It is time for food activists to embrace the need for effective regulation, rather than to acquiesce to the libertarian philosophy that we do not need regulation if we buy locally. In recent years, the FDA's SWAT team–like raids on small farms selling raw milk or goat cheese has caused a backlash against all food regulation. Most of us would agree that this is an outrageous misappropriation of resources. In a country facing so many food-related dangers, the FDA should not be policing small operations selling to consumers willing to take the risk because they know the producer and are confident of the products' safety. Yet the agency takes a pass on the real health hazards, such as the chemicals and residues in the millions of tons of imported foods.

We must demand that the USDA spend its resources wisely and protect all Americans from the hazards in the industrialized and globalized food system. We must build the political power to give members of Congress and the executive branch the backbone to stand up to the selfish economic interests of those that put their quarterly profits before the health of the American public. To have a safe food system that serves everyone, food activists must add food safety and effective regulation to the good-food movement's agenda.

PART V

The Story of Factory Farms

Over the last two decades, small and medium-size livestock farms have given way to factory farms that confine thousands of cows, hogs, and chickens in tightly packed facilities. Large numbers of livestock farmers have been driven out of business, while a small percentage has adopted factory farming practices largely at the behest of the largest meatpackers, pork processors, poultry companies, and dairy processors. Despite ballooning in size, however, many livestock operations are just squeezing by, because the real price of beef cattle, hogs, and milk has been falling for decades. The largest meat processors, which operate essentially as monopolies with unchecked power, are the beneficiaries of the cheap grain policies that have made factory farming profitable. Ending industrialized meat production is a critical component of reforming the food system.

8

COWBOYS VERSUS MEATPACKERS:
THE LAST ROUNDUP

Commerce is entitled to a complete and efficient protection in all its legal rights, but the moment it presumes to control a country, or to substitute its fluctuating expedients for the high principles of natural justice that ought to lie at the root of every political system, it should be frowned on, and rebuked.
—James Fenimore Cooper, *The American Democrat* (1838)

The burly cowboy roping cattle on the western range, an American icon immortalized in western movies and country songs, has just about disappeared from the national landscape. Long marked by violence and lawlessness—from the range wars of the nineteenth century to the land-grabbing exploits of the western cattle empires—the U.S. cattle industry has been devastated in more recent decades by a type of economic violence. The titans of beef have eliminated the cowboy and created a system that pushes independent ranchers out of business, drives cattle off the range, and creates huge profits for some of the largest corporations in the country.

Mike Callicrate's blog, *No-Bull Food News*, is an apt description of the outspoken opposition to big agribusinesses. The owner of Callicrate Cattle, his vocation is fighting for the independent cattle producer. Callicrate says he was "blacklisted" by the monopolistic beef packers because of his advocacy. In 1996, he was one of ten ranchers who filed a class-action lawsuit against IBP, the giant meatpacker that merged with Tyson, for its unfair, deceptive, and discriminatory cattle-buying practices. The case ended when the Supreme Court refused to hear the case against IBP. Callicrate continued to criticize the industry voraciously for its market power, including Farmland National Beef, which retaliated by refusing to purchase his cattle. Without a market for his cattle, he was forced to close down.

Callicrate, energetic and entrepreneurial, invented a widely used castration device that has provided the resources for him to circumvent the market power of the meat industry. Undeterred by being driven out of business, he invested a "few million" into remodeling an existing processing plant and opened Ranch Foods Direct. A local source of high-quality meat, the company primarily does business in Colorado Springs, where he distributes beef to more than a hundred restaurants.

Callicrate concedes that most ranchers do not have this option, because they have no way to slaughter and market their beef. Unless a rancher can sell a quarter or half of its beef to a consumer, it is hard to make a profit selling direct. Most consumers do not want to store this much beef and they are only interested in steaks and good-quality roasts, not all of the pounds of hamburger that come from purchasing part of a cow.

To mitigate this problem Callicrate has been involved in developing and promoting mobile slaughter units that can be used on-site at farms. But he believes the only way that independent ranching can continue is if the government begins enforcing antitrust laws. He explains that ranchers face untold obstacles to making a living. Cattle used to be sold at a competitive live auction, where the big meatpackers had to compete with smaller firms. Today there is no place for an independent rancher to take his herd to market and get a fair price.

Eric Schlosser, author of the best seller *Fast Food Nation*, has an apt description of the industry.

> Over the last twenty years, about half a million ranchers sold off their cattle and quit the business. Many of the nation's remaining eight hundred thousand ranchers are faring poorly. They're taking second jobs. They're selling cattle at break-even prices or at a loss. The ranchers who are faring the worst run three to four hundred head of cattle, manage the ranch themselves, and live solely off the proceeds. . . . Ranchers currently face a host of economic problems: rising land prices, stagnant beef prices, oversupplies of cattle, increased shipments of live cattle from Canada and Mexico, development pressures.[1]

Just how do cattle become burgers? Increasingly, the whole process—from farm to plate—is industrialized, beginning with the vial of semen that is used for artificial insemination. Production of semen is dominated by three corporations—World Wide Sires, Cooperative Resources International, and

ABS Global—that you have likely never heard of but whose progeny you have eaten if you eat beef.

Beef cattle are raised successively in different types of operations before they go off to slaughter and processing, known in the industry vernacular as "packing." In the first stage of production, cow/calf farms have breeding heifers and raise calves that have a nine-month gestation period, weigh sixty to one hundred pounds at birth, and are weaned at between six and ten months. Cows can reproduce for seven to nine years, but they are often plagued with birthing problems that can cause the loss of calves and high veterinary bills. Calves graze on pasture or rangeland alongside their mothers until they are weaned, lowering the risk of disease that happens under crowded indoor conditions. Raising calves is done by smaller family farms, who are willing to do the nurturing required and take pride in raising the animals.

When calves reach about 400 to 650 pounds, most are sold to stock-feeding operations, also often run by family farmers, that raise the cattle on pasture or range as they mature and gain weight. Sometimes this stage is skipped, and calves are kept longer at the cow/calf operation before they are sent directly to a feedlot. In 2008, half of all beef cattle were raised on 675,000 farms and ranches with fewer than one hundred head of cows.[2] But most of these cattle ultimately end up on feedlots.

Feedlots fatten cattle using a grain-based diet until the animals are just under two years old or weigh around twelve hundred pounds. Feedlots increase rates of weight gain by the use of pelletized natural or synthetic sex hormones that are implanted in the ear skin of cattle. According to Dr. Samuel S. Epstein, professor emeritus of environmental and occupational medicine at the University of Illinois Chicago School of Public Health, the hormones used in beef production are associated with an increased risk of reproductive and childhood cancer. He says that "residues of these hormones in meat are up to twentyfold higher than normal" and "still higher residues result from the not uncommon illegal practice of implantation directly into muscle." Unfortunately, the USDA does not monitor the meat for hormone residues.[3]

Feedlots have become larger and more profit-driven over the past two decades, as meat production and packing have become more monopolized. But cattle have not always been raised this way. Until the 1960s the animals were raised on open rangeland or pasture located on ranches around the nation, especially in Texas and the western states. Changes in farm policy and the development of hybrid grains and irrigation encouraged the production of large amounts of grain that could be used to feed cattle, even though their digestive systems are designed for grass, not grain. By the 1970s, feedlots became the preferred method of fattening cattle, a process called "finishing" in the industry. Over the following decades, research and technology also facilitated the creation of new, specialized grain-based feeds and hormone injections or implants to increase weight gain.

Until recently, feedlots were run as family-owned operations that housed fewer than a thousand cows. Today, many thousands are housed in acres of steel pens, each corralling around two hundred cattle that are fed from a mechanized feed delivery system or a trough that can be filled by a tractor. The cattle stand without shelter, shade, or grass and sleep on their own waste, which forms a concrete-like surface that is dusty when it's dry and sewer-like if it rains. The cattle are squeezed together tightly so that no calories are lost to unnecessary movement.

Cattle spend up to six months eating a feed mixture that contains corn by-products derived from ethanol production, animal by-products, cottonseed meal, grains, and alfalfa. The purpose of the high-calorie diet is to create fat deposits known as marbling—the source of flavor and a tender texture in steak.

As feedlots have gotten much larger, they have formed partnerships with or are owned by meatpackers such as Tyson Foods and JBS, the Brazilian meat giant. They marketed most of the nation's beef cattle.[4] Now the largest beef feedlots finish the vast majority of beef cattle. In 2008, the largest 12 percent each finished more than sixteen thousand and marketed nearly three quarters of all cattle.[5]

In a 2010 analysis of the latest agricultural census data, Food & Water Watch found that the number of beef cattle on feedlots larger than five hundred head had risen to 13.5 million in 2007, adding about 1,100 every day for five years. The five states with the largest inventories on feedlots all have more than a million factory-farmed beef cattle. In Texas the average is over twenty thousand head. In California, Oklahoma, and Washington, it is over twelve thousand.

Unfortunately, more than a hundred years after Upton Sinclair wrote *The Jungle* the meatpacking industry's disregard for workers, consumers, and livestock has not changed. In 1906 Sinclair wrote his novel to expose the mistreatment of immigrants in Chicago, using the stockyards as a backdrop. He is famously quoted as saying, "I aimed at the public's heart and by accident I hit in the stomach." Sinclair had spent several weeks doing undercover work at a meatpacking plant, where he experienced the horrible working conditions and witnessed the abusive labor practices, the exploitation of children, and unspeakable filth. One of the most attention-grabbing scenes was when workers fell into rendering tanks along with parts of animals and were ground up and sold as lard. As a result of the public outcry, the Pure Food and Drugs Act of 1906 was passed.

However, the market power of the five companies that controlled 60 percent of the industry at that time was a problem. In 1919, the FTC completed a report originally requested by President Woodrow Wilson that showed there was no competition in the meat industry and that it needed to be dramatically restructured. It showed there was a lack of competition in both the purchase of livestock and the sale of fresh meat.

In 1921 Congress passed the Packers and Stockyards Act (P&SA), a law designed to restore competition and fair trade practice to meatpacking. At that time, the big five—Swift, Armour, Cudahy, Wilson, and Morris—had become the big four because Morris merged with Armour. The Packers and Stockyard Administration (PSA), now called the Grain Inspection, Packers & Stockyards Administration (GIPSA), was formed after passage of the law to regulate livestock marketing activities at public stockyards and the operations of meatpackers and live poultry dealers.

The USDA-enforced PSA was designed to prevent meatpackers and processors from using unfair, deceptive, or unjustified discriminatory practices against producers. PSA also bars anticompetitive actions such as manipulating or controlling prices, creating monopolies, and conspiring to allocate territory or sales. USDA's enforcement of the PSA against meatpackers and processors has been uneven and limited, and some provisions of the act have never been implemented—almost one hundred years later.

Although the power of the meatpacker was weakened for about twenty years, during which more than two thousand small meatpackers operated, after World War II many small retail food stores with their own butcher shops were forced out of business as large grocery store chains emerged. Then, as fast-food chains became the largest purchasers of beef, the meatpacking industry began to consolidate again.

By the 1980s the stampede to consolidate had greatly intensified. President Reagan's Department of Justice was anxious to rewrite and weaken antitrust law. The agency filed a friend-of-the-court brief in support of Cargill/Excel, which was attempting to merge with Spencer, the nation's third-largest beef packer. Monfort, at that time the fifth-largest beef and lamb packer and distributor, tried to stop the merger, and the company's president was quoted in the *New York Times* as saying that it "scared the hell out of him."[6] Cargill appealed to the Supreme Court, winning in a 6 to 2 landmark decision that restricted companies like Monfort from filing private antitrust suits to block competitors from merging. By 1990, three companies dominated the meatpacking industry: IBP (now owned by Tyson Foods), Excel/Cargill, and ConAgra.

As consolidation increased, meat recalls became common. In July 2002 ConAgra recalled 19 million pounds of *E. coli*–contaminated beef in twenty states—then the second-largest recall in history. The giant food processor now concentrates on prepared foods, grain milling, and industrial seasonings and flavors. And in the last twenty years, the consolidation has continued to keep up with retail consolidation.

One of the other major changes in the beef industry is boxed beef. Pioneered by IBP, the process involves cutting up the carcass at the packing plant and delivering it in vacuum-sealed bags to wholesalers, supermarkets, and fast-food restaurants. This so-called technical advance has resulted in the large grocery chain driving further consolidation of the meat industry. Today, the big-four grocery chains that make more than 50 percent of sales are interested in dealing with only a few wholesale suppliers. Walmart, with its 20 percent market share, wants a large, steady stream of product from as few sources as possible. The same is true of fast-food restaurants, which buy large amounts of beef. They prefer to deal with large meatpackers that can supply high volumes of meat.

The Meat Price Investigators Association, a legal action group formed by five hundred operators of feedlots, was involved in several lawsuits against meatpackers and grocery stores in the late 1970s and early 1980s. The association found that "the supermarkets control the ultimate consumer demand for beef by the specials they feature in their meat departments. If the wholesale price of beef threatens to move higher they lessen the demand for beef by either raising their prices to a level where consumers hesitate to buy or feature other items, such as poultry, ham, etc."[7]

Of the roughly sixty-seven pounds of meat that each American consumes per year (on average), 65 percent is eaten at home and 35 percent is consumed at restaurants and from food service businesses. Midwesterners lead the nation in consumption, at seventy-three pounds, followed by Southerners and Westerners, at sixty-five pounds each. Rural consumers eat more than urban and suburban consumers, and low-income consumers eat more beef than households of other incomes.[8]

Ground beef is the only type of beef eaten more often at restaurants, with McDonald's, Burger King, and Wendy's representing 73 percent of all fast-food sales that include hamburgers. McDonald's, as the single largest buyer of beef, purchases a billion pounds a year at a cost of approximately $1.3 billion.

The market power of these fast-food chains, too, has led to more consolidation in the beef industry.[9]

The end result is that beef packing is the most concentrated industry in the livestock sector. Feedlots are getting larger in order to sell to an increasingly consolidated meatpacking industry, with just four firms—by market size, Cargill, Tyson Foods, JBS, and National Beef—slaughtering more than 80 percent of cattle today.

But the further processing of beef is becoming more concentrated as well. The top three players in beef also dominate this part of the industry, with Tyson leading and followed by JBS and Cargill. Processing includes the production of ready-to-eat meals, frozen hamburgers and marinated beef, refrigerated prepared foods, and canned products. For instance, JBS, with its international portfolio, manages a variety of brands, such as Friboi, Sola, Swift, and Anglo, and makes various companies' private-label brands.[10]

What are these corporations that control such a huge portion of our meat production and processing? Cargill is not only one of the largest players in the meat industry, it's also one of a handful of powerful corporations that control the entire global agricultural system. As of 2007, Cargill's feedlot business was the third largest in the United States, feeding seven hundred thousand head each year. In 2010, Cargill operated three cattle feedlots in Texas, one in Kansas, and one in Colorado. It holds at least as much power in the Canadian beef industry: according to the Canadian National Farmers Union, after XL Foods acquired Tyson's Canadian beef operations in 2009, Cargill and XL Foods controlled over 80 percent of beef slaughter in the country.[11] Cargill also has beef operations in Argentina and Australia.[12]

Tyson Foods, best known for producing poultry, is also a major player in beef. If the entire amount of hamburger the company produces were shaped into quarter-pound hamburgers, it would circle the globe ten times. Arkansas-based Tyson Foods was founded in 1935 by John W. Tyson, who began hauling chickens from Springdale, Arkansas, to market in Kansas City and St. Louis. In 1936, he delivered his first load of five hundred chickens to Chicago, netting a profit of $235. He wired the money home with a message to pay his debts and buy more chickens. By the late 1940s, Tyson had entered the feed and hatchery business. By 1958, with the addition of a slaughtering facility, the company had become fully vertically integrated. John W. Tyson also originated the idea of attaching chicken coops to a flat-bed truck for long-distance hauls, a system that is still used today.[13]

Tyson acquired dozens of other companies through the 1990s, and in 2001 it entered the beef business in a big way by acquiring IBP, the world's largest supplier of beef and pork at that time. Today, as the second-largest meat producer in the world, Tyson Foods has 114,000 employees in four hundred facilities. It says it is in the business of providing protein, and each week it slaughters approximately 450,000 cattle.[14]

JBS's founder, Brazilian Jose Batista Sobrinho, began with a small farm and slaughtered one or two cattle every day to sell to butcher shops. In 1953, he acquired a small slaughtering plant that killed five head every day, and in 1968 he began acquiring other facilities.[15] Today the company has 128,000 employees and operates on every continent. It began expanding in Brazil in the 1970s, and in 2005 it acquired its first foreign company, Swift Armour, Argentina's largest beef processor, followed by the American company Swift Foods in 2007. It went on to acquire companies in Australia, Europe, and Latin America, and bought the U.S.-based giant Pilgrim's Pride, the largest poultry company, at a time when it was facing bankruptcy.[16]

JBS used an alliance with Brazil's development bank and an aggressive acquisition strategy to make the company the largest purveyor of meat in the world. It went public in 2007, raising $5 billion, and went on a buying spree when cattle operations in the United States were struggling. Today, six of Jose Batista Sobrinho's children are in management, including his son Wesley Batista, who is chief executive of the company. Their strategy is global, focusing not only in the United States, but in Russia, China, and the Middle East.[17]

While significantly smaller than Cargill, Tyson, or JBS, National Beef is the fourth-largest processor, slaughtering 3.7 million cattle in 2010. Based in Kansas City, Missouri, the company also has operations in Kansas, California, Pennsylvania, and Georgia. It processes and markets fresh and case-ready beef, which is packaged for sale at retail stores; beef by-products; and leather. It is owned by U.S. Premium Beef, a vertically integrated company that produces cattle and processes and markets beef under a number of brand names.

These four companies enjoy a level of concentration that allows them to exert vast power over the market, resulting in tremendous leverage over independent cattle producers. The pressure to sell to larger meatpackers has encouraged independently owned feedlots to get bigger, in part to compete with the large meatpacker-owned ones. Because the large beef packers now also own their own cattle and operate feedlots, thus controlling supply through all stages of production, they have reduced their need to buy from independent and small operators. Packer-owned feedlots enable the meatpackers to drive down cattle prices, keep consumer beef prices high, and push down the prices paid to producers. Because meatpackers own and slaughter cattle, they can be sellers or buyers at an auction. They are sometimes on both sides of a sale, such as when the slaughterhouse they own buys cattle from one of the company feedlots. They distort or manipulate prices through these relationships. For example, the meatpackers can slaughter their own cattle when the price is high and buy at auction when prices are low, driving down prices for other independent cattle producers.

Company-owned feedlots can be immense. The world's largest beef processor, JBS, owns the Five Rivers Cattle Feeding company, which in 2010 had a capacity of 839,000 head on thirteen feedlots in Colorado, Idaho, Kansas, New Mexico, Oklahoma, Texas, and Wisconsin. The average Five Rivers feedlot has about a 65,000-head capacity, but the largest, in Yuma, Colorado, has a capacity of 125,000. In July 2010, JBS announced that it intended to buy McElhaney feedlot, one of the country's twenty-five largest feeding operations, with 130,000 head of beef.[18] This continued consolidation is dramatically increasing the market power of the Brazilian company, which is headquartered in São Paulo.[19]

Corporate-owned feedlots are much bigger than independently owned ones, and they lack roots in their local communities. Cargill, an international company headquartered in Minnesota, has feedlots in Texas, Colorado, and Kansas. Decisions made by JBS originate far away in Brazil. While farmers and ranchers drink the same water and breathe the same air as their neighbors, the corporate owners of these largest feedlots are located thousands of miles from any environmental problems they may create.

Most cattle feedlots are located in rural counties, yet the large numbers in these areas produce the same amount of waste as some of America's largest cities. The manure is stored on-site until it is spread onto nearby farm fields. But feedlots can flood or generate polluted runoff, and overapplied manure can leach into groundwater or leak into nearby waterways.

According to Duke University's Center on Globalization, Governance & Competitiveness, only about 60 percent of the larger feedlots have any formal written guidelines for environmental issues in general, and only about half have manure-management programs. Most feedlots use waste lagoons to capture runoff, and many use berms and fencing or landscaping to control runoff and minimize erosion.

Runoff from spray fields and lagoons pollutes waterways and drinking water with heavy metals, pathogens, antibiotics, and ammonia. Effects on human health occur from exposure to bacteria, viruses, or other toxics in the animal waste. Water pollution around feedlots raises nitrate levels and can cause "blue baby syndrome" in infants, a disease that can be fatal or cause developmental problems. The nitrogen pollution also causes algae to grow uncontrollably, choking out sunlight and nutrients needed by fish and plants. Feedlots in Texas are partially responsible for the seven-thousand-square-mile dead zone in the Gulf of Mexico. Numerous lakes in the state have been seriously polluted by feedlots.[20]

By the last decade of the twentieth century, ranchers could see that the future for independence was bleak, given the ever-increasing power of the meatpacking industry. Although the existing laws could be used to curb coercive market power, the U.S. Department of Justice and USDA have taken a laissez-faire approach to agricultural market power in recent decades. Pressure mounted on lawmakers to force GIPSA to use the power that was given to them when Woodrow Wilson was president to level the playing field and allow small producers to compete with large-scale factory farms. The "fair farm rules" have been waiting to be written by the agency since 1921.

An informal coalition of farm organizations—the National Family Farm Coalition, the Rural Advancement Foundation International (RAFI), the Ranchers-Cattlemen Action Legal Fund (R-CALF), and the National Farmers Union—were joined by the Western Organization of Resource Councils, Food & Water Watch, the Organization for Competitive Markets (OCM), the Missouri Rural Crisis Center (MRCC), the Iowa Citizens for Community Improvement, and others to fight for a new section in the 2008 Farm Bill that would provide a more competitive market for livestock growers. While the alliance did not achieve all of its goals, the massive piece of legislation included a provision that directed the little-known USDA department GIPSA to write rules for enforcing the 1921 law.

GIPSA's mission, as part of the USDA's Marketing and Regulatory Program, is to ensure a competitive marketplace for the benefit of American agriculture and consumers. However, the agency has been very lax in executing its mission, because agribusiness wields its enormous economic and political power to intimidate the political appointees at USDA and the career staff that report to them.

In defiance of the meatpacking industry, President Obama promised during his first campaign that, if elected, he would help fix the rules that allow the meat industry to take advantage of the people who raise the animals Americans eat. Subsequently, the meatpackers applied political pressure on President Obama and on members of Congress to prevent the rules that emerged from stopping noncompetitive practices, manipulating the livestock market, and abusing poultry growers through unfair contracts.

In May 2011, 147 members of Congress proved once again that the meat industry can buy public policy. The trade associations for Big Meat—major campaign contributors in congressional races—wrote to Secretary Vilsack in an effort aimed at obstructing the GIPSA rules. The letter, initiated by Representative Jim Costa, a Democrat from the Central Valley of California, was typical of the sabotage routinely directed at antitrust laws. The letter asked for Vilsack's "prompt response to the concerns that have previously been raised on this matter" and his "commitment to conduct a more thorough economic analysis."[21]

Unfortunately, while the administration got off to a good start in keeping its promise, the president lacked the courage to move forward rules that would address the market manipulation and uncompetitive behavior of the meatpackers and capitulated to industry pressure. In the winter of 2011, toothless rules were finalized that will do little to curb the market power of the meat industry.

No one knows better the lengths to which the food industries will go to get their way than Dudley Butler, the former administrator of GIPSA, who was appointed to the office by President Obama. Soon after the administration's surrender to the meat industry, Butler resigned. He is a Mississippi trial lawyer and rancher who has represented contract poultry growers in disputes with the chicken titans, and he made no bones about his intention to protect smaller growers against agribusiness. He was attacked viciously and repeatedly by the meatpackers, the industry trade association, and their lobbyists during the two-year process of investigating and writing the rules.

One of the ferocious attacks on Butler was for speaking at a 2009 meeting of the OCM, an organization focused on enforcing antitrust law and establishing competitive markets. As John Howell, the publisher of Butler's hometown Missisippi paper *The Panolian*, wrote:

> Also not surprising have been the attacks from the farming corporatocracy. "GIPSA's J. Dudley Butler threatens U. S. Livestock Production," an August 2010 headline in *Beef Magazine* stated. The story that followed likened him to a "fox guarding the henhouse." Butler has also undergone pummeling in hearings before the House Agriculture Committee whose members enjoy the largesse of campaign donations from the beef and meatpacking industry. He has been derogatorily described as a "notorious plaintiff's attorney," and "one of the 'Johnnie Cochrans' of ag law" in an Internet publication.[22]

Beef Magazine, a pro-industry trade journal, used a partial response to an audience member's question at the OCM meeting to make it seem as if Butler was intending to encourage more lawsuits against the meatpacking industry with the GIPSA rules. R-CALF CEO Bill Bullard later called the beef trade journal out: "Meatpackers' apologists, some of whom claim to be journalists, are openly engaged in an unethical smear campaign targeted at Dudley Butler. . . . The purpose of the proposed GIPSA rule is to prevent monopolistic meatpackers from capturing control of the livestock supply chain." Fred Stokes, former executive director of OCM, charged that Butler's comments at the meeting were taken out of context and "unethically twisted" by the meatpackers. He went on to say, "I carefully reviewed the taped conversation . . . and there is no way any ethical person could draw such a perverse conclusion."[23]

Advocates had been cheered by Butler's appointment, because he made clear that he was going to enforce the PSA, and the meat industry knew that they were in for a fight. He was instrumental in starting a process to make GIPSA regulations more protective of livestock growers and for holding hearings on agriculture and antitrust issues. During 2010 five hearings were held in Iowa, Alabama, Wisconsin, Colorado, and Washington, D.C. Thousands of farmers, ranchers, and contract poultry growers traveled long distances to tell USDA and DOJ officials—at the first-ever joint agency hearings—about the need to implement the GIPSA rules.

Ranchers testified repeatedly about how the concentration in the industry had pushed down the price that beef producers receive at auction because there is no competitive pressure to bid up prices. In many cases only one, sometimes two, of the major beef packers will attend a feedlot auction, and sometimes only one buyer actually bids. Nearly three out of five feedlots sell auction cattle to a single beef packer, which keeps prices low.[24]

USDA-commissioned studies in the 1990s also found that concentration drives down the price beef producers receive. These prices have fallen steadily over the past two decades. Cattle producers receive only a small and declining amount of the auction price: the net return has fallen from $36 a head between 1981 and 1994 to $14 a head between 1995 and 2008.[25]

Meatpackers supply their slaughterhouses with a combination of cattle they buy at auctions, some they already own, and others secured with contracts with feedlots or producers. The contracts, which are drawn up by the meatpackers, are known as "captive supply arrangements." Beef producers or feedlots will agree—via a marketing contract—to deliver cattle to meatpackers in the future. Often the agreements allow meatpackers to lower the agreed-upon price upon delivery. The USDA estimates that two fifths of slaughtered cattle are obtained through captive supply.

These cattle producers receive lower prices than they might get at auctions and receive worse terms than a more favored supplier. The meatpacker often manipulates the price, since it is dominant in both buying and selling. Captive supply arrangements also favor some producers with special premium prices and terms. They are often the giant feedlots that can provide many heads. This disadvantages smaller sellers, who must rely on the cash market that meatpackers dominate. And these arrangements are all confidential, creating a market that is so opaque that one supplier has no idea what prices others are receiving.

Ranchers said loudly and clearly that the captive supply reform proposal, which has been sitting at the USDA since the 1990s and has been part of debates during the last two Farm Bills, should be adopted. They would allow contracting only if they were based on prearranged, set prices and firm dates of delivery, and if the contracts are transparently and publicly offered. Meatpackers would be prohibited from using a pricing system that could provide unfair advantages to some producers and disadvantage others.

Allan Sents testified at the Fort Collins, Colorado, DOJ/USDA hearing about the power and leverage of the packers. Sents operates a ten-thousand-head commercial feedlot in central Kansas and has been working in the industry for forty years. He explained to agriculture secretary Tom Vilsack that one of the packers he deals with had offered captive supply arrangements to some of his competitor feedlots even though he is closer to the slaughter facility, and they would often take no cattle from his operation. He went on to describe the situation:

> So I told the buyer [packer], I said, "Well, if you're going to discriminate . . . against using that way, I'm going to allow the other packer buyers first opportunity to buy our cattle." I thought it was a turn about—a fair play–type of thing. Well, the original buyer then didn't like that kind of response and told his buyer [representative] to quit coming into our yard. So for three months we didn't get a representative from that major packer into our yard just because we had tried to play ball the same way the packer was trying to deal with us; a very evident sign of intimidation, and why you hear these stories of why producers are afraid to stand up.[26]

At the hearings independent ranchers such as South Dakotan Bob Mack discussed how the meatpackers have distorted the market and presented evidence why Congress should ban packer and processor ownership of livestock.

During the debate over the GIPSA rules, the National Farmers Union's president, Roger Johnson, called for a farmers' bill of rights, arguing: "The

Packers and Stockyards Act has been around for ninety years. It's time to start enforcing it. The Farmer and Rancher Bill of Rights will protect farmers and ranchers from anticompetitive behavior by packers and processors. Industry will no longer be allowed free rein to abuse livestock producers who have limited market power."

During the 2011 attempts to defund implementation of the GIPSA rules, the president of the Wisconsin Farmers Union, Darin Von Ruden, similarly stated that the rule "will help level the playing field for independent family farmers raising livestock by creating a fair market environment in the industry, making sure small producers have equal access to the market as large producers." He went on to say that the big meatpackers "have tremendous power to dictate not just what meat is available, but how that meat is raised." [27]

R-CALF, an organization that represents thousands of U.S. cattle producers, has been one of the most steadfast advocates for antitrust reform and for implementing rules to establish a competitive livestock market. The organization's slogan, "Fighting for the U.S. Cattle Producer," is exactly what it has been doing through lawsuits, lobbying, and advocacy.

In a briefing paper on the need for regulation, R-CALF does not hold any punches, stating: "The Packer Lobby wants to chickenize the fed cattle industry, relegating it to an industry where the terms of production and terms of marketing are controlled by the packers themselves. . . . Once the beef packers achieve their goal of forcing independent cattle feeders out of the cash market, they will dictate and control the terms of production and terms of marketing for fed cattle through contracts, just as they now do in the poultry and hog industries." [28]

Former Nebraska state senator Cap Dierks, hailing from the hills of north-central Nebraska, is the type of legislator who could never be bought by the industry. Raised in Ewing, where he still lives today, his family has been ranching in Nebraska since the 1880s. His youngest son now manages the farm, although his other three children maintain an interest in farming. Dierks has seen the massive changes to agriculture, and he really hopes to see in his lifetime that independent ranchers and farmers are able to compete fairly.

In 2010 Dierks introduced a bill advocating for landowners along the proposed route of the Keystone XL pipeline, a project that, if implemented, will adversely affect farms and ranches. The bill would have required TransCanada, the company building the pipeline, to set aside funds for leaks, repairs, and other mitigation costs. Dierks also introduced a bill in 2006, and again in 2009, that would restrict non–family farm corporations from acquiring or obtaining an interest in any farming or ranching operation in Nebraska. (In

November 2010 a twenty-four-year-old conservative activist, Tyson Larson, who campaigned against Dierks's age, defeated him. Although now retired, Dierks continues to lobby the Nebraska legislature on food safety issues and large animal medicine and to stop the "chickenization of beef.")

Dierks is representative of many of the individuals who have been fighting long and hard to stop the abusive practices of the meat industry. Even though industry won this round, the coalition fighting for fair competition is determined to use the next Farm Bill to reopen the processes. It will take a broad-based coalition effort to prevail over an industry willing to use dirty tricks to maintain their market power over livestock producers and to drive the independent rancher out of business.

The Panolian's Howell sums up the situation: "Anyone who has stirred that much reaction from big lobbying and industry groups might be doing something right." [29]

9

HOGGING THE PROFITS

There is a sufficiency in the world for man's need but not for man's greed.
—Mahatma Gandhi

Industrial hog operations are a very nasty business. Thousands of pigs are packed together tightly in giant warehouses, where they generate tons of liquid and solid waste, posing health hazards to the surrounding community and degrading the environment. The feces and urine produced in hog factories—containing ammonia, methane, hydrogen sulfide, cyanide, phosphorus, nitrates, heavy metals, antibiotics, and other drugs—fall through slatted floors and into a catchment pit under the pens. Giant exhaust fans pump the toxic fumes out of the warehouse-like buildings, twenty-four hours a day, to prevent the hogs from dying. The accumulated waste is then pumped into enormous lagoons that can cover six to seven and a half acres and hold as much as 45 million gallons of wastewater. Leaking and flooding lagoons pollute local waterways, and the fumes from the waste spread on neighboring fields choke and sicken the local community.

Raising pigs, which are social animals with high cognitive abilities and excellent memories, under these harsh conditions is cruel. In his landmark *Rolling Stone* article "Boss Hog," Jeff Tietz describes the heartbreaking life of the factory-farmed hog: "Smithfield's pigs live by the hundreds of thousands in warehouse-like barns, in rows of wall-to-wall pens. Sows are artificially inseminated and fed and delivered of their piglets in cages so small they cannot turn around. Forty full-grown 250-pound male hogs often occupy a pen the size of a tiny apartment. They trample each other to death. There is no sunlight, straw, fresh air or earth." [1]

Pigs have not always been raised this way. One of the unusual factors about the rise of the factory hog farm is that it happened very quickly. In 1992 less than a third of U.S. hogs were raised on farms with more than two thousand animals, but by 2004 four out of five hogs came from one of these giant operations, and by 2007, 95 percent were.[2] An analysis by Food & Water Watch found that between 1997 and 2007, 4,600 hogs were added to a factory farm every single day, increasing the total to more than 62 million. As with most industrialized animal production, the industry has selected certain parts of the country to host hog factories, with Iowa, North Carolina, Minnesota, Illinois, and Indiana producing two thirds.

This transformation was facilitated by federal policy changes pushed by the pork packers and other agribusinesses. Chief among these was the overproduction of corn, soybeans, and other crops used for hog feed that made the feed price artificially low—below the cost it took to raise the crops. Permitting these crop prices to fall below their cost of production and then paying farmers some of the difference with taxpayer dollars subsidized the growth of hog factories.

These artificially low feed prices also encouraged livestock producers to buy feed rather than pasture their livestock or grow their own. Since producers no longer needed land for these purposes, it became economically feasible to confine large numbers of animals together in factory farm facilities without an enormous amount of land.

The Environmental Protection Agency's disjointed, toothless, and lax oversight of factory farms further enabled hog operations to increase in size. Weak oversight of waste disposal, a major expense of hog operations, reduces the costs of factory farming and encourages the development of larger and larger operations. Adequate oversight was blocked repeatedly by the livestock industry, which opposed any regulation of these pollutants.

Further facilitating the growth of hog factories has been the failure of the Department of Justice to prevent the largest meatpackers from merging into a virtual monopoly. The wave of mergers and acquisitions has concentrated the pork-producing sector into the hands of a few powerhouses that employ heavy-handed tactics that minimize the prices they pay for livestock.

This has meant that independent hog operators who sell their livestock on open markets have nearly disappeared in the face of massive consolidation in the industry. Two out of three hogs are now slaughtered by the four largest pork processors. These companies not only slaughter and process the hogs, but they exert tremendous control over farmers through production and marketing contracts. As pork processors have come to increasingly own the hogs they slaughter, the vertical integration and control of the hog sector has pushed prices down and encouraged operators "to get big or get out."

This trend has been documented by the USDA. In 1993, almost all hog sales (87 percent) were negotiated purchases between farmers and packers or processors (known as "spot market" sales). Because the packers relied on family farmers, it gave these producers some negotiating power with the large corporations. But by 2006 nearly all hogs (90 percent) were controlled by the packers, either by owning their hogs outright (20 percent) or through production-contracted hogs (70 percent).[3]

These arrangements result in a market supply held captive by packers and depress the spot price, making it completely unfeasible to have a small hog farm. During the period from 1989 to 1993, before the massive growth of factory hog farms and contract production, the average monthly price was $75 per hundredweight (i.e., one hundred pounds, a standard measurement of weight for some livestock). During the 2004 to 2008 period, average monthly hog prices were $52 per hundredweight, a 31 percent decline. A USDA-funded study found that a 1 percent increase in the use of packer ownership or contract production causes the spot market to fall by nearly the same amount (0.88 percent).[4]

In a report called "Killing Competition with Captive Supplies," the Minnesota-based Land Stewardship Program (LSP) found that "packer control of the market is pervasive" and that "farmers reported facing daily what they call a mind game, which they describe as pressure from agricultural leaders to conform to the new factory farm system of hog production." Among the report's findings: "Packers' practice of acquiring captive supplies through contracts and direct ownership is reducing the number of opportunities for small- and medium-sized farmers to sell their hogs. With fewer buyers and more captive supply, there is less competition for independent farmers' hogs and insufficient market information regarding price. Lower prices result."[5]

Today pork processors, like Smithfield, by far the largest pork producer and packer in the United States and the world, engage in abusive contract relationships with factory farm operators. These operators assume economic risk by taking out large loans to build the warehouse-like facilities and equipment that have automated feeding systems. Contracts can require farmers to build or upgrade facilities, which can require significant investments. For a typical hog-finishing operation, six eleven-hundred-head hog houses typically cost between $600,000 and $900,000.[6] In 2005, three out of five hog operators (61 percent) were required to make these capital investments.[7]

1 **Cargill**
2 **Tyson**
3 **JBS**
4 **National Beef**

80% of all US cattle

1 **Smithfield**
2 **Tyson**
3 **JBS**
4 **EXCEL**

66% of all US hogs

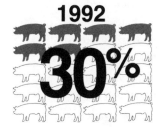

1992
30%

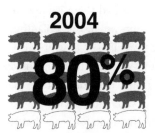

2004
80%

2007
95%

change in percentage of hogs that are raised on **factory farms** (more than 2,000 animals)

This debt makes the hog factory operator extremely dependent on the cash from the operation to pay back the loans they have taken out for building and upgrading facilities. Operators that rely on a steady contract relationship with a packer or processor are unable to complain about shoddy treatment or unfair terms for fear of retaliation that could end their business.[8]

The terms for hog production contracts can significantly disadvantage producers. Some provide a strict management manual for growers, eliminating the producer's autonomy to make decisions.[9] Operators are also responsible for securing permits for manure disposal and for taking on the environmental liability associated with it.[10] Some contracts even have a provision that allows the pork packer to evict farmers from their own hog barns and force them to hire company-selected managers to finish if the packer decides that the farmer was not properly caring for the livestock.[11] In many cases the companies use the contracting process to force factory operators to increase the scale of their operations or lose the sales.

Unfortunately, a window of opportunity for tackling contract abuses was lost in the fall of 2011, as described in the previous chapter, when the Obama administration capitulated to the meat industry and failed to reform the rules.[12]

Reforming these GIPSA regulations has been at the top of the agenda for many hog producers for many years. Chris Peterson, a farmer from Clear Lake, Iowa, and president of the Iowa Farmers Union, views the packer ownership and control of hogs through contracts as the major impediment to the return of a level playing field for farmers. At the 2010 USDA/DOJ joint hearings held in connection with revising the GIPSA rules, Peterson testified in support of them. He said that when he started raising hogs in the 1970s, "Iowa had tens of thousands of independent hog producers." But now, he said, "that market has been ruined. . . . It basically no longer exists." Peterson explained:

> Out in the countryside you had them calling you on days they needed hogs. And I remember times when I was planting corn or whatever and, my gosh, I don't want to really sell hogs today, you know. But then the packers start calling, and you play them against each other. And lo and behold, there comes a time when you shut the planter off and you make an extra $5 or $10 a hog. You load up a load of hogs. And I tell you what, one thing we've been trying to do for years that I stand for, if we want to solve this problem: ban packers from owning livestock.[13]

The pork industry, along with the rest of the meat industry, went into high gear lobbying vociferously against the rule. They used all of the economic and political power they had in Washington, D.C., to pressure the Obama administration into capitulating, including running strident ads in cities that Secretary of Agriculture Tom Vilsack visited, saying that the rule would kill jobs.[14]

A number of meat-related trade associations, including the National Pork Producers Council, which is heavily influenced by packers like Smithfield, hired an agribusiness-friendly economics consulting firm to report on the "economic impacts analysis" for use in their lobbying campaign. Their analysis supported the assertion that the pork industry would experience $69 million in upfront impacts, $79 million in ongoing impacts, and $259 million in indirect costs. The reason given for these losses: if GIPSA required changes in marketing agreements and contracts.[15]

Unfortunately, the industry won this round. And Smithfield was one of the big winners.

As the largest of the four corporations that together control 66 percent of the U.S. hog market, Smithfield is politically and economically powerful.[16] It is also the largest hog producer and processor in the world, and each year it slaughters approximately 26 million hogs; company-owned farms produce 16 million of the animals it slaughters.[17] Smithfield now owns more hogs than the next eight largest pork producers combined (among them Tyson Foods, the giant poultry multinational; JBS/Swift, the largest beef packing corporation; and Excel, a processor of beef, pork, and prepared meats).[18]

The company's roots are in the small town of Smithfield, Virginia, where in 1936 Joseph W. Luter and his son Joseph W. Luter Jr. opened the first packing plant. Joseph Luter III sold the firm in 1969 to develop a ski resort in Virginia, but Smithfield was failing by the mid-1970s and management lured Luter to rejoin the company.[19]

Under Luter, the company went through what in corporate speak is called "a thorough business reorganization,"[20] which consisted of buying out other investors' shares, firing managers, and embarking on a high-speed growth plan centered on gobbling up other meat companies. In 1978, Smithfield purchased a plant in Kinston, North Carolina, and by 1981, when it bought out local rival and longtime pork competitor Gwaltney of Smithfield, it had doubled in size. Next Smithfield expanded into the Midwest, when it bought the Wisconsin-based company Patrick Cudahy. Baltimore-based Esskay was purchased two years later. But the real turning point came in 1987, when Luter launched a fifty-fifty partnership with the then fifth-largest pork producer, Carroll's Foods. Now, Smithfield was not just slaughtering hogs but also raising them.

By the 1990s, Smithfield had become truly vertically integrated, control-ling the hogs from birth all the way through processing, and even controlling their genes. Luter made an exclusive deal with the British firm National Pig Development Company to develop a "genetically perfect" pig that was lean and easy to process.[21] Today Smithfield owns several specific genetic lines of hogs. The company is known as Smithfield Premium Genetics, and it markets the genetically manipulated products under the label Smithfield Lean Gen-eration Pork.

Smithfield continued devouring competitors in both the production and processing industries, and it ventured into other meats. The number of acquisitions provides a sense of the company's voracious appetite for control-ling the industry. They include: Valleydale; John Morrell; Lykes Meat Group; North Side Foods; Moyer Specialty Foods; Packerland Packing; Stefano Foods; Farmland Foods; Cumberland Gap Provision; Cook's; Armour-Eckrich; But-terball; Murphy Farms; Vall, Inc.; Alliance Farms; MF Cattle Feeding; Five Rivers Feeding; and, finally, Premium Standard Farms.

It is known for its dirty tactics. When it announced the intention to pur-chase Murphy Farms, then Iowa attorney general Tom Miller sued Smithfield for violating the state's corporate farming law, which strived to prevent mo-nopoly ownership.[22] Rather than delaying the merger, Smithfield and Murphy Farms blatantly concocted a sham transaction to transfer ownership. Murphy sold its Iowa assets to former manager Randall Stoecker, who in turn estab-lished a corporation called Stoecker Farms. As the sole officer and shareholder, Stoecker received a loan of more than $79 million from Murphy Farms to acquire the Iowa assets. His only payment was in the form of two promissory notes, which deferred most of the debt for ten years. Immediately afterward, Murphy transferred all of its remaining, non-Iowan assets to Smithfield—including control of Stoecker's finances—in effect giving Smithfield complete control of what had been Murphy Farms' assets.

Through the use of two lawsuits, Smithfield successfully overturned Iowa's corporate farming law, which prohibited vertical integration through the ownership of livestock by packers. In 2005, during the time the second lawsuit was being appealed, Smithfield reached a settlement with the state's attorney general.[23] Miller agreed not to enforce the law against Smithfield, and Smithfield in turn gave $1 million a year over ten years—a paltry amount that a major corporation considers the cost of doing business—to fund a ten-year environmental training program and $240,000 over four years to fund scholarships at Iowa State University. Smithfield also agreed to pay $100,000 annually for ten years to fund "innovative" hog production. So it goes with companies that have tremendous resources and political power.

In 2006, when Smithfield announced its intention to buy the nation's second largest pork company, Premium Standard Farms, the alarm bells started ringing. A September 19, 2006, press release from Iowa Republican senator Chuck Grassley said that the United States needs family farmers and independent producers to make the free and open market system work, and that expanded packer ownership of hogs, exclusive contracting, and captive supply are all adversely affecting their ability to compete in the marketplace. Grassley stated:

> I cannot fathom how Smithfield, which is the largest and fastest-growing integrator, can continue to be allowed to purchase hog operations across the country. Over the last several years Smithfield has made it perfectly clear that it intended to purchase its competitors to assert its dominance in the pork industry. This is alarming. I expect the Justice Department to take a serious look at this merger.[24]

The Department of Justice Antitrust Division, the agency responsible for enforcing the nation's competition laws, did take a look, but unfortunately it rarely finds the courage to say no to a merger. In 2007, the agency announced it would allow the merger to move forward. Grassley concluded: "It looks like nobody's going to stand in the way of all this vertical integration until we've just got one meatpacker in the country. Maybe then the Justice Department will figure out we've got a problem."[25]

Butt Smithfield was not getting a free ride in Iowa: Iowa Citizens for Community Improvement (ICCI), based in Des Moines, has been fighting back. Four Catholic priests, including Joe Fagan, now a former priest, originally formed ICCI in Waterloo, Iowa, in 1975 as a secular social justice organization. Purportedly retired, Fagan is still a crackerjack activist and was recently lambasted by the right-wing blog *Uncoverage: The Right Idea* for questioning a Republican presidential candidate at a public meeting about his views on privatizing Social Security.

Fagan, who was born on a farm in 1940 and attended a one-room school in Dubuque County, was an associate priest at a Waterloo Catholic Church when he decided to help start ICCI and become a community organizer. At its inception, the organization was funded by eight churches, and it still receives some faith-based funding, as well as foundation money. Nowadays, much of its money comes from its 3,300 members, who not only pay dues but also are active in the group's grassroots campaigns. ICCI has members in ninety-eight of Iowa's ninety-nine counties, and a staff of sixteen.[26]

Originally, the organization focused on community issues like removing abandoned properties, funding neighborhood improvements, and preventing utilities from gouging customers. But during the farm crisis the group started working on agriculture issues, such as renegotiating mortgages for more than two hundred farms and helping to secure $32 million in loans for small and midsize farms. ICCI's current director, Hugh Espey, is the high-energy, strategic mastermind behind many of the tactics that have made ICCI so successful, such as building a strong coalition that links farmers, environmental, labor, consumer, and immigrant and civil rights under a single social justice banner. Under Espey's leadership ICCI has built a statewide membership that has the power to win some good legislation and stop bad legislation in the Iowa statehouse that agribusiness wants.

Espey grew up in the Mississippi River town of Quincy, Illinois, where his father was a doctor and his parents were Goldwater Republicans. When he left home and went to college at Drake University in Des Moines, he was exposed to a bigger world and began questioning the values he had grown up with. After receiving a BA in sociology and going on to the University of Iowa for an MA in sociology, he saw a newspaper ad for a community organizer position with ICCI that would help make a difference in Council Bluffs, a town located along the Missouri River in western Iowa. Espey worked on neighborhood issues and recruited member activists for ICCI. He went on to work for ICCI in Sioux City and eventually became ICCI's executive director.

Espey explains that ICCI started working on agriculture issues because the farm crisis of the 1980s was devastating for the state. Banks simply weren't willing to work with farmers, who were about to lose their homes and livelihoods. In the early 1990s ICCI began to organize to stop factory farms, but in 1995 they jumped in full force, with Missouri Rural Crisis Center (MRCC) and other organizations in the Campaign for Family Farms and the Environment (CFFE), an effort still going strong today.

CFFE has been an important factor in the Midwest in helping local groups succeed in fighting factory farms; the cooperation between MRCC, ICCI, and the Land Stewardship Program (LSP) is a strong multistate effort with a progressive and inclusive agenda. The coalition bridged racial and urban-rural divides and was able to include environmental, labor, faith-based, and consumer organizations in the fight against factory farms.

In Iowa, the strategy was to organize at the local level to build political momentum for better enforcement of state and federal laws. ICCI has an impressive track record in Iowa, including stopping the construction or expansion of over six dozen factory farms. They have helped neighbors of factory farms successfully reduce their property taxes when their homes were devalued by odor and pollution. ICCI helped mount litigation when they created a nuisance that severely affected citizens' quality of life. The organization has won numerous battles over local control of factory farms by challenging the permitting process of Iowa's Department of Natural Resources (DNR) and forcing the agency to issue stiff fines and penalties to factory farm polluters. ICCI's campaign resulted in the agency referring pollution cases to the Iowa attorney general for stronger enforcement action. Despite opposition from extremely powerful agribusiness opponents, ICCI won Iowa's first clean air fight by obtaining a standard for hydrogen sulfide, a chemical emitted from factory hog farms, in September 2004.

When Republican governor Terry Branstad announced that he wanted to double the output of Iowa's factory farms by 2050, ICCI went on the offensive. ICCI member and independent livestock farmer Garry Klicker told the governor at a listening session in February 2011: "We don't need less environmental regulations. We need local control, stronger permitting standards, and tough new fines and penalties to crack down on factory farm polluters."[27]

Klicker, a longtime ICCI leader and factory farm fighter, makes no bones about how he feels about it. He told the Burlington *Hawk Eye*, "My feeling is, if a farmer is raising hogs in a confinement, they probably need to be visited by [Iowa's Department of Human Services], because they probably have kids in the closet."[28]

Although Iowa has the largest number of hogs in the country—almost 18 million were counted in the last agricultural census—ICCI is continuing to successfully oppose the construction of new and expanding facilities.[29] A five-thousand-head Cargill operation was stopped in Adair County in August 2011; the ICCI staff worked with their membership to "educate" and "influence" the Board of Supervisors to recommend to the DNR that the permit be denied. ICCI won 2 to 1. Next it invited DNR staff to the county and organized a meeting that included many of the neighbors of the proposed facility. In an unusually quick win, the DNR denied the permit without the Board of Supervisors appealing.

Iowa is not the only state where activists are fighting factory hog farms. Rhonda Perry, executive director of the MRCC, has been fighting the same types of battles in her state. She and her husband, Roger Allison, one of the founders of MRCC, have spent the last twenty-five years facing down the pork industry and some of the other largest corporate agribusinesses in the world, and against all odds they have won a significant number of rounds. Both were raised on family farms in Missouri, coming of age during the farm crisis of the mid-1980s. They met and married during the early years of their struggle for family farm justice.

In 1985 Roger Allison's parents faced foreclosure on their family farm. At that time, it was the federal government that was keeping the prices farmers received much lower than the cost of doing business, even though the cost of fuel and other inputs had increased dramatically. USDA's Farm Home Administration, the lender of last resort for farmers (since renamed the Farm Service Administration), was holding hundreds of auctions on the steps of local courthouses, selling farms for pennies on the dollar.

Allison decided to fight back through litigation and political organizing. His family sued the USDA for the illegal foreclosure and won; the lawsuit was eventually part of a large class-action suit against the agency, which finally halted the unlawful foreclosures.

Eventually Allison and Perry were able to buy Allison's parents' farm from the USDA to add to their own. Today, besides continuing to fight for family farmers at MRCC, they raise row crops and seventy-five cattle in a cow-calf operation, and also help run Patchwork Family Farms, a co-op of fifteen independent family hog farmers who sell directly to consumers.

Perry says that before the onslaught of factory hog farms, the state had been well suited to small-scale and diverse family farms. Hogs were a good supplement to farm income, because they mature quickly, and if prices dropped, a farmer could just grow fewer hogs the next time around. When Perry was a girl, hogs were ready to go to market frequently, and her family would call the five places they could sell to, and then choose the one with the best price. Today there is almost no place in Missouri that an independent farmer can take hogs to market: the giant integrator Smithfield has all but taken over.

Today Missouri is a tragic case study of the impacts of corporate hog farming—a sad story reflected statistically in a loss of farms that happened overnight. According to the USDA, the number of small, family-farmed hog operations in Missouri dropped from 23,000 in 1985 to 3,000 large ones in 2007, a decline of 87 percent. While the number of farms has declined dramatically, the number of hogs grown in the state has remained constant, with 95 percent raised in operations that have more than two thousand animals. Meanwhile, the consumer price of pork has increased 71 percent between 1985 and today, while the farmer's share has decreased 49 percent.[30]

Perry goes on to say that the profits from these higher prices are not flowing into the rural communities of Missouri. Local agriculture is the economic engine for rural areas. When big, vertically integrated corporations move in, they do not use the local fencing company or buy feed grain, farm equipment, or other necessary materials needed for farming from local businesses. So not only are farms consolidated and farm families displaced, but the economic well-being of the community is compromised.

While many of the issues are the same today as when MRCC was founded, Perry says "the enemy is different." The USDA was the main lobbying target in the mid-1980s, but today it's more complex, involving government policies and the economic power of agribusiness. One of MRCC's major missions since the early 1990s has been fighting factory hog farms.

Premium Standard Farms (PSF) was the first company to invade Missouri and develop a completely vertically integrated model. They built the first hog factory, with eighty thousand sows, in 1995 and built a processing plant at the same time. They described the model as being "squeal to meal." Because Missouri prohibits corporate-owned farms, the company executives claimed they were just some "good ol' boy" hog farmers from Delaware, coming to help make the business more efficient. They bought land; built giant facilities; provided their own breeds, piglets, and feed; and packaged the meat under their own brand name. When PSF went public it could no longer be considered a "family farm" operation in Missouri. So in 1993, in the dead of night on the last day of the legislative session, a three-county exemption to the anticorporate farming law was snuck into a larger bill—of course, these were the counties where PSF operated.

But the company was losing money, even though it claimed that its model was the future of hog farming. MRCC investigated PSF's Securities and Exchange Commission filings and proved that the company was losing money, and that the model was a failure. PSF went bankrupt, was restructured, and merged with Continental Grain Company. Eventually the new entity merged with Smithfield.[31]

However, MRCC's early entry into the battle to stop corporate hog farming has resulted in Missouri having fewer hogs than many Midwestern states—it is seventh nationwide. It was in the 1990s that MRCC began its valiant fight against factory hog farms, using a combination of grassroots organizing, litigation, legislative campaigns, and electoral work. One of the group's early feats was a 145-day demonstration in front of the USDA office in Chillicothe, Missouri, protesting the agency's policies. Every day, farmers and regular citizens joined the procession of tractors, and on Monday nights

MRCC held a meeting with farmers at the site. It is through these types of actions that the group has developed strong ties with the farming community throughout the state and has built the political power to successfully stop hundreds of factory farms.

Perry attributes much of MRCC's success and staying power to the fact that they have farmer members who are involved in the local organizing, because it's in their self-interest. The local control issue necessitates broad-based coalitions that sometimes include local elected officials, since MRCC has to go up against the biggest agribusiness corporations and their trade associations, including the Farm Bureau.

These regressive interests say they believe in property rights, but what they really mean is property rights for corporations. MRCC organizes around the issues related to property. People at the local level have the right to say if a hog farm should be permitted. This has enabled them to win many legislative battles, including passage of the Good Neighbor Act of 1996, which created the first state standards for factory farms. Every year since 2003 MRCC has successfully defeated attempts by corporate agribusiness to take away local control.

Sometimes MRCC has stopped hog factories at the local level before they are built, and sometimes it used state environmental laws or federal enforcement. More recently it has been passing health ordinances at the local level. It is able to do this because Missouri is one of the few states to have local control—meaning citizens can decide locally if they want to live next to a hog factory or a toxic waste dump. This has enabled them to organize at the county level to both stop hog farms and pass local health ordinances.

Meanwhile, as family farmers were fighting Smithfield in the Midwest during the 1990s, the company was seeking to find sources for hogs in the East. It was also refining its strategy for becoming vertically integrated—for controlling production from genetic research and breeding to packaged pork chop. In 1992, Smithfield entered into a joint venture with the North Carolina–based Carroll Foods, the fourth-largest hog producer in the United States. In 1999 it acquired the company outright, becoming the largest hog producer in the world.[32]

This sealed North Carolina's fate as the number two hog-producing state in the country today. North Carolina's 11 million hogs create the stench, sewage, and sickness that are the trademarks of the industry. In one animal factory located in the state, 2,500 pigs produce 26 million gallons of liquid waste, 1 million gallons of sludge, and 21 million gallons of slurry (a thin mixture of water and manure) per year.[33] The waste is then sprayed onto nearby agriculture fields, widening the area affected by the toxic chemicals.

These manure cesspools emit noxious odors into the surrounding communities. The stench has been known to nauseate pilots at three thousand feet in the air.[34] The odor affects the quality of life for people living in the rural communities near these facilities, many of whom can no longer hang out their laundry to dry, sit on their porches, or even open their windows. Even worse, residents experience a wide range of health problems, including asthma, allergies, eye irritation, and depressed immune function, along with mood disorders, such as heightened levels of depression, tension, anger, fatigue, and confusion.[35]

The burden of these facilities is concentrated in some of North Carolina's most impoverished areas. Nearly two thirds (61 percent) of the factory-farmed hogs are located in five eastern counties.[36] One study found that the state's industrial hog operations are disproportionately located in communities of color and in communities with higher rates of poverty.[37]

Hog farming has always been an important part of North Carolina's agriculture. North Carolina, where intense hog production increased significantly during the 1980s, embodies the risks created by the rapid rise of big pork-packing companies and factory farms. In the 1990s, lenient environmental regulations and local zoning exemptions attracted corporations like Smithfield and Premium Standard Farms, which transformed the state into a pork powerhouse. After Smithfield and PSF merged in 2007, Smithfield controlled an estimated 90 percent of North Carolina's hog market.[38] The state now has more hogs than people.[39]

But as the hog population surged, the number of farms plummeted more than 80 percent, as factory farms drove traditional farms out of business. In 1986, North Carolina had 15,000 hog farms, but by 2007 just 2,800 remained.[40] The state's hogs produce 14.6 billion gallons of manure every year.[41]

North Carolina's waters have been polluted repeatedly by waste from hog factory farms. The public first became aware of problems with the lagoon and spray field system in 1995, when a lagoon burst and released 25 million gallons of manure into the New River.[42] Lagoon spills were responsible for sending 1 million gallons of waste into the Cape Fear River and its tributaries in the summer of 1995,[43] 1 million gallons into a tributary of the Trent River in 1996,[44] and 1.9 million gallons into the Persimmon Branch in 1999.[45] Hog waste was also the likely culprit for massive fish kills in the Neuse River in 2003, when at least 3 million fish died within a two-month span.[46]

Perhaps the most infamous example of the environmental dangers of hog factories occurred in 1999, when Hurricane Floyd hit North Carolina. The storm flooded fifty lagoons and caused three to burst, leading to the release of millions of gallons of manure and the drowning of approximately 30,500 hogs, 2.1 million chickens, and 737,000 turkeys.[47]

In 1997 North Carolina established a moratorium on building new hog waste lagoons, and in 2007 the legislature made the ban permanent.[48] Unfortunately, this doesn't affect existing lagoons. Watchdog groups that have been tracking the industry for years note that, despite the moratorium, these existing lagoons continue to expand.[49]

And Smithfield's slaughterhouses are no better for the environment than its factory farms. The company's plant in Tar Heel, North Carolina, is the second largest in the world, slaughtering some 34,000 hogs each day.[50] It pulls 2 million gallons of water from the local aquifer daily and returns about 3 million gallons of wastewater to the Cape Fear River.[51] Like Smithfield's plants elsewhere, the facility has been cited for several environmental violations.[52]

In 1997 the company received one of the largest Clean Water Act fines in U.S. history after officials found that Smithfield and two of its subsidiaries in Virginia failed to install decent pollution equipment and to treat its waste, resulting in five thousand violations of the company's permitted limits for phosphorus, fecal coliform, and other pollutants. The pollutants flowed into the Pagan River, the James River, and the Chesapeake Bay for more than five years.[53] Judge Rebecca Beach Smith stated that Smithfield's violations "had a significant impact on the environment and the public, and thus in total their violations of the effluent limits were extremely serious."[54] The company was fined $12.6 million,[55] which amounted to 0.035 percent of its annual sales.[56]

But it is not just the plant's environmental record that makes it so infamous. It's also Smithfield's labor practices.

According to "Packaged with Abuse: Safety and Health Conditions at Smithfield Packing's Tar Heel Plant," a report commissioned by the United Food and Commercial Workers union (UFCW), Smithfield has engaged in abusive labor practices in several ways. The report alleges that Smithfield employee Vanessa McCloud's job for seven years was to cut the skin off of frozen pork as it came down the line at breakneck speeds. One day, Vanessa slipped a disk in her back while on the job. She was not able to return to work immediately and was fired, according to the report. She received no worker's compensation and has since applied to Medicaid in hopes of paying her medical bills. She has no idea how she will support herself and her children because of her debilitating injury.[57]

The report discloses that Vanessa's experience is typical of workers at the plant. To meet production goals, the processing lines move extremely fast, and workers who fall behind have reported being verbally abused or even fired.

Others do their best to keep up, but very few work in the plant for more than a few months before experiencing an injury from the grueling work.[58] The list of injuries reported at the plant is lengthy—the report claims that repetitive motion disorders, such as carpal tunnel syndrome, contusions, blunt traumas from slipping and falling on wet floors, cuts and punctures, infections causing the fingernail to separate from the finger, fractures, amputations, burns, hernias, rashes, and swelling are all potential dangers to workers—and injuries are on the rise. From January to July 2006, 463 injuries were reported at the Tar Heel plant, up from 421 during the same period the previous year.[59]

The report goes on to say that instead of helping the wounded workers, Smithfield uses intimidation to prevent them from reporting their injuries. Even when they do report them, they are often denied workers' compensation. Then, because of their disabilities, they frequently cannot find gainful employment again.[60]

Smithfield is engaging in a legal vendetta; it is suing the UFCW, the research group that wrote the report, and others—no surprise, given the company's resources and access to top-notch law firms. The lawsuit argues that

> as part of the Defendants' ongoing scheme to extort money and property from Smithfield, Defendants' intentionally and maliciously caused to be published in the Report false, misleading and baseless information about the working conditions at Smithfield's Tar Heel plant. . . . [T]he entire basis for the Report was to facilitate the Defendants' collective desire to portray Smithfield's Tar Heel plant as an unsafe workplace in which safety laws and standards were habitually and intentionally violated, as part of their effort to damage Smithfield's business reputation.[61]

Both the UFCW and the research group have denied these allegations and motioned the court to dismiss Smithfield's complaint.[62]

The UFCW counters Smithfield's allegations by saying that the company "picks up in the middle of the story that began in the early 1990s," when Smithfield was cited for illegal conduct in two union elections.[63] Workers at the Tar Heel plant had attempted to form a union through the UFCW in both 1994 and 1997. The elections were initially lost, but the results were overturned in a decision by a National Labor Relations Board judge, who charged Smithfield with several violations of federal labor law that inhibited a free and fair union election.[64] The NLRB ruled that the violations included threatening employees with plant closure or job loss because of their union activities, unlawfully interrogating employees about union activities, offering to remedy employee grievances and improve benefits in attempts to dissuade employees from selecting the union, and even firing employees because of union activities.[65]

In December 2004 the NLRB upheld the order for a new union election at the Tar Heel plant, but Smithfield appealed. Meanwhile, the UFCW charged Smithfield with exploiting racial divides as a tactic to prevent plant workers—most of whom are African American and Latinos—from organizing. The UFCW claims that Smithfield kept these workers at separate stations and attempted to turn them against one another during the 1997 elections. The company held separate meetings for the two groups, telling the Latinos that if they voted for the union they would be deported and telling African Americans that if they voted for the union the Latinos would replace them.[66] Since 1997, the share of Latino workers at the plant has increased dramatically. Former supervisor Sherri Buffkin testified to the U.S. Senate that Smithfield liked immigrant workers because they were "easy targets for manipulation."[67]

From 2000 to 2005 Smithfield took advantage of a special state law to maintain its own private police force to patrol the plant and intimidate workers from standing up for their rights, according to the UFCW. The force carried concealed weapons on and off duty and arrested workers and detained them in an on-site cell. The Smithfield police force arrested at least ninety workers and charged them with a variety of crimes. Ultimately, many of the charges were dropped by the county court, but the arrested employees were stuck with the court costs and attorney fees.[68]

In 2004 the NLRB issued a complaint against Smithfield, its company police, and Smithfield's sanitation subcontractor for violations of labor law. The charges included physically assaulting and falsely arresting employees, firing workers for union activity, threatening employees with arrest by federal immigration authorities, and threatening employees with bodily harm.[69]

In the winter of 2008 Smithfield workers were given the opportunity, after years of fear and intimidation, to vote on whether they wanted to be represented by the UFCW. In 2009, after an intense and hard-won seventeen-year battle for fair treatment and better working conditions, workers at the Tar Heel plant voted for a four-year contract that increased wages.[70]

Separately from "Packaged with Abuse," the UFCW claims that Smithfield operates a facility on the premises of the Tar Heel plant where workers are sent after an injury. This clinic is responsible for approving time off and compensation claims. Numerous employees have reported that they are given cursory exams and sent back to work.[71] A 2005 report by Human Rights Watch about workers' rights in U.S. slaughterhouses found that "workers at Smithfield . . . often described [the company clinic] as a disciplinary arm of management, denying claims and benefits and often failing to report injuries."[72]

From 2003 to 2006, Smithfield's Tar Heel plant was forced to pay $550,000 worth of workers' compensation claims against the company. Workers were forced to hire attorneys to recover their medical costs and lost wages due to being injured on the job. Nearly all of these workers were fired at some point after they were injured.[73]

In 2003 a twenty-five-year-old employee climbed into a tank to clean it and was quickly overcome with toxic fumes and died. The North Carolina Department of Labor Occupational Safety and Health Administration's investigation found that the young man had been improperly trained and supervised, and that the tank was not properly labeled as a dangerous confined space. The agency fined Smithfield just $4,323.[74]

The safety issues continued after this incident. In March 2005 the federal Occupational Safety and Health Administration conducted a walk-through and safety inspection of the plant and found more than fifty violations of safety and health laws, with most of them categorized as "serious." These included a lack of safety training, unguarded blades, missing guardrails, blocked exits, illegible signage, and improper safety procedures.[75]

And now, although U.S. activists continue to fight hog factories, another battle with Smithfield has emerged on the global stage. Having created so much anger from environmentalists, farmers, and consumers at home, Smithfield began looking overseas to increase profits. The company began its Eastern European venture by lobbying and courting public officials in Poland and Romania, making friends in high places, and "fending off" local opposition groups to create a conglomerate of feed mills, slaughterhouses, and climate-controlled barns housing thousands of hogs.[76]

In 1999, Smithfield bought out Poland's leading processing company, Animex, and began exploiting the country's lax environmental regulations and cheap labor. The takeover was supported by a $100 million loan from the European Bank for Reconstruction and Development and its partners.[77] Not too bad for Smithfield, considering it acquired Animex for just $55 million when then CEO Joseph Luter valued the company at $500 million. "Just ten cents on the dollar," he boasted.[78]

Smithfield's philosophy was clear. The current CEO, C. Larry Pope, says of their operations in Eastern Europe: "Politically, it is acceptable, and we've got people in Western Europe who make twenty euro an hour when you've got people in Eastern Europe who make one or two euro an hour. You've got land in Western Europe, very hot place. Land in Eastern Europe, they will virtually give you. Plants in Western Europe are very expensive. Plants in Eastern Europe, they will virtually give to you for small dollars."[79]

By 2009 Smithfield had five hundred farmers raising hogs for the Polish plant and had become the largest pork producer in Romania, with forty farms and cropland covering fifty thousand acres. The company's profits have been enhanced by European agricultural subsidies at the same time that it has overproduced hogs, driving small farmers out of business. It has also created environmental disasters from seeping hog waste, water pollution, and odor— the same problems that plague Smithfield's operations in the United States.[80]

Smithfield's assault on Poland and Romania has done more than just cause environmental and human health problems. Anna Witowska-Ritter, who has been organizing against Smithfield in Eastern Europe for Food & Water Watch, knows firsthand the devastation. Witowska, who received a PhD in sociology from Jagiellonian University in Krakow in 2010, has been battling the onslaught of Smithfield since 2004. She spent several years in the United States working in the nonprofit sector but eventually moved back to Poland with her American husband. Now a mother of two young girls, she still finds the time to educate citizens, the media, and policy makers about Smithfield's deplorable practices.

Witowska-Ritter says, "Smithfield took advantage of the desperation related to unemployment in Poland and Romania. I have seen factory farms in both countries. Some of them use the cooperative facilities left over from the Communist era." She goes on: "They represent the victory of corporate capitalism over communism, but they still make quite a pitiful sight. Smithfield was successful in silencing people. They give money to schools and other community activities as a way to shut people up."

Citizens of Romania were shocked by Smithfield's methods. In one 2007 incident, in the town of Cenei, hundreds of carcasses of hogs killed by a heat wave at a Smithfield operation were left lying around for about ten days. "We couldn't breathe anymore," said an adviser at Cenei's town hall. "I live a kilometer away from the farm, and at night I had to close the windows to sleep. The Americans have made our village a hotbed of infection."[81]

The swine flu that hit Romania in 2007 was also a source of embarrassment. Officials had asked Smithfield to stop breeding pigs and transferring them from one farm to another, but Smithfield paid no attention. Then it was discovered that of the thirty-three factory farms in the country at the time, eleven were never authorized by sanitation authorities and had to be closed down. When veterinary doctors attempted to inspect the operations, guards with bulldogs refused to let them in.[82] Csaba Daroczi, assistant director at the Timisoara Hygiene and Veterinary Authority, explained: "Smithfield proposed that we sign an agreement that obliged us to warn them three days before each inspection. These people have never known how to communicate with the public authorities."[83]

Smithfield's communication with the public is not much better. Its response to the plague that ultimately halted the country's pork exports to the European Union (EU) was arrogant at best. "We have nothing to say to the press; the swine plague is under control; journalists can just publish our communiqués," said the director of Smithfield's Timisoara operation, after receiving orders from corporate headquarters to refuse all contact with journalists.[84]

Witowska-Ritter says of Smithfield's operations in Poland that when a TV reporter went undercover and filmed unhygienic practices, such as sending old meat back to stores with new expiration dates and washing moldy meats to make them look fresh, they just blamed it on "Polish sloppiness." She is continuing her battle against Smithfield and working to reform agriculture in the EU, because Eastern Europe is being exploited in a way that would never be allowed in Western Europe.

As could be expected, as in the United States, the influx of corporate agriculture has crippled local communities and markets. Smithfield's expansion into Eastern Europe resulted in an overproduction of pork, to the point that the prices offered for meat were below the cost of production for independent hog farmers. The outcome has been the loss of the traditional farms that have been the hallmark of these countries.

Iowa hog farmer Larry Ginter, a longtime ICCI member and opponent of factory farms, made the connection between the plight of American farmers and the struggles of farmers in other countries when he testified at the Iowa hearing on revising the GIPSA regulations. He noted: "Labor, family farms, democratic rights are in a pitched battle against the dictatorship of capital. We've got to understand that this is an international struggle. Those Mexican workers coming up here are family farmers. Those Sudanese workers in the packing plants are family farmers and workers being driven off by the big dictatorship of capital. We have to understand that we are not alone in America." Urging his fellow farmers to action, Ginter concluded, "Nothing can happen on the farms unless farmers turn the wheel and plant the seed."

Smithfield shows no sign of slowing down its destruction anytime soon. Nor does it appear that the titans of meat have given up on even further consolidation. During the summer of 2010, Smithfield's stock rose 10 percent because of rumors that the giant Brazilian beef and pork meatpacker JBS/Swift was interested in buying the company. But, fortunately, around the country a large cadre of factory farm fighters like Ginter are working to protect their communities and to change federal farm policy.

Kendra Kimbirauskas, who serves as president of the Socially Responsible Agriculture Project (SRAP), notes, "In our work, we have learned that once people are educated about the terrible effects of factory farms, many of them become passionate about changing the system." SRAP is a national organization working with a network of community groups around the United States to fight factory farms and regenerate the critical infrastructure necessary to restore local food systems that are based on family-scale agriculture. Kimbirauskas goes on to say: "In this movement, there are many necessary approaches to regenerate our food system. SRAP is not just working to build a vast activist network that reaches into every state, but also to provide the on-the-ground tools that are vital in supporting the growth of a nation of localized food economies."

10

MODERN-DAY SERFS

Corporation: an ingenious device for obtaining individual profit without individual responsibility.

—Ambrose Bierce, *The Devil's Dictionary*

Most Americans have no idea that the chicken they eat is raised by a farm family that makes an average of $15,000 a year—that is, if none of their back-breaking labor is counted. Contrast this with the amount of profit made by KFC or a company like Tyson Foods.

According to Mike Weaver, a contract grower who is the president of the Contract Poultry Growers Association of the Virginias, when a consumer buys "a twelve-piece bucket of chicken at KFC or other chains, it costs around $26.99 [price depends on the store location] or thereabouts. Of that $26.99, the integrator [poultry company] receives around $3 to $5 depending on the price of chicken when it was bought. KFC gets around $21, and the grower who spent at least six weeks raising that chicken gets 25 cents!"[1]

Weaver says that the industry doesn't like the growers to talk to one another, but "they don't intimidate me." He has been growing poultry for ten years, beginning with turkey after he retired as a special agent with the Fish and Wildlife Service's Office of Law Enforcement. After three years, he switched to broilers just before Pilgrim's Pride merged with JBS. He charges that the industry lies to people and will not listen to grievances.

This unfair relationship exists because the food industry has amassed tremendous political and economic power through the massive consolidation that has transpired since antitrust regulation was eviscerated by the Reagan administration. This is especially true in the poultry industry. Over the past twenty-five years, as larger companies acquired smaller, regional processors and cooperatives, the industry has become increasingly concentrated. In the past decade, the five largest poultry producers—Tyson Foods, JBS/Pilgrim's Pride, Perdue, Sanderson Farms, and Koch Foods—now sell 70 percent of all chicken consumed in the United States.[2]

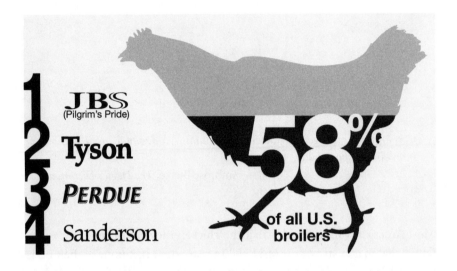

The top two companies, Tyson and JBS/Pilgrim's Pride, sell half of the one billion chickens Americans eat every year. Demonstrating how concentrated meat production has become, Tyson is both the largest chicken producer in the country and the second-largest chicken, beef, and pork processor in the world. Brazilian-owned JBS/Pilgrim's Pride is the second largest chicken producer in the United States but the largest meat producer in the world.

As a result of its power, the poultry industry is free to cheat the men and women who grow the birds it depends on for its profitability. The industry has created an animal production system that has been described by investigative journalist Dave Mann as "a story of modern-day sharecropping and indentured servants."[3]

Mann is referring to the fact that the large poultry-processing companies, known as integrators, control every step of chicken production ruthlessly—from the bird's genetic material to the cutlets at the meat counter. At one time there were dozens of types of chickens raised for meat. But since the 1950s chickens have been bred to grow increasingly large breasts and to convert feed "efficiently" into flesh during the six weeks they live. For instance, Tyson Foods, the largest chicken producer in the country, owns Cobb Breeding Company, the developer of the Cobb500 chicken, the most popular factory farmed chicken, and one that is used widely in the industry.

As Don Tyson, the late president of Tyson Foods, said, "If breast meat is worth two dollars a pound and dark meat is worth one dollar, which do you think I'd rather have?" This analysis is at the heart of the U.S. poultry industry's investment in bird genetics.[4]

Over the past several decades the poultry industry has developed just a few genetic lines that are now used predominantly by the entire industry. These birds are uniform in size and reach their optimum at approximately five to six weeks. The uniform size and shape of the birds facilitates processing, allowing slaughter lines to move quickly, thus reducing costs. It also reduces hatching costs, because one hen can produce a large number of offspring in a very short time.[5]

The poultry giants raise birds in the hatcheries they own and provide them to growers when they are a day old. They also own or control the mills that produce the drug-laced feed necessary for growing chickens in extremely crowded and dirty conditions. It contains antibiotics to prevent the chickens from getting sick and also often contains arsenic, which is intentionally added to give the meat a rosy color.

Today, as a result of the control that the industry wields all along the production chain, 98 percent of the chicken eaten in the United States is produced on contract.[6] This means that the giant poultry integrators pay growers for the service of raising the birds, not for the actual chickens.

Dr. Robert Taylor, a tenured professor of economics at Auburn University, has spent twenty-three years studying and writing about the poultry industry. He explains how the industry works: "Integrators generally own or control the breeding flock, hatcheries, chicks, assignment of baby chicks to growers, feedmills, feed ingredients, transportation of feed, and processing (slaughter) plants. These companies, integrating all decision making affecting poultry production, direct the course of action in all key areas of production: placement of baby chicks, the number of chicks placed with each grower, what birds are fed, and when birds ready for processing will be picked up from the grower."[7]

The integrators (the equivalent of "packers" in the beef and pork industries) use production contracts to keep the cost of raising chickens low and to manage the supply of birds needed in the slaughter plants they own. The companies have structured the industry so that they do not have to invest any capital in the factory farm facilities or equipment necessary for growing birds. Growers assume all of the debt and financial risk for building the warehouse-like barns, while taxpayers assume the risk on defaulted loans. This is because the USDA's Farm Service Administration (FSA) guarantees the loans, providing an economic benefit for both the integrator and the bank that makes money on the loan's interest, whether it is defaulted on or not.

A poultry grower from Alabama, Scott Hamilton, explains how the loans work: "A grower does not even see a written contract until after they've gone to the bank to get the loan to build the houses on their land. The bank often makes the loan based on a letter of intent from the poultry company."[8]

Growers accept this arrangement because they are usually desperate to make a living on their farms. Potential growers are hoodwinked by the company into believing that it is a viable business opportunity. The FSA has played a role in legitimizing the abusive contracting practices of the industry by encouraging farmers to get into the business and facilitating the loan process. For farmers in certain areas of the country, like the Deep South or the Delmarva Peninsula (shared by Delaware, Maryland, and Virginia), growing poultry is often a last-ditch effort to save the family farm.

Once the farm is put up for collateral, the grower is in "debt bondage," according to the Rural Advancement Foundation International (RAFI), a public interest organization that provides assistance to poultry growers. RAFI says that growers sign one-sided, abusive, take-it-or-leave-it contracts.

> A typical poultry farmer takes out a bank loan for $1 million or more to invest in specialized poultry houses and machinery. This debt is financed over a decade or more. The farmer's house and land are collateral for the loan. The contract, however, may be good only for the life of one flock— four to six weeks. Farmers must accept any new contract that the company presents, even if the terms are unfavorable. The alternative is termination, bankruptcy, and the loss of their farm and their home. Unfair and abusive contract terms are now commonplace in the industry.[9]

Taylor concurs, based on the many contracts he has analyzed. He says that it took growers about ten years to trust that he wouldn't tattle to the integrators, but eventually he began to find poultry contracts slipped under his office door, providing him with an important window into the industry's economics. He says that once growers pay down their loans, the integrators force them to make expensive upgrades. Keeping growers in debt is part of the leverage that companies have over growers. Growers who decline to upgrade are almost always terminated as growers.[10]

According to Taylor, the structure of the industry "is plantation-like," and poultry growers are "serfs with a mortgage." He explains that the poultry industry is the most "vertically integrated" sector in all of agriculture—a system in which the companies have ownership and control over every step of the chicken supply chain, from genetics to the branded store package.[11]

He says poultry production is "lopsided and deceptive," the most asymmetric of any industry. Integrators have learned to put nothing in writing. They tell growers about gross revenue, but not all of the associated costs. Because the companies have all of the data on the economics and the growers have none, the growers are paid unfair prices—basically raising chickens for free.

Taylor says most growers have little knowledge about how much or how little they make on their poultry operations. The only information on costs and returns is garnered from the Alabama Farm Analysis Association, an association that charges growers an annual fee of $1,200. They are told to save all of their receipts and other information from the integrator, which are reviewed by young agricultural economists who go through the information and allocate costs. If proper managerial accounting is done, and even a portion of their labor is included at a modest charge, they are losing money.[12]

Taylor is no stranger to controversy. When he was at Texas A&M in the 1980s and evaluating corn ethanol for the U.S. Office of Technology, he was lambasted by Texas state legislators, but he stood his ground. A tenured professor today, he is helping growers face down an extremely powerful industry that he says is profiting by lying and taking advantage of growers. Integrators control all of the information—monthly information on pay for growers, production, mortality, and the weight of birds.

One of the ways the industry cheats growers is by paying them based on a ranking system that makes them compete with other growers in putting weight on the birds—a process known as "feed conversion." The catch in all of this is that the company provides the necessary inputs: the day-old chicks and drug-laced feed. However, the companies often rank growers that have been given chicks from different hatcheries or feed from different feedmills, creating an unfair situation. Even though the company is responsible for the quality of the chicks and feed, the company ranks growers as if it is a straight-up fair competition.

Hamilton explains: "The difference between a top ranking and a bottom ranking can mean many thousands of dollars to a grower for a seven-to-nine-week flock. The irony is that while the company portrays this system as a competition, there is really only one winner, and that is the company."[13]

According to Hamilton, "Some growers have seen through that smoke screen to understand that only through working together will they gain the leverage to demand better contracts." He says that the ranking system has often been used as a tool to retaliate against growers who speak out against contract abuses. Hamilton told a U.S. Senate subcommittee that he himself experienced this drop in ranking when he became active in the Alabama Poultry Growers Association.

> I saw my ranking fall and was put on a probation-like program. I had sick birds, through no control of my own. . . . In a more extreme example, a breeder-hen grower in Georgia, Chris Burger, was the victim of severe retaliation by his poultry company when he tried to organize a breeder-hen grower group in his area. The company deliberately targeted him and delivered chickens with cholera to his farm. He was able to sue, and years later he won his case after it was proven that the company deliberately targeted him with the bad birds because of his organizing efforts. But his victory in court paled in comparison to the loss of his farm and the loss of his family to divorce related to the stress of those years.[14]

Valerie Ruddle, a poultry grower for JBS/Pilgrim's Pride in West Virginia, has experienced the full range of integrator abuses, including retaliation. Usually poultry growers are too fearful of losing their property to speak out. But no one is going to keep Valerie Ruddle from speaking the truth about the industry. She says that growers are really afraid, but she will not be intimidated. She once challenged an executive at Pilgrim's Pride: "You've already hurt me financially as much as you can. You've done it all. There's nothing else you can do."[15]

Ruddle grew up in the Washington, D.C., metro area and "never in a million years" thought she would raise poultry. But in 2003, her elderly in-laws, who raised turkeys for Cargill, were forced to either upgrade the facility, taking on decades more of debt, or face being dropped as a grower.[16] Ultimately, they were shut down, and after two years of no income they faced foreclosure on the property.

Valerie and her husband, Russell, could not stand by and watch their family lose the farm, so they decided to take over the poultry operation. However, they knew that Cargill's upgrade requirement was unreasonable, because the turkey facility was in good shape, so they decided to grow chickens for Russell's employer, Pilgrim's Pride, now owned by JBS.

Although the Ruddles had a healthy savings account, and both had good jobs (Valerie at the local propane company and her husband as an electrician at the Pilgrim's Pride processing plant), they had never planned to go into debt for $812,000 by assuming the mortgage on the 147-acre farm and to build new warehouses and other equipment for growing chickens. They were not allowed to use the existing turkey warehouses but had to invest in new structures built to Pilgrim's Pride's specifications.

Based on the cash flow and profit estimates of Pilgrim's Pride, the Ruddles' business investment should have paid off. But it did not. The debt on the farm and chicken operation depleted the Ruddles' savings; without their off-farm income, they would be bankrupt. Their economic situation really worsened after Pilgrim's Pride declared bankruptcy in 2008 (it was acquired by JBS in 2009). The company was run so inefficiently that it cut costs in ways that hurt its bottom line and negatively affected growers.

For instance, the company fired some of the chicken catchers—the people who literally enter the warehouse when the chickens are mature and catch and load them in trucks. This cost the Ruddles money, because the remaining catchers could only remove birds from two barns in one day. Their three barns are considered "a flock," and growers are ranked on an average weight gain for their flock. The math for the feed conversion is done based on the first day chicks are placed in a barn and the last day birds are picked up—which, if the pickups occur on different days, works against the grower, since it appears that it costs more to feed the birds.[17]

Also problematic for the Ruddles is that they have very little say over when and if they get birds. The integrators often promise a prospective grower a five-year contract, with guaranteed birds. Valerie says that people should read the fine print, because it is the company that decides how many flocks a grower receives. Even when a grower has a contract, the company decides "flock to flock" when the grower will get more chicks. Growers that do not "cooperate" with the company may not get another flock, even though they have assumed the tremendous financial debt for the facilities.

The retribution they have experienced for being outspoken about company practices, including Valerie's position as secretary of the advocacy group Contract Poultry Growers Association of the Virginias, has caused further economic hardship. In May 2010, Russell was fired from his job as an electrician at the Pilgrim's Pride processing plant, a job he had held since 1995. He was told that he took too much time off for his "side business" of raising chickens to be processed at the plant. He had only taken his vacation time and a few hours of unpaid leave over the course of the year, to be present when birds were removed from the warehouses for slaughtering—a Pilgrim's Pride requirement. The fact that the company often delivered and removed flocks over more than one day not only hurt the Ruddles financially, because of the way costs are calculated, but it required Russell to take off more time.

In December 2010, Valerie was asked to testify at one of the hearings organized as part of the process to revise the GIPSA regulations. On December 3, the day the press release about Ruddle's testimony was posted, Jeff Bushong, assistant live production manager at Pilgrim's Pride, visited her at work to say that the company would not be providing any more birds until contractual obligations had been met. He also said to her that he understood she would be out of town the following Wednesday, December 8. This was the day she was to testify at the hearing in Washington, D.C.[18]

Ruddle laid out in her testimony how integrators have taken advantage of growers.

> You talked about asymmetry. They have a lot of that in the poultry in-dustry. It's called vertical integration. . . . [T]hey supply everything to us. They supply our birds. They supply our feed. They supply their medica-tions that they need. Everything comes from them. However, we supply 100 percent of the labor. We supply 100 percent of the utilities. We supply the mortgage for the buildings that these birds are grown in. . . . We work very hard. The labor is not simple and it is very time consuming.[19]

Valerie went on to explain that a grower must simply trust the integrator about the weight of the feed, because it is not weighed on delivery. And the grower must trust the company on the weight of the flock at the slaughter facil-ity, because the grower is discouraged from being present to verify the weight. Growers are paid a base pay, but the price per pound ends up varying depend-ing on the average daily weight gain of the flock. The integrator, who also determines when a flock is sent to slaughter, can manipulate these numbers by picking up birds early or catching them in the multiple barns on different days. The Ruddles lost a penny per pound on a flock that weighed 379,000 pounds— losing out on $3,790—because of manipulation by Pilgrim's Pride.[20]

The Ruddles have also experienced problems related to ranking. Valerie says that this metric is based on the feed conversion rate for each flock of birds—an unverifiable figure formulated by the integrator. Growers do not know who they are being ranked against or whether the growing conditions or feed quality may have differed significantly.[21]

She says that sometimes she gets sick chicks and inferior feed from the company. The company provides her with three types of feed for different aged chickens over their short life span: starter, grower, and withdrawal. No doubt the chickens are "withdrawing" from the antibiotics and other feed ingredients—such as the arsenic used as a growth promoter. She notes, "In-puts from the integrator, such as feed and chick quality, are 80 percent of making a good chicken."[22]

Although Ruddle has three chicken warehouses, today growers are encouraged to invest in five or six houses that each have the capacity for 35,000 birds. Each bird has a quarter of a square foot for its short and miserable life of about seven weeks. Not only is the size of the warehouses increasing, but the scale of poultry farms has grown rapidly, as growers try to eke out a living by increasing the volume of birds they produce on contract. The median-size poultry operation increased by 15 percent in four years, rising from 520,000 birds annually in 2002 to 600,000 birds in 2006.[23]

An analysis by Food & Water Watch found that at the time of the last agricultural census in 2007, over one billion broiler chickens were raised on large farms in the United States. This nearly 88 percent increase in production means that 5,800 chickens were added to larger farms for every hour of the last decade.

It has not always been this way. Until the early 1940s, poultry was raised mostly on small farms for local consumption. Demand for poultry increased dramatically during World War II, because beef was rationed, and poultry filled the void. Most of the chicken was raised and sold locally, and during this period many farmers got into the poultry business. The birds sold for meat did not produce eggs; they were older hens and roosters that were allowed barnyard freedom, making the meat flavorful and fibrous. Sold as whole chickens, they had to be either roasted whole or cut up for frying.[24]

After the war, the new interest in poultry brought federal dollars pouring into genetics research to develop birds that were larger and tender, more like the young roosters, called "spring chickens." Up until this time, birds had been bred primarily for laying eggs. In 1946, A&P, the nation's first supermarket chain, seeing the new market for chicken, developed a national contest called the Chicken of Tomorrow. The goal was to create a grassroots competition for a high-yielding broad-breasted chicken that would reach maturity in a very short period of time.

In 1948, the contest culminated in chicken breeders from twenty-five states submitting more than 31,000 eggs that were hatched and fed under the same conditions. Judging was done by staff from the University of Delaware's poultry research substation.[25] The winning chicken, the Vantress, became the standard chicken raised by growers for the next decade, and it is a distant relative of the Tyson-owned Cobb500 chicken grown on factory farms today.

After this development, the new era of nutrition research led to improved feed formulas containing vitamin B12; antioxidants; and, as discussed previously, antibiotics. Vaccinations were developed for disease control. Buildings with automatic feeding mechanisms and ventilation equipment meant that large numbers of birds could be raised.[26]

Small growers could not make the large capital investments necessary to build the new-style buildings and equipment necessary for scaling up. This was particularly true because the technological advances increased production, causing a decline in prices. This provided an opening for the large feed companies to expand their role in producing broilers. They saw the economic benefit of encouraging an industry that was highly dependent on large amounts of feed. Their access to capital allowed them to enter into the hatchery business and to build or acquire processing plants. By 1955, nearly all chicken was grown under a contract arrangement.

Moreover, similar to today, another driving force in creating the consolidated and vertically integrated poultry industry was the demand from grocery stores for low-cost supplies of meat in high volumes. According to the USDA's Economic Research Service, during the ten-year period following World War II, retail chicken prices dropped 30 percent to 40 percent, compared to price increases of 75 percent to 90 percent for red meats during the same period.[27]

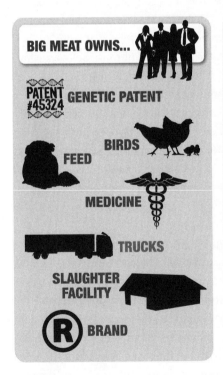

One tactic for lowering the cost of production was to operate in economically disadvantaged parts of the country. The industry came to be centered in two regions: the Delmarva Peninsula and the South. Delmarva had hosted a small chicken industry since the 1920s, when an entrepreneurial woman from Ocean View, Delaware, started a small brood of 500 chicks and by 1926 was raising 25,000 meat chickens. Other farm families saw her success, and soon a small industry grew up in the region that has flat land and a good climate for growing the grains needed for feed.[28] By the time of the A&P contest, Delmarva had a thriving poultry industry, and the University of Delaware had developed a poultry program.

In the South, by the 1950s, the cotton industry had moved westward and soybeans had become the crop of choice. Mild weather and the "excess labor" due to the Eisenhower-era agriculture policies made it an ideal location for the growth of the industry. The South had a ready pool of farmers willing to put in long hours for low pay and workers desperate enough to work in the hazardous and unpleasant slaughter facilities.

Passage of the Poultry Products Inspection Act of 1957, which required inspection of broilers sold across state lines, further increased pressure on small operations. Large capital investments were needed to meet the requirements of the new law. Only the companies that could increase production by installing automated processing systems could cover the costs of the new food safety investments.

By the 1970s companies with the primary business of selling feed had gotten out of the broiler business, and integrators like Tyson, which focused on all stages of production, came to dominate the industry. With the advances in bird genetics, feed conversion, and other new technologies needed for confinement, the stage was set for highly capitalized firms to acquire their competitors.

But even with the technological advancements, one of the primary ways that these vertically integrated operations became profitable was by shifting costs to the grower. So over the course of several decades, the contracts became more ruthless. The first type of contract used created an arrangement where the integrator provided credit to the grower, who supplied the enclosed barn, equipment, labor, fuel, and feed. Once the broiler was sold, the grower paid back the company and kept whatever profit was left. Different types of contracting arrangements followed, including flat-fee contracts that required the grower to provide housing and labor, while the integrator provided feed, medicine, and chicks and retained title to the broilers. Variations of the flat-fee contract quickly came into use with the principal goal of moving risks and costs of production onto the grower.

In addition, about half of the growers have only one or two integrators located near their farm. And once a grower is associated with one integrator, it is very difficult for them to sign a contract with another company, since there is an informal blacklisting of growers who advocate for themselves. This means that growers have little choice but to accept whatever terms the companies offer.

Today, with most contracts being flock-to-flock, there is no assurance that a grower will ever get another one to raise. Even when flock-to-flock contracts are automatically renewed, growers are dependent on the companies to keep providing them with new ones, and they are paid via the unfair and nontransparent ranking system.

Kate Doby, a grower from Cameron, North Carolina, lives on a third-generation farm. She and her husband built two broiler houses and raised poultry until October 2008, when their contract was terminated. Initially she was given a ten-year contract, which was the length of her loan. But before the end of the contract, it was changed to a one-year term. When she was told to make an expensive upgrade before the current loan was paid off, she refused. She testified at the GIPSA hearings: "The company tells the grower, if they make these upgrades that they're going to get paid more under this ranking system. I asked the company managers, I went to them, I said: 'You want me to do this. Show me on paper where I'm going to make this money back to justify borrowing more money, when I still owe money on these houses?' They couldn't do it." [29]

Doby says that many growers "have borrowed a lot of money to make upgrades demanded by the company" and "precious farmland and homes were put up" for collateral. These growers are now finding themselves in financial trouble, because they are forced to go flock-to-flock, putting them at the mercy of the integrator.

Taylor explains that the integrator serves as an "overlord," directing the production process. After the company delivers the feed and the chicks, the grower only gets paid for what is returned to the company in meat. Dead birds are the property of the grower, and as the system has evolved, the tremendous amount of waste generated on the farm has become the grower's liability as well. Taylor and a co-author explain, "Once one enters the life of a grower, the trap is closed: high capital costs and large debt to enter the business, no input or product price, no market in which to sell goods and no way out except bankruptcy if the integrator 'dumps' the grower." [30]

The average on-farm income for poultry growers—not counting any of their labor—was $10,000 for a small operator with less than 266,000 birds a year and $20,000 for medium-size operations with between 266,000 and

660,000 birds per year. These meager earnings barely make a dent in the debt for the poultry-house upgrades. Poultry growers lost money ten of the fifteen years from 1995 and 2009.[31]

Once in a while there is justice. In 2010 a jury in Oklahoma returned a $7.3 million verdict against Tyson Foods for its "deceptive and coercive business practices." Growers charged that Tyson used economic clout to coerce them into growing chickens below the cost of production. The company attempted to strong-arm them into making expensive upgrades, used a secretive system for calculating pay, and refused to allow them to verify the weight of feed or mature birds. Originally, fifty poultry growers sued the company, but the suits have been broken up into smaller ones.[32] In March 2012 the Oklahoma Supreme Court overturned the jury verdict and ordered a new trial based on Tyson's claim that juror questionnaires were filled out improperly and that a mistake was made in interpreting the law.[33]

According to Taylor, "Contracts allow growers to subsist, but not to grow, profit or prosper. There is no 'wealthy' subset of chicken growers. . . . Often the grower's ranking changes more because of factors controlled by the integrator than by the grower's management."

The concentration and economic power in the broiler industry is matched in the egg-producing segment of the industry. There is a good chance you have never heard of the giant companies that produce eggs, though there is an excellent chance you have eaten their products many times. A handful of egg companies produce a large share of the eggs most Americans consume. The largest four, Cal-Maine Foods, Rose Acre Farms, DeCoster, and Moark LLC, raise more than 257 million chickens.[34]

The average size of egg-layer operations has grown by half, increasing to 614,000 at the time of the last agricultural census in 2007. In 2009, the four largest firms owned 30.2 percent of the laying hens in production.[35] When a few firms dominate the marketplace, the major players can collude and manipulate prices and drive practices that are more intensive and larger scale. Some of the largest companies have been implicated in a scheme to manipulate the price of eggs at the grocery store by allegedly colluding to artificially reduce egg production and drive up retail prices.[36]

For example, in 2009 Land O'Lakes and its egg supplier, Moark LLC, agreed to pay $25 million to settle a price-fixing class-action suit that alleged Land O'Lakes companies had conspired with other industry partners to reduce the egg supply and drive up the retail price. The suit contended that producers lowered hen cage space (which reduces egg production), coordinated practices among firms to reduce flock size, and exported eggs below their cost, all in an effort to reduce supply and raise prices. Land O'Lakes agreed to provide documents related to other companies' participation in the alleged conspiracy.

The same type of abuse marks the plight of workers who "catch" the chickens once they reach maturity and load them into trucks. Catchers are pressured to work quickly in the stinking, dirty, and dusty environment that is full of antibiotic-resistant bacteria from the chicken dung. The faster they work, chasing after and catching chickens, the more dust rises for them to choke on.

Carole Morison, a former contract grower, was one of the founders of the Delmarva Poultry Justice Alliance, along with Jim Lewis, an Episcopal minister. She explains that chickens are caught by hand when they reach a "marketable age," at around seven weeks, and weigh 5.5 pounds. The birds are crated and loaded onto trucks by forklift.[37] As part of their campaign to publicize Perdue's treatment of catchers, they handed out leaflets at grocery stores saying that chicken catchers have the dirty job of catching birds and loading them onto trucks for slaughter. They are paid so poorly they have to catch 2,500 birds "before they can bring home one Perdue chicken from the grocery store." This is one more way the poultry industry increases its bottom line at the expense of the people who work for them.

The situation for poultry-processing-plant workers is no better. This type of work requires being in a plant with a temperature at under fifty degrees and standing for hours on end at a conveyor belt gutting, cutting, or deboning poultry. Conveyor-belt work is awkward, because the machines are built to a standard size, but people are different heights. So tall workers stoop and short workers stretch. The straight-handled knives used for deboning also cause serious injuries—sometimes crippling.[38]

An immigrant worker explains: "If we are not done with the truck full of chickens, we cannot leave work at the end of our shift. We are slaves. . . . You just have to be very fast. You're not always working safely because you have to keep up with the production line. The managers always want more production in less time."[39]

Chickens whiz by at a breakneck pace, and since workers have no control over line speed, they cannot rest. Line speeds have increased dramatically, from 143 birds a minute a decade ago to 190 today. The repetitive nature of the work creates a range of painful and debilitating injuries. Twenty percent of poultry workers are injured on the job. Rates of repetitive stress injuries are now the third highest in the industry because workers make the same cutting motion from ten thousand to forty thousand times a shift.[40]

Processing plants are a miserable work environment—bone-chilling cold, wet, and slippery. A worker explains: "You have to be careful with the knives and the machines, because everything is so slippery. A lot of fat falls on the machines and the floor. There's fat everywhere. Everything's greasy. So when there's a disk cutter with a rotating blade, your fingers are in danger."[41]

One in five workers in the meatpacking and processing sectors is in the United States illegally.[42] Nationwide, poultry workers are 50 percent Latino, and more than 50 percent are women. The companies pay low wages, and an average worker qualifies for Head Start, food stamps, the National School Lunch Program, and the Low Income Home Energy Assistance Program.

A 2000 survey of poultry companies by the Department of Labor (DOL) found that over 60 percent of plants violated basic wage and hour laws. A majority of poultry plants illegally force employees to work off the clock by not compensating workers for job-related tasks before and after their shifts, and for brief breaks during the workday. The DOL survey also confirmed that over half of poultry plants, mainly nonunion plants, illegally force workers to pay for their own safety equipment by deducting the costs of required gear from workers' paychecks. Tyson and Perdue have been sued and have had to pay millions of dollars for cheating workers out of wages and failing to compensate workers for time spent taking off protective gear.[43]

In 1991 twenty-five workers died from smoke inhalation in a fire at a poultry plant in North Carolina. Rescue workers found that the owner had ordered the fire exits to be locked to prevent stealing. Twenty years later, working in a poultry plant is still one of the most dangerous jobs in America. Because the integrators mix antibiotics in the feed of chickens to prevent infection from overcrowding, the birds harbor antibiotic-resistant bacteria, which is passed on to poultry workers. In a study of poultry workers, researchers found they were thirty-two times more likely to be resistant to the antibiotic gentamicin than people not working with poultry. Poultry workers are also exposed to dust and droppings from poultry, causing respiratory illnesses like pneumonia and asthma.[44]

Speaking of her years interacting with poultry workers, Morison reflected on her community's experience with Perdue: "Publicly they speak of doing the 'right thing,' and privately within the company realm they do the complete opposite. The worst [incident] in my mind was of a pregnant processing-plant worker who was forced to urinate on herself because to allow her to use the bathroom would mean that her spot on the line would have been empty and it would have slowed processing speed down." She went on to say, "In this same Perdue processing plant, workers were continually harassed by a supervisor who constantly threw pieces of chicken, hitting workers if it was deemed they weren't moving fast enough."

It is not only the workers in the poultry industry who suffer—the conditions in which the birds live and die can only be described as cruel, from their birth to slaughter. Broiler operations use birds that have been bred to have large breasts, and they grow so fast that their size does not keep pace with their hearts or lungs, causing heart attacks and other health effects.

Thousands of birds are crowded together for their brief lives in extremely crowded conditions in the filthy warehouses where they live in their own waste. At the end of their short lives, the birds are roughly loaded in tractor trailers, denied food and water, and sent to processing facilities where they are hung by their feet and slaughtered by the truckload.[45]

Laying conditions for egg-producing hens are no better than for meat chickens—just different. Eggs are produced in large-scale operations with hundreds of thousands of layer hens held in each facility. A handful of firms own multiple farms or contract with a number of large layer operations, most of which house their birds in small cages that are stacked from floor to ceiling. Sixty-nine million eggs were produced in the five states where production is concentrated: Iowa, Ohio, Indiana, California, and Pennsylvania.

In October and November 2010 the Humane Society placed an undercover activist as a worker at an egg farm owned by Cal-Maine, producer of 8 billion eggs a year. The living conditions were extremely dirty and overcrowded, with each bird having a space the size of an 8″ x 11″ sheet of paper. When the hens' wings, necks, legs, and feet became caught in cage wire, they could not reach food or water and suffered injuries that led to a miserable death. Eggs were covered in feces.[46]

A video, shot as part of the Humane Society investigation, shows what appear to be employees cutting the toes off young turkeys before tossing them down a chute to a bloody conveyor belt. It also seems to show employees scooping up a handful of turkeys and tossing them into a bin, dropping some on the floor and leaving them there, as well as an employee pulling a cart of injured animals over to a grinder and throwing them in.[47]

Beyond the cruelty that marks the poultry industry is the truly massive pollution associated with producing large numbers of chickens in one area.

Poultry facilities generate tremendous volumes of stinking waste that emit dangerous gases, such as ammonia, methane, and hydrogen sulfide.

For instance, layer hens can produce as much manure as the sewage generated in medium-size cities in the counties where the poultry farms are concentrated. The 13.8 million layer hens in Mercer County, Ohio, produce as much untreated waste as the entire population of greater Dallas–Fort Worth, as do the 20.1 million broiler chickens on factory farms in Shelby County, Texas. The 7.7 million layers in Sioux City, Iowa, produce as much manure as all the sewage in Seattle. And the 17.5 million broilers in Franklin County, Georgia, produce as much waste as the greater Philadelphia metro area.

A typical broiler factory. Photo by author.

As with the hog industry discussed earlier, the waste from laying hen operations is sometimes drained into lagoons that leak and overflow. Other egg factories and most broiler operations store dry litter (manure and bedding) in sheds or outdoors until it is spread on farm fields. After it is applied to the soil, much of the waste runs into local waterways, causing people to become sick from breathing harmful gases or from drinking water polluted with nitrates, dangerous microbes, antibiotic-resistant bacteria, and viruses.

Scott Edwards, a former attorney with the environmental organization Waterkeeper and now a litigator with Food & Water Watch's Justice program, knows firsthand the environmental devastation caused by factory farms. Wa-terkeeper brought a lawsuit against Perdue in 2010.

Edwards says of the situation in Maryland: "It's almost downright impres-sive when, a week after being sued for violating the Clean Water Act, you can walk into the governor's office and get him to do a public relations dance for you by having him hand you an award for environmental stewardship. And how almost remarkable is it that Perdue has the ability to call state senators—some of whom they make campaign contributions to—and get them to intro-duce into the state budget a hold on public funding for the environmental law school clinic that is representing the Bay's interests in that suit?"

Edwards has used the Clean Water Act, passed under the Nixon admin-istration, to rein in factory farming. He explains that this sweeping set of comprehensive environmental laws was designed to bring some degree of corporate responsibility to industrial polluters by holding them accountable for their waste discharges through a transparent permitting and monitoring system. Agricultural waste is known as nonpoint source pollution, a technical term for pollution that does not originate from one place, like an industrial facility or manufacturing plant.

Edwards comments on Perdue: "They've managed to stick all responsibil-ity for their billions of pounds of waste on someone else: local chicken growers, 'chickensitters,' who have neither the means nor resources to properly dispose of even a fraction of the manure produced by Perdue's birds. Perdue owns the chickens. It owns the feed given and the drugs administered to them."

He goes on to explain: "It's all orchestrated through unconscionable adhe-sion contracts that leave their growers burdened, broke, and in violation of the law. No other industry in the country has managed to game the system in quite the same way."

Although the poultry companies own the chickens and the feed that goes into them, the farmers are responsible for the management of the manure. Poultry litter—chicken manure and manure-laden bedding (usually rice hulls or straw)—is stored on farms, where it is applied to farmland as fertilizer. In many dense poultry-production areas, the volume of poultry litter greatly exceeds the fertilizer need and capacity of nearby farmland. With so many birds and so much manure, the accumulated litter can pose a significant en-vironmental risk.

Unfortunately, although factory farm activists were hopeful that the Obama administration's EPA would take action to address the problems caused by the waste from industrialized animal facilities, the agency has caved in to industry pressure. In 2010, as a result of a lawsuit brought by several environmental groups, the EPA agreed to conduct an inventory of the largest factory farms that would include the location of facilities, the quantity of manure produced, and its use. The agency had never previously tracked the data that would provide it with the information to assess the full scope of the waste problem. But the agency capitulated to the industry. It has announced that no new information will be sought from industrialized animal facilities.

Unfortunately, the failure of the Obama administration to take action against the poultry industry to protect the environment or to strengthen GIPSA regulations to curtail contract abuses suffered by growers must be seen as part of a long-standing policy of both political parties. The environmental degradation and unfair treatment of growers and workers in the poultry industry have been well documented over the past several decades. The abusive behavior of the industry had so intensified by the mid-1990s that each year, during the appropriations process, USDA officials would ask for new GIPSA authority and more funding to improve investigations and regulation of the poultry industry.

For Every KFC 12-Piece Chicken Bucket
($19.09 in Manhattan)

25¢ to grower

$3–$5 to JBS

Remainder to KFC

In March 1999, citing a series of articles in the *Baltimore Sun* outlining the chicken growers' plight, Michael V. Dunn, the USDA's undersecretary of marketing and regulatory programs, said, "We desperately need those additional funds, and we need some more teeth [in the law] to do something to assist these producers out there."[48]

Yet more than a decade later, the situation has only grown worse, because of the increased political power and influence of the poultry industry. Confronting this industry is not just a matter of justice for growers, chicken catchers, and slaughterhouse workers; it goes beyond addressing the horrendous pollution that is destroying waterways and leading to the largest estuary in the country, the Chesapeake Bay, becoming a dead zone.

Changing the behavior of the poultry industry requires building the political power to reestablish antitrust regulation and enforcement. It means dealing head-on with an industry that is willing to sacrifice the environment, people's health, and the lives of its workforce to increase profitability. We must take this challenge on—because it is a matter of survival.

11

MILKING THE SYSTEM

*And so it appears that most and perhaps all of industrial agriculture's mani-
fest failures are the result of an attempt to make the land produce without
husbandry.*

—Wendell Berry, *Bringing It to the Table:
Writings on Farming and Food* (2009)

One of my fondest memories is of my father coming in the kitchen door carrying a bucket of milk from our Jersey cow with the cream already rising to the
top. Jerseys, originally bred in the Channel Islands off the coast of England,
have a sweet disposition and, as small cows, cost less to feed. As someone who
had been a dairyman in his twenties, my dad always said that Jerseys produce
a little less milk, but it is the best, because of the high butterfat content. My
mother and I would wait for gravity to work its magic, causing the heavy
cream to rise to the top. Naturally impatient, I would usually stick my fingers
in for a taste. We'd scoop the cream into a jar and shake it hard until it turned
into butter—a low-tech version of the butter churn. My father had a strong
relationship with our Jersey, named Bossy, whom he milked twice a day and
considered part of the family.

John Kinsman, age eighty-six, feels the same way about the thirty-six cows
that he milks on his farm in Lime Ridge, Wisconsin, providing organic milk
for Cedar Grove, a specialty cheese maker near his home. Spry and fit, Kinsman still milks and does heavy farm work on his 150-acre organic farm, where
eighty acres are devoted to rotational grazing and hay production and the
other seventy acres are woodland. He comes from a farming family (his name
having originated in thirteenth-century England from Kynneman, meaning
"keeper of livestock"). His grandparents farmed in Vermont in the nineteenth
century and came to Wisconsin in a covered wagon.

His father purchased the Kinsman farm after World War II, shortly after John came home from Guam, where he served during the war. Kinsman says, "The farm was like Stonehenge, so rocky that weeds wouldn't grow." The family used pesticides produced by Monsanto until the early 1960s, when John became seriously ill with a neurological disorder. His doctor diagnosed the cause of his illness as nerve damage from agricultural chemicals, and he never used chemicals on the farm again.

Today his organic dairy herd grazes in a rotational system that provides the nutrition best suited to the digestive system of ruminant animals. No grain is fed on the Kinsman farm. Cows, along with goats, sheep, and other related species, have difficulty digesting grain, which they are often fed on factory farms. Every twelve hours, John moves the herd to new pasture; he says that the cows complain loudly if he is late in moving them to fresh grazing. In the winter he feeds them hay that he has produced on his farm.

Kinsman is as much an activist as a farmer. In the 1970s he helped organize a program that brought African American children from Mississippi to Wisconsin for home stays, and he became very active on civil rights issues. He was a leader during the farm crisis of the 1980s and has fought the use of artificial hormones in milk. In 1994, he founded Family Farm Defenders (FFD), a Wisconsin-based advocacy organization whose leadership is primarily dairy farmers. He says they are fighting for sustainable agriculture—a term Kinsman defines broadly as encompassing animal welfare, workers' rights, consumer safety, and environmental stewardship.

Two decades ago there were only two factory farms in Wisconsin, and now there are more than 150. In 2010 FFD and several citizen organizations marked the opening day of the annual World Dairy Expo in Madison with a protest against taxpayer-subsidized factory farm expansion in the state. According to their September 28 press release, they also presented the First Annual Wisconsin "Land of 10,000 Lagoons" Awards to winners in the following categories: most manure in one location, most cows held in confinement, and most taxpayer subsidies. FFD stated that factory farming has "been fueled by lackadaisical regulation, millions in subsidies, no environmental liability requirement, flagrant farm labor abuse, as well as blatant political corruption and corporate influence peddling."[1]

In 2006 the Wisconsin legislature passed legislation against local control of factory farm siting and placed decision making in the hands of an industry-dominated task force appointed by the governor. The state's Department of Natural Resources (DNR) has an embarrassing record of always siding with factory farms in the siting process.

"The relaxation and nonenforcement of laws designed to protect drinking water in Wisconsin is appalling. The fact that no meaningful DNR monitoring and enforcement exists speaks for itself," commented Jennifer Nelson of Sustain Rural Wisconsin Network at the award-giving ceremony.

Dairies have been transformed into milk-production factories over the last twenty years. They sell their milk to a company, usually a giant cooperative, that collects the fluid milk at the farm. The corporation-like co-op sells milk to a processor to be pasteurized, bottled, and distributed. Food manufacturers convert fluid milk into cheese, ice cream, and industrial dairy ingredients for other processed-food companies. Large retail stores sell the finished products and, in many cases, dictate to processors and manufacturers the wholesale price of the product.

And very little of the money paid by consumers for milk ends up in the hands of farmers. The growing spread between what consumers pay and what farmers receive is captured by the dairy processors and retailers that dominate the industry. The Utah commissioner of agriculture and food noted in June 2009 that consumers are not gaining from the declining milk prices that farmers receive, saying, "We are concerned that retailers have not reduced the retail price of milk to reflect the huge reduction in the wholesale level."

The massive changes in the dairy industry are in large part a response to changes in the retail market. Regional and local supermarket chains have disappeared in recent decades and national supercenters and discounters have emerged as new grocery retailer powerhouses that exert power over the food suppliers that are invisible to consumers. Rural sociologists Mary Hendrickson and William Heffernan, known for their groundbreaking work on consolidation in agriculture, reflect on how the dairy industry has been "restructured" to serve large retail grocery stores: "For decades, most dairy farmers seemed immune to the consequences of restructuring because through their local and regional cooperatives, they were also a dominant processor of milk for their local or regional markets. National markets did not exist. Even most of the investor-owned firms operated at a community level. . . . Vertical integration, which formally connects the dairy processing stage to the retail stage, is probably the major driver of the restructuring at this time."[2]

They go on to say that the dairy industry changed in response to large retailers' power and influence in the marketplace, which meant that they can dictate terms to food manufacturers and processors "who then force changes back through the system to the farm level."[3] The food processors and manufacturers have embraced consolidation in retailing because it cuts down on the transaction costs of dealing with large numbers of customers.

And as one would expect, the largest dairy processor, Dean Foods, has a customer list that matches the biggest retailers: Walmart, Kroger, Supervalu, Costco, and others. As a large supplier of organic dairy and soy milk, it also does business with Whole Foods. Consolidation in the grocery industry has dramatically driven consolidation in all sectors of the dairy industry, beginning with the farms and including the megacooperatives, fluid milk processors, and dairy product manufacturers.[4] They have found it most profitable to form powerful alliances along the chain of production, which begins at the source of production: giant factory farms.

The increased scale and production by farms, processors, and manufacturers is causing the country to hemorrhage dairy farms and farmers, with 52,000 dairies lost in just a decade.[5] As late as 1998, the majority of milk was produced on small farms with fewer than two hundred cows, but today more than a quarter of all milk comes from industrial dairies that have over two thousand cows. These new megadairies can house ten thousand cows or more, crowding them into high-density feedlots with no access to grass.

Milk produced on the 65,000 remaining dairy farms is funneled through a handful of powerful buyers and retailers that use their market power to push down the prices farmers receive for milk.[6] Production has remained fairly constant, because the scale of the farms has increased significantly.[7] Dairy cows are treated like milk-producing machines, milked in round-the-clock shifts. Until the advent of factory dairy farms, dairy cows produced milk for as long as twenty years. In contrast, cows are kept on factory farms for up to three years, after which they are sent to slaughter, usually for fast-food hamburgers.

Today dairy cows produce milk for about ten months after a pregnancy and birth of the calf—a necessary precondition for lactation. On factory farms, they are impregnated through artificial insemination, and after the calf is born it is taken from the mother almost immediately. Female calves are usually kept as part of the herd to eventually produce milk, while male calves are used for veal, which is really a by-product of the dairy industry. Veal is usually produced using inhumane crates that prevent the calf from exercising. They are also fed a diet that induces anemia—all for the purpose of producing the pale-colored meat that veal is famous for.

Dairy herds also spend their short lives in intensive confinement, where they are fed grain and watered. They are milked with machines as many as four times a day to relieve their udders of the large amount of milk that they have been bred to produce. They are also sometimes injected with a genetically engineered hormone, recombinant bovine growth hormone (rBGH), that increases milk production by 8 percent to 12 percent. Almost 43 percent of large-scale dairies with more than five hundred head use the hormone, as do 29 percent of midsize dairies and 9 percent of small dairies.[8]

It is well documented that rBGH causes high rates of the painful bacterial udder infection mastitis. The drug containing the hormone Posilac, originally developed by Monsanto, has been sold to pharmaceutical giant Eli Lilly, which is promoting sales in the developing world. Significant concerns remain about use of the drug, because it increases another powerful hormone, IGF-1, that is linked with increased rates of colon, breast, and prostate cancer in humans. The hormone cannot be used in the European Union, Japan, Canada, or Australia, and its use in the United States has diminished because of a consumer backlash.

Factory dairies create enormous amounts of waste from the thousands of dairy cows, thereafter polluting groundwater, contributing to airborne particulate pollution, and producing excess phosphorus and nitrogen runoff to streams and rivers. Small dairies, in contrast, generate less manure, and they usually apply it to cropland or incorporate it into pasture as fertilizer. Because big dairies generate far more manure than they can use as fertilizer, they must either store it in giant lagoons or apply it to cropland at excessive rates, where it leaches into groundwater and runs off into nearby rivers and streams. Many factory-farmed dairies have caused significant manure spills and environmental hazards in recent years.

There are many examples of this pollution around the country. In 2010, at a 1,650-cow Randolph County, Indiana, dairy operation, a manure lagoon liner detached, floated to the surface of the lagoon, and became inflated with gases from decomposing manure.[9] The manure bubbles were large enough to be seen from satellite photos, but the operator, who had declared bankruptcy after milk prices collapsed, could not afford to repair the liner.[10] After the county shut down local roads and banned school buses from the surrounding area, because of the risk posed by potential noxious gas releases or explosions, Indiana environmental officials deflated the bubbles.[11]

A one-thousand-cow dairy operation in Frederick County, Maryland, reimbursed the county and a local city $254,900 for providing emergency water supplies, testing, and other costs after a 576,000-gallon manure spill in 2008 polluted the town's water supply, which had to be shut off for two months.[12]

In 2009 a 250,000- to 300,000-gallon manure spill from a 660-head Pipestone County, Minnesota, dairy leaked into a tributary after a pipe between manure basins clogged and overflowed. The spill killed fish and closed a state park to swimmers for Memorial Day weekend after heightened levels of fecal coliform were found in the park's waters.[13]

The rise of the polluting factory-farmed dairy industry—4.9 million cows—has been more pronounced in western states and has transformed the national dairy landscape over the past decade. An unbelievable 650 additional dairy cows were added every day between 1997 and 2007.[14] There are more than 2.7 million cows on factory-farmed dairies in California, Idaho, Texas, and New Mexico.

The largest factory-farmed dairy counties produce as much untreated dairy waste as the sewage produced in major American metropolitan areas, which is required to go to treatment plants. The more than 464,000 dairy cows on factory-farmed dairies in Tulare County, California, produce five times as much waste as the population in the greater New York City metropolitan area.[15] The nearly 240,000 dairy cows in Merced County, California, produce about ten times as much waste as the city of Atlanta, Georgia.

The emergence of western factory-farmed dairies has contributed to the decline of local dairy farms in the Southeast, Northeast, Upper Midwest, and parts of the Midwest.[16] Although traditional dairy states like Wisconsin and New York added 340,000 dairy cows to the largest operations over the decade, the growth in these states paled in comparison to the growth of factory-farmed dairies in western states.

Beginning in the 1800s, and continuing into the twentieth century, most milk was produced on small farms and sold locally. As the country became more urbanized, farmers began selling milk to processors for distribution in the city. After the advent of refrigerated railroad cars and trucks, marketing and distribution for milk became more regional. Eventually, many farmers banded together to form cooperatives, so that they could use collective bargaining with buyers. From just after World War II and into the late 1960s, milkmen delivered a large percentage of the milk people drank to their homes. But as grocery store chains appeared, power shifted to the retail sector and those companies that serve retail chains.

Only a few dairies dominated processing and distribution from the 1920s to the 1970s, including Borden Dairy Company (later Kraft), Beatrice, Carnation, and Pet. According to a report by the USDA's Economic Research Service, "Acquisitions by corporations were at an all-time peak in the late 1920's when the National Dairy Products Corporation and the Borden Company started their growth. After dropping off during the Depression, acquisitions of more than 1,000 companies were recorded during World War II, a level never again reached. The Federal Trade Commission (FTC) brought a virtual halt to acquisitions by the eight largest dairy companies in the mid-1950's."[17]

By the 1960s these companies had become part of large diversified firms created through mergers and acquisitions. However, up until the 1990s, a large number of medium-size milk processors were still local, family-owned businesses that bought milk from local dairies and supplied local consumers and retailers.[18] The extreme perishability and constant production of milk makes dairy farmers especially dependent on their buyers. They have to move their milk while it is still fresh, which gives buyers substantial leverage over farmers. This has become especially true as the milk-processing industry has consolidated and specialized, farmers have fewer and fewer options in their area, and small dairy farms cannot survive.

Joel Greeno milks forty-eight cows on his 160-acre farm in Kendall, Wisconsin, and rotates them on 160 acres of pasture. He began his dairy operation in 1993 and a year later helped found the Family Farm Defenders. He recently testified on the state of the industry and about the problems faced by dairy farmers.

> I'd just like to start with the Greeno family [that] has farmed in Monroe County, Wisconsin for 150 years now. Mom and Dad's twenty-ninth wedding anniversary present was a farm foreclosure, and their thirtieth anniversary present was a sheriff's auction on the courthouse steps. That forced me to move my dairy operation out of their facility. . . . You know, my dad battled through polio, a debilitating back injury, cancer, but he couldn't beat low milk prices, and something has to be done. In mid-January, a New York State dairy farmer shot fifty-one of his cows and then himself. We know of nearly one hundred dairy farmers that have committed suicide to date since the '08 crash. It's got to stop. . . . Dairy farmers deserve dignity. They deserve justice. They deserve cost of production plus a profit.[19]

Dairy farmers, especially those with small herds, are being driven out of business because a tiny handful of companies control the industry. The mantra in dairy, like other areas of agriculture, is "get big or get out." A few companies buy the majority of their milk from farms and process it into dairy products and industrial food ingredients. Most dairy farmers market their milk through cooperatives. These cooperatives allow producers to pool the product and participate in the pricing set by the federal dairy marketing order that was first established in 1937, when the USDA began administering the federal milk marketing orders that divide the country into regions. They establish the definitions for different classes of milk, allegedly based on the cost of production in different regions.

Today the corporate-like cooperatives determine how to distribute the milk payments among the membership, and the cooperative is not required to pass any price premiums for the highest-value products to its members. In many areas the cooperative is the only buyer, forcing farmers to endure this discriminatory treatment by the cooperative because they do not have other viable marketing alternatives.[20]

Consolidation also slashed the number of dairy cooperatives by half in twenty years, but the smaller number market a larger share of milk. In 1980, there were 435 dairy cooperatives that marketed 77 percent of the fluid milk; by 2002, there were only 196 cooperatives, but they marketed 86 percent of the milk.[21]

Dairy farmers are extremely vulnerable because of the alliances between corporate-style cooperatives and the milk handling, processing, and manufacturing industry. Dairy Farmers of America (DFA), by far the largest cooperative, controls 30 percent of milk production in the United States, according to the group's Web site. The top four dairy co-ops control 40 percent of fluid milk sales: DFA; California Dairies, Inc.; Land O'Lakes; and Northwest Dairy Association. DFA is a marketing "cooperative" with more than eighteen thousand members and ties to big processing companies that collect and market milk. DFA was created in 1998 out of the merger of four large cooperatives. Dairy farmers effectively are required to market their milk through DFA to access the marketplace, and they take whatever price DFA offers.

Warren Taylor has a lot to say about DFA and the rest of the dairy industry. And he has firsthand experience. His father, Bert Warren, was a well-known dairy technology expert, and he followed in his dad's footsteps. In 1974, Taylor received a degree in dairy technology from Ohio State University and started his career with Safeway Stores as a project manager for the nation's first computer-controlled milk plant. He went on to work for Safeway as a corporate staff engineer specializing in processing design for thirty milk and ice cream plants in the United States and Canada. At that time, between 1977 and 1987, Safeway had the world's largest fluid milk bottling operation. After leaving Safeway he worked on engineering and design projects for DFA, Dannon, Land O'Lakes, and other processors.

Taylor says that in the 1950s, when his father was in the industry, milk was fresh and alive, produced from grazing cows, and processed and sold locally. Taylor and his wife, Victoria, wanted to provide the same kind of healthy dairy products for their community. Today they operate Snowville Creamery in Pomeroy, Ohio, located in the southeastern part of the state. His milk is the freshest that is available on the grocery store shelf—minimally pasteurized and not homogenized. As an engineer with years of experience in the dairy industry, he was able to design and build a state-of-the-art creamery. The milk exceeds FDA pasteurization standards and often makes it to the store shelf in a day. The dairy broke ground in 2006 on Bill Dix and Stacy Hall's three-hundred-acre dairy farm, where 130 cows eat very little grain, spend most of their time outside, and are grazed rotationally on a mostly grass diet.

Taylor says that over his thirty years of association with the industry "it has become so consolidated that it's dominated by two giants: DFA and Dean Foods." He laments that even organic milk is not a real alternative, because in most cases, the cows are milked on factory farms, and Dean Foods now owns one of the largest organic brands, Horizon.

Taylor's assessment of the industry is correct. Dean Foods is the largest dairy company in the country, at the top of the Dairy 100, which ranks companies by sales of dairy products and also lists companies engaged in any aspect of the dairy industry. DFA, ranked eleventh in size, is also the primary—and in some regions the exclusive—supplier of milk to Dean Foods.[22] In turn, Dean Foods processes and markets around 40 percent of the nation's fluid milk supply,[23] 60 percent of all organic milk,[24] and 90 percent of soy milk.[25]

Dean Foods began buying strong regional milk brands in the 1980s: between 1987 and 1998, Dean bought fourteen fluid milk companies.[26] Dean Foods also merged with Suiza Foods, which was a merger of two of the largest fluid milk processors. Dean Foods is the most common source of milk in the dairy case, but consumers rarely see a Dean label. Dean or one of its subsidiaries owns or sells more than fifty brands, including Alta Dena Dairy, Berkeley Farms, Borden, Country Fresh, Garelick Farms, Horizon Organic, Land O'Lakes, Lehigh Valley, Mayfield Farm, Meadow Brook, Meadow Gold, Reiter, Shenandoah's Pride, Silk Soymilk, Swiss Dairy, Verifine, and several dozen others.

Ranking the size of dairy industry players is confusing, both because the industry is global and because there are so many steps in the chain of production from the milk to manufacturing products such as cheese, butter, ice cream, and other foods. Because dairy companies can engage in many aspects of the industry, from processing milk to manufacturing products, it can be difficult to rank companies by product—so instead, this is done by gross earnings. Dean ranks first in terms of size, Nestlé is second, Saputo (headquartered in Quebec) is third, Kraft is fourth, and Land O' Lakes is fifth. Many of the dairy companies have business alliances that are not obvious to the consumers who buy the products.

Like the lack of transparency in the ownership of brands, pricing for milk

the top four dairy cooperatives control 40% of all fluid milk sales

1. **Dairy Farmers of America**
2. **California Dairies, Inc.**
3. **Land O' Lakes**
4. **Northwest Dairy Association**

MILK

40%

TOP 4 CO-OPS

Dean

MILK

80%

Together, the top 4 co-ops and Dean Foods control 80% of all fluid milk sales

MILK **40%**

MILK (USDA ORGANIC) **60%**

SOY **90%**

Dean's market share of conventional fluid milk, organic milk, and soy milk can be a mystery to those not familiar with the industry. Today's pricing system has its roots in the early decades of the twentieth century. Beginning in 1925, processors were paying farmers for milk according to its use: fluid milk (for drinking) or milk for manufactured products, such as cheese, butter, and ice cream. This concept, known as "classified pricing," is still in use today. Milk to be sold in cartons to drink is priced higher than milk for manufacturing dairy products.

Milk prices for farmers have been on a roller-coaster ride. In the summer of 2007 prices reached a record high, and as prices rose, large-scale dairies added more cows to capitalize on the higher price of milk. Over the following two years, overproduction caused prices to fall by half. Although milk prices fell, production costs did not: during 2008 the cost of feed rose 35 percent and the cost of energy rose by 30 percent.[27] Many dairy farmers were losing between $100 and $200 per cow every month in 2009.[28]

According to Taylor:

Milk pricing has become increasingly unstable and erratic. For dairy farmers who have no control over what they are paid, this has been disastrous. For the large milk processing firms who buy milk from farmers, and sell to grocery stores there has been unprecedented profits and monopoly control. The instability in milk pricing is due to increasing concentration to the point of near monopoly. America's dairy farmers were recently paid as little as $12 per 100 pounds of milk while their cost of production was $17. They had to borrow and take on new debt.[29]

Fluid milk prices are especially vulnerable to manipulation by commodities traders. The market that dairy farmers have to contend with is the Chicago Mercantile Exchange (CME), where commodity futures market prices are determined by digital trading that takes place at lightning speed. The spot price of cheddar cheese at the CME is the basis for the federal government's milk price formulas. While this price setting should be transparent, it tends to be secretive and understood only by insiders. The cheese commodity futures trade occurs for half an hour a week, is estimated to involve forty or fewer traders working for half a dozen firms, and covers 80 percent of the cheese marketed in the United States.[30]

In 2008 the Dairy Farmers of America and two of its former executives were fined $12 million for attempting to manipulate the price of fluid milk through cheddar cheese purchases at the CME (two other executives paid smaller fines).[31]

The very small number of traders representing huge dairy companies can actually influence the price of cheese at the CME (and thus the price paid to farmers for their milk) by holding or selling cheese at strategic moments. The Government Accountability Office (GAO) determined that cheese prices at the CME were prone to manipulation.

[T]he spot cheese market remains a thin market which in combination with the presence of a small number of traders that make a majority of trades and the spot cheese market's pricing structure contributes to questions about the potential for price manipulation. . . . Academic analyses and cheese and dairy industry participants have raised a number of concerns associated with thin markets being susceptible to manipulation. These concerns include the following: Dominant traders may be able to attempt to manipulate prices more easily in thin markets. Prices in thin markets may not reflect supply and demand, even without manipulative behavior by dominant traders.[32]

As Joel Greeno explained:

The crisis in dairy is real. The problems in dairy are man made. . . . The Chicago Mercantile Exchange decides everything in dairy prices. . . . The CME cheese trading is a highly leveraged, thinly traded market with few players. Currently, two players mainly control CME block cheddar trading. . . . Two players do not make a market, and at least not a true market indicator. All of the industry marches in lock step with CME prices. With a history of price fixing, the CME block trading cannot be the deciding factor in milk pricing.[33]

The GAO documented the small number of traders dominating the market. Between January 1, 1999, and February 2, 2007, two market participants purchased 74 percent of all block cheese, and three market participants sold 67 percent of all block cheese. During the same time period, four market participants purchased 56 percent of all barrel cheese, and two market participants sold 68 percent of all barrel cheese.[34]

John Bunting, an activist dairy farmer from Delhi, New York, explains that, before the 1980s, "farmers were paid a price based on the cost of inputs plus profit." He says that with the stroke of a pen, Reagan did away with parity (fair pricing), and he removed resources from antitrust enforcement. Bunting has written extensively on these issues. In his report "Dairy Farm Crisis 2009:

A Look Beyond Conventional Analysis," he writes: "Dairy markets are run by an oligarchy—a few elite players—with little or no governmental oversight. As such, the current financial situation provides an opportunistic moment for key players to unduly depress farm milk prices and reap both profits and market power." [35]

Bunting has experienced firsthand the market power of the big players in the industry. DFA, the only purchaser of milk in his area, will not buy from him. He says that "his politics put him on the wrong side" of finding a market for his milk and he had to cull his herd. He has only a few cows now that produce milk for his family, and the rest he feeds to the pigs.

He goes on to say that ever since the Reagan era, the price farmers are paid for milk has fallen almost every year, but the price consumers pay for milk has continued to rise. Besides the manipulation on the CME, Bunting says, the other factor in the low prices is globalization and the importation of milk protein concentrates (MPC) that replace milk in cheese.

MPCs are dumped on the market way below the price of milk and are used in foods like Kraft Singles—which they label a processed cheese product to circumvent an FDA rule on labeling—and other cheese products and processed foods. They are created by putting milk through an ultrafiltration process that removes all of the liquid and all of the smaller molecules, including the minerals that the dairy industry touts as being essential for good nutrition.

What is left following the filtration is a dry substance that is high in protein and used as an additive in products like processed cheese, frozen dairy desserts, crackers, and energy bars. Because MPCs are generally produced as a dry powder, exporters can ship it long distances very cheaply; almost all of the dry MPCs used in America are imported.

MPCs can be the product of cow's milk or they can come from animals like yak or water buffalo that are used for milk in some exporting countries, such as India and China. With such imprecise definitions and vague standards, it is hard to imagine how regulatory agencies manage to keep track of where this ingredient comes from and how it is used. As it turns out, they don't have to, because MPCs are largely unregulated.

Bunting writes: "Importing MPCs into the U.S. is the same as importing milk, except that MPCs are loaded into box trailers at the dock for transportation to plants. No one notices. USDA does not count MPCs as milk. If the imported MPCs for December 2008 could be converted back to milk hauled in tanker trucks, the convoy would be nearly 65 miles long, bumper to bumper." [36]

Bunting is not the only dairy farmer furious about the situation. Around the country, MPCs have been identified as one of the factors causing the price of milk to plummet for farmers. Joaquin Contente is president of the California Farmers Union. He and his brother own a farm that has been in the family since the 1920s. In a July 2009 congressional hearing, Contente called upon the federal government to include "the cost of production, not the costs paid to producers more than 30 years ago." Among the causes he named for the "devastation of the dairy industry" were rising concentrated dairy imports, a lack of competition in the marketplace, consolidation, and rising input costs. He went on to sum up the problem: "Dairy producers have fought for years to pass legislation to regulate dairy imports. . . . So far, dairy processors and food manufacturers, with their well-funded lobbying firms, have fought off any regulation. To end the dairy crisis, lawmakers need to direct their attention to the dairy imports that are flooding our market and forcing so many operations to the brink of financial disaster."[37]

Imports are one more nail in the coffin of family dairy farms. Some short-term actions could be taken to address this issue, like transparent labeling of MPCs and other types of imported dairy ingredients in processed foods and "cheese food," alerting consumers to the inferior nature of these products. However, in the long term, to address the problems associated with corporate-driven globalization, trade policy must be reformed. This is the only way to prevent cheap, low-quality products from undercutting the viability of local and regional production of milk and other foods. As long as MPCs and other cheap dairy inputs destabilize the market, it will be impossible for dairy farmers to make a living or for consumers to have access to safe and healthy food.

Stopping the heartbreaking impoverishment of dairy communities and the hemorrhaging of family dairy farms also means addressing the arcane and complicated way that dairy farmers are paid. The federal government's formulas for calculating the price of milk should be based on the cost of producing it and enough profit to provide them a livable wage. But just a few large companies now use these formulas to lower farmers' prices nationwide.

Moreover, key to achieving reform in what dairy farmers get paid is addressing the rampant consolidation that marks the industry. From the cow's udder to the milk carton, a few large players control every step of the dairy supply chain. They set U.S. dairy policy and control the venues where farmers' milk prices are determined. Dean Foods, the largest milk processor in the country, benefits enormously from policies that encourage overproduction and low prices. Likewise, Walmart and the other grocery giants also profit from having the market power to strike sweetheart deals with the largest milk suppliers for high volumes of cheap milk.

While these companies duck criticism by declaring their commitment to sustainability and family farmers, they cynically use their economic and political power to maintain their privileged position in the marketplace at the expense of farmers and consumers. It is time for the USDA and the Department of Justice to open an investigation into industry giants like Dean Foods' practices, including its relationship with the grocery industry.

DFA, and the handful of other corporate-like cooperatives that operate as virtual monopolies, must also be investigated. They are using their tax-exempt status and the special privileges it brings to hold dairy farmers hostage to a market with one buyer. The decisions that DFA makes about how to distribute milk payments to its member producers are unfair and unethical, especially since the cooperative does not pass on any price premiums for the highest-value products. While DFA is exempt from antitrust law based on size, it is not exempt from antitrust rules based on the use of predatory practices against its own membership. An investigation must be opened on the collusion that this corporate cooperative engages in with milk processors like Dean Foods.

If we truly want to rebuild a regional food system, we must collectively build the political power to break up dairy monopolies. Rebuilding a local food system requires enabling dairy farmers to make a living, while providing consumers with a high-quality product. To do so means we must focus our energies on revitalizing antitrust protections and creating farm policy that allows farmers to get paid according to logical rules in a fair marketplace.

PART VI

Corporate Control of the Gene Pool: The Theft of Life

Over the past forty years, scientists in search of personal gain and prestige have pursued manipulating the genetics of life-forms ranging from bacteria and seeds to fish and animals. Largely at the expense of taxpayers, corporations have gained control over the basic building blocks of life, threatening the integrity of the gene pool and our collective food security. Essentially unregulated, biotechnology is moving far into the realm of science fiction, as scientists actually seek to create life. Understanding the history and the intentions of the prophets and proponents of biotechnology is critical to reining in these dangerous technologies and creating a sustainable food system for the future.

12

LIFE FOR SALE: THE BIRTH OF
LIFE SCIENCE COMPANIES

*Whenever science makes a discovery, the devil grabs it while the angels are
debating the best way to use it.*

—Alan Valentine

A new generation of scientists, who had enormous faith in their own abilities and the role of technology, came of age after World War II. One such man was Herbert W. Boyer—the son of a coal miner and a railroad worker. Born in 1936 and raised in Derry, Pennsylvania, Boyer was more interested in football than biology—not the most likely candidate for changing the face of science. But his football coach also doubled as Boyer's science teacher, and he encouraged Boyer to become interested in science. Boyer became fascinated at a young age by the 1953 discovery of DNA. He eventually earned a PhD in bacteriology at the University of Pittsburgh and a postdoctoral fellowship at Yale in enzymology and protein chemistry. In 1966, at the height of the sixties counterculture movement, he went to San Francisco to be a professor at the University of California. Pictures of him at that time show a man with longish hair and a mustache. And contrary to current stereotypes of pro-GMO scientists, he protested on behalf of civil rights and at anti–Vietnam War rallies.[1]

Meanwhile, while Boyer pursued his work, scientists at both Stanford and Harvard were researching gene splicing and recombining. Paul Berg had spliced genes on segments of the DNA molecule of a virus. Stanley Cohen, located two floors below him at Stanford, was isolating small rings of genetic material from the *E. coli* bacterium. Cohen heard Boyer discuss his work at a conference in Hawaii, and at a fateful meeting afterward, held in a Honolulu deli, they decided to join forces.[2] This legendary event is captured in a sentimental sculpture that graces the campus of the corporation Boyer helped found, Genentech.

The two men, using Berg's work on gene splicing, figured out the basics of genetic engineering—essentially meddling with billions of years of evolution by moving the genes of one species into another. The result of their historic collaboration was published in 1973.

J. Michael Bishop was on the faculty at the University of California during this period, and he wrote an introduction to an oral history describing the birth of biotechnology. He says that some members of the University of California faculty were concerned about the contamination that could occur from Boyer's work, and that Boyer was moved to "inferior quarters, bereft of virtually all professional amenities." Boyer destroyed some of the expensive equipment that he borrowed for the lab, perhaps as "modest revenge for his ostracism." Bishop goes on to describe the events that transpired after Boyer and Cohen's work became public.

> It was nice, it was powerful, and for some, it was worrisome. Alarms sounded. What mischief might ensue such meddling with the natural order? Before long, the research community had implemented a voluntary moratorium on research with recombinant DNA until the issue of safety could be resolved. In December of 1974, concerned parties met at the Asilomar Conference Center in Pacific Grove, California, and drafted the first of several sets of guidelines by which recombinant DNA could be pursued. . . . For Herb Boyer, the Asilomar conference was a "nightmare," an "absolutely disgusting week," and the biohazard committees that the conference sired, "an incredible waste of time and money."[3]

The moratorium on genetic engineering was short-lived—which shows how these types of decisions often go unmarked in the public at large and move forward with no debate. Change was coming fast and furiously.

In 1976 Robert Swanson, a venture capitalist in San Francisco, heard the scientific buzz about the new breakthroughs. With a BS in chemistry and a master's degree in management, he also saw the lucrative potential in commercializing biotechnology. Swanson approached Boyer, and after a short time, they launched Genentech—and a new industry was born. Genentech was focused on replacing chemical treatments with genetically engineered (GE) drugs made from living cells that would act as drug factories. The company developed the first genetically engineered human therapeutic drug by taking a human insulin gene and inserting it into bacteria that would continue to reproduce itself, then extracting the protein and purifying it for use as a medicine.

An academic commercializing his discovery was almost as revolutionary as the GE technology itself. In the mid-1970s, it was not yet common for scientists to benefit financially from scientific achievements made in a university laboratory. So perhaps Boyer should also be recognized as one of the first scientists to focus on advancing this concept—a practice that has dramatically changed academic institutions—and in many ways sullied scientific integrity.

Boyer has been actively engaged in Genentech since the beginning, serving as vice president from 1976 to 1990 and as a director of the board since the company was formed, while simultaneously serving as a professor of biochemistry and biophysics at the University of California at San Francisco until 1991 (he is now an emeritus professor).

At the company's inception, Swanson rented space, found investors, and breathed life into the company, while Boyer hired and oversaw young scientists. The culture has been described as "macho" because "biotech began as a guy thing."[4] A poster of a topless woman surfing graced one of the labs until the end of the 1980s, when a female staffer finally complained. The employees of Genentech, who call themselves GenenExers, have gone on to form dozens of biotech companies. Tom Abate, a journalist writing for the *San Francisco Chronicle*, captures the company's culture.

> Genentech has been stained by scandals involving its inventions, its sales force and its chief executive. Company veterans say Genentech's accomplishments and excesses both stem from the same traits: a fierce competitiveness, a drive to succeed, a sense of making history. . . . To blow off steam, Genentech created the Friday afternoon Ho-Hos, beer and pizza busts that evolve over time. Departments would compete to put on the best Ho-Ho. Once a year, Swanson and Boyer would don grass skirts, order a pig and stage a Hawaiian extravaganza.[5]

An intense and aggressive race had begun to establish biotech companies and to push new products out the door ahead of competitors. California, with its culture of cutting-edge thought—from the free speech movement and counterculture of the 1960s to the entrepreneurialism of Silicon Valley—was ground zero for the new industry. But then again, before the millions could begin rolling in, someone had to bankroll the start-ups. And biotechnology is a very expensive proposition that requires not just initial funding but also frequent and large infusions of cash.

Swanson and Boyer funded Genentech originally by convincing Tom Perkins of the Kleiner Perkins venture capital firm to turn on the money spigot. He started out with a few hundred dollars and ended up investing millions. Venture capitalists line the famous Sand Hill Road in Menlo Park, California. It is the Wall Street of California, whose money guys have provided the jump-start for waves of technology, including biotechnology and the dot-com revolution.

A key to the birth of this heavily financed industry is what Martin Kenney, a professor at University of California–Davis, calls "the university-industrial complex," a phenomenon that predates even Boyer. Biotechnology was born in the labs of academia, but "the multimillion-dollar, multiyear contract between a single university and a single company" became the engine that drove its growth. Indeed, the 1974 contract between Harvard Medical School and Monsanto was one of the arrangements that became a model for the university-corporate relationships that gave birth to biotechnology.[6]

Ray Valentine, a university professor at Davis, followed in the footsteps of East Coast universities. In 1975, as a young professor in the Department of Agronomy and Range Science, he was the agricultural representative to the meeting at Asilomar debating the future of biotechnology—the one that had so angered Boyer. Valentine said it was a big step for a farmer to be with all those Nobel Prize winners. In his talk he suggested that perhaps the new technology could be used for fixing nitrogen in plants and he immediately saw the potential for commercializing the technology in agriculture.[7]

In the late 1970s he wrote a paper proposing an institute to spin off commercial products from campus research that drew a cool response from the university. But as a proponent of academic cooperation with corporations, he procured a research grant from Allied Corporation for a research project for biological nitrogen fixation, and it gave Allied an exclusive, royalty-bearing license to any patents. This created quite a stir at the university—which was already known for its ties to agribusiness. According to Kenney, "Shortly after securing the Allied Corporation grant for UCD, Ray Valentine founded and became vice-president of the first agricultural biotechnology firm, Calgene. Then one week after UCD secured the Allied Corporation grant, Allied purchased 20 percent of Calgene's stock for $2 million."[8]

Valentine had other supporters in high places. He spoke of how Nixon personally intervened in securing a research grant that was in danger of lapsing, and that without the money for his lab, Calgene would never have been born. He claimed that he had taken an oath not to speak of it until after Nixon's death.[9]

Another opportunity arose when Valentine joined forces with Norman Goldfarb, who was looking to quit his job at Intel and start his own company. Goldfarb's family had bankrolled companies ranging from Genentech to Apple, and he had money to invest. After Calgene was established, Valentine kept his day job at the university, but he also became vice president of research and founded a science advisory committee.

In the 1980s Calgene began the long process of working on genes for herbicide resistance—the research that eventually led to the development of Monsanto's weed killer Roundup. Calgene was viewed by its founders as a "force for revolutionary change" that would follow in the footsteps of Genentech and Apple, according to one of the company's first scientists, Vic Knauf. Calgene had some big-time investors, including Procter & Gamble and the French chemical giant Rhône-Poulenc. But what caught investors' imaginations was a "discovery that would solve every vegetable shopper's pet peeve and turn Calgene into a jolly green gene giant." The idea was to make a better tomato—a flavorful one with an extended shelf life that would also make good tomato paste. The science advisory committee at Calgene was not impressed with the idea, which inspired Dave Stalker, a senior researcher at Calgene, to make the tomato his personal cause. Campbell's Soup was convinced to fund the research. Calgene isolated the gene that makes tomatoes soft and decided to construct another gene to keep the first gene from doing its job.[10] Amazingly, it worked.

After years of research and development the tomatoes went on the market in 1994, three days before receiving approval from the FDA. The first genetically modified food approved for humans, they were marketed with the trademarked name MacGregor tomatoes and grown from Flavr Savr seeds. Genetically engineered tomatoes were a big flop, in part because the company did not start with seeds from flavorful tomatoes and use traditional breeding methods to improve them. While the tomatoes had a long shelf life, they did not remain firm if they were picked after turning red and so still required harvesting while green. Former Calgene scientist Belinda Martineau writes: "Instead of taking a generous bite out of the $4 billion U.S. retail market for fresh tomatoes, Calgene suffered negative gross margins in its tomato business that led to losses to the tune of tens of millions of dollars a year. The Flavr Savr gene was, in the end, of only marginal value to a fresh market tomato business. Less than two years after its national rollout, the Flavr Savr tomato forever disappeared from the supermarket shelves."[11]

GE tomatoes for food processing and paste were a failure as well. Calgene and the British company Zeneca, which was also funded by Campbell's

Monsanto's Acquisitions

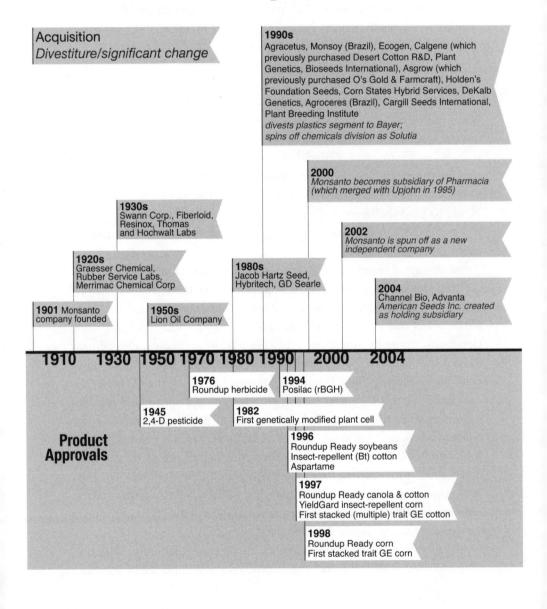

Acquisition
Divestiture/significant change

1990s
Agracetus, Monsoy (Brazil), Ecogen, Calgene (which previously purchased Desert Cotton R&D, Plant Genetics, Bioseeds International), Asgrow (which previously purchased O's Gold & Farmcraft), Holden's Foundation Seeds, Corn States Hybrid Services, DeKalb Genetics, Agroceres (Brazil), Cargill Seeds International, Plant Breeding Institute
divests plastics segment to Bayer; spins off chemicals division as Solutia

2000
Monsanto becomes subsidiary of Pharmacia (which merged with Upjohn in 1995)

1930s
Swann Corp., Fiberloid, Resinox, Thomas and Hochwalt Labs

2002
Monsanto is spun off as a new independent company

1920s
Graesser Chemical, Rubber Service Labs, Merrimac Chemical Corp

1980s
Jacob Hartz Seed, Hybritech, GD Searle

2004
Channel Bio, Advanta
American Seeds Inc. created as holding subsidiary

1901 Monsanto company founded

1950s
Lion Oil Company

1910   1930   1950 1970 1980 1990   2000   2004

1976
Roundup herbicide

1994
Posilac (rBGH)

1945
2,4-D pesticide

1982
First genetically modified plant cell

Product Approvals

1996
Roundup Ready soybeans
Insect-repellent (Bt) cotton
Aspartame

1997
Roundup Ready canola & cotton
YieldGard insect-repellent corn
First stacked (multiple) trait GE cotton

1998
Roundup Ready corn
First stacked trait GE corn

Through Time

2008
Aly Participacoes (Brazil), Marmot (Guatemala),
De Ruiter Seeds, EarthMap Solutions
Posilac (rBGH) sold to Eli Lilly

2009
WestBred, Pfizer's Chesterfield
Village Research Center,
Seminium (Argentina)
Sunflower assets sold to Syngenta

2006
Diener Seeds, Sieben Hybrids, Kruger Seed, Trisler
Seed Farm, Gold Country Seed, Heritage Seeds,
Fielder Choice Direct, Campbell Seed (m&s)
*Monsanto and Dow agree on global cross-licensing
of intellectual property, the first of many such
agreements*

2010
ANASAC corn & soy
plant (Chile)

2011
Divergence, Beeologics,
Panon Seeds (Hungary)

2007
Delta & Pine Land Company, Agroeste Sementes (Brazil),
Western Seed, Rea Hybrids, Jung Seed, Moweaqua Seed,
Bo-Ca Enterprises, Hawkeye Hybrids, Lewis Hybrids,
Hubner Seed, ICORN Inc., Poloni Semences
*divests Monsanto Choice Genetics (swine), divests NexGen
& Stoneville cotton brands to avoid DOJ antitrust suit,
launched International Seed Group as holding company*

2005
Seminis Inc., NC+ Hybrids,
Stewart Seeds, Trelay Seeds,
Stone Seeds, Specialty Hybrids,
Fontanelle Hybrids, Emergent
Genetics, Stoneville Cotton

2005	2006	2007	2008	2009	2010

2005
Roundup Ready alfalfa and
sugar beets approved
Both withdrawn in 2007

2011
Roundup Ready alfalfa
reissued, Roundup
Ready sugar beets
partially reissued,
drought-tolerant corn,
fatty-acid soybeans

Soup, worked out a legal arrangement for developing GE tomatoes. While Calgene developed those for the fresh market, Zeneca created tomatoes for use in commercial food processing. After FDA approval, the tomatoes were sold at a lower price and advertised as genetically engineered, but consumer acceptance was low, and as a result neither tomato is on the market today.[12]

The cost of the failed tomato project was high for Calgene. As Monsanto began to turn itself into a life sciences company, shedding its chemical businesses, it acquired its first stake in the company in 1996 and bought the remaining shares in 1997. Monsanto wanted ownership of Calgene's research into genetically modified oils, cotton, and fresh produce.

This important acquisition by Monsanto was part of the biotech merger mania of the 1990s. The industry grew and developed through mergers, acquisitions, and alliances that involved biotech research firms, seed distributors, and chemical manufactures. The pressure to grow and become more profitable also led to an alignment with pharmaceutical companies. Bryan Bergeron and Paul Chan explain in their book on the industry: "The primary means of growth is acquiring seed companies in order to directly access the seed market. Because seed companies ultimately decide which biotechnology to incorporate into their product lines, without a captive seed company, there is no guarantee that an agricultural biotech company will be able to bring its technology to market."[13]

Monsanto, the world's largest producer of genetically engineered seed, with ownership of 674 agricultural patents, has become the dominant player and the symbol of the industry. It was founded in St. Louis, Missouri, in 1901 by John Francis Queeny, who was a buyer for Meyer Brothers Drug Company, and thus he was familiar with the drug and chemical industry. He married well, naming the company after his wealthy father-in-law, Mendes de Monsanto, a sugar financier from St. Thomas. According to a 1936 article in *Time*, "He figured he had sold enough remedies and condiments to be able to make money manufacturing them. . . . He gave the company his wife's name because he planned to keep on peddling drugs under his own."[14]

Monsanto first produced saccharin—a derivative of coal tar that had been discovered accidently by a researcher at Johns Hopkins University and was later in the twentieth century linked to bladder cancer. Monsanto began as the major supplier of saccharin, caffeine, and vanillin to Coca-Cola—a prophetic alliance in hindsight.

While Monsanto had begun marketing agricultural chemicals in 1945, it got into agrochemicals in a larger way with the introduction of 2.4-D and, as a result, created an Agricultural Division in 1960. This jump-started the company's production of herbicides—Ramrod in 1964 and Lasso in 1968. Monsanto's cell biology research program, the start of its research into genetic engineering, began in 1975 in reaction to the publicity around Boyer and Cohen's discovery of how to splice genes.

The herbicide Roundup was commercialized in 1976, and by 1981 Monsanto had changed its strategic focus to concentrate on biotechnology and established a molecular biology group. The company started pouring money into research on the use of recombinant DNA technology to transfer genetic material from one organism to another to add desirable traits.

As a result, Monsanto became the first company to genetically modify a plant cell. Monsanto's search for their golden apple was based on its desire to develop herbicide-tolerant seeds that would allow Roundup to kill weeds without harming the crops in the same field.

Biotechnology challenges traditional breeding methods for desirable crop and livestock traits. It is different than the traditional selective breeding of plants over many generations within the same plant species until the desirable trait manifests itself. At this time, splicing genes is a very expensive business that requires state-of-the-art laboratories and highly paid scientists. Monsanto needed a great deal of money to execute its strategic plan.

Monsanto's search for a company with guaranteed profits resulted in the purchase of the maker of NutraSweet, G.D. Searle, in 1985. It seemed like a good match, because the drug company also had an experienced sales and marketing staff. But by 1988, Searle was the focus of hundreds of lawsuits and a flood of bad publicity because of its intrauterine device (IUD), the Copper-7, which killed or injured hundreds of women. Incidentally, Donald Rumsfeld, who was secretary of defense in the 1970s under Gerald Ford and again in 2001 under George W. Bush, ran Searle between 1976 and 1985. According to Washington, D.C., attorney Jim Turner, Rumsfeld said in a 1981 sales meeting that he would push to get the artificial sweetener aspartame (NutraSweet) approved within the year, using his political clout. While serving as a member of Ronald Reagan's transition team, he recommended Dr. Arthur Hull Hayes Jr. to lead the FDA, which soon thereafter approved the sweetener.

In the early 1980s Monsanto began testing genetically manipulated crops. Jerseyville, Illinois—a town in the southwestern part of the state not far from St. Louis—became the front line for the first biotechnology field trials in 1987, first with tomatoes and then with Roundup Ready soybeans.

In 1993 the FDA approved Monsanto's first biotechnology product—the artificial recombinant bovine growth hormone (rBGH), which was created to increase milk production in cows. Although dairy cows produce the hormone naturally, artificially elevating its levels increases milk production in a way that can lead to mastitis in cows and may have detrimental effects on human health.

The NewLeaf potatoes, manipulated through the addition of a gene that repells the potato beetle, was introduced in 1995. The GE potato flopped, however, and is no longer marketed. The seed was too expensive for farmers, and consumer pressure on McDonald's, the largest single user of potatoes, to stop using it in its french fries sealed its fate. People really didn't want to be guinea pigs for genetically engineered food.

Monsanto responded to the consumer backlash by focusing more on commodities hidden in processed food—canola; soy; corn; wheat; and a nonedible crop, cotton. Roundup Ready canola and cotton were introduced in the mid-1990s and quickly took over market share. In 1996, Roundup Ready soybeans were approved. Commercializing GE soy had been on the Monsanto agenda since its biotechnology program was initiated. Manipulating the genetic material of the soybean so that it could survive being doused with the herbicide presented opportunities for vast profits from both products. Well-known Indian activist Vandana Shiva called the control of the entire production process of plants from breeding to cultivation and sale the hijacking of the global food supply.

Soy is a versatile crop grown widely in the United States for protein and oil that is used in a broad range of food products, animal feeds, and industrial products, so controlling its production was a harbinger of the changing landscape for farming. Today, 80 percent of soybeans are genetically modified, and Roundup Ready is responsible for generating 40 percent of Monsanto's profits. Shiva laments, "Soybeans and soybean products are being pushed as global substitutes for diverse sources of foods in diverse cultures. They are being promoted as substitutes for the diverse oilseeds and pulses of India and for cereals and dairy products worldwide."[15]

Next, Roundup Ready corn was approved and commercialized in 1996. And Monsanto was the first company to introduce stacked traits in corn—the combination of more than one trait like resistance to corn borer insects and Roundup Ready.

In 1998 Monsanto won a bidding war to acquire the DeKalb Genetics Corporation, enabling the company to quickly put genetically engineered seeds into the hands of a large number of farmers, since DeKalb controlled 11 percent of the corn seed market at that time. The purchase of the large corn seed company Holden, also in 1998, gave Monsanto a dominant position in the corn seed market by the end of the century.

Continuing its marathon of biotech deals, Monsanto and Empresas La Moderna, a Mexican biotechnology company, signed a collaborative agreement with Mendel Biotechnology enabling the two companies to obtain exclusive access to Mendel's technical capabilities and genomics for corn, soybeans, and several fruits and vegetables. Mendel's expertise in identifying the function of genes and patenting the corresponding DNA sequences was critical for Monsanto's work on the next generation of biotech products.[16]

In a deal fraught with problems, including a lawsuit against both companies over wrongfully obtained genetic material, Monsanto bough Cargill's international seed business in 1998, excluding the North American seed business. Cargill's massive international marketing infrastructure combined with Monsanto's biotechnology capabilities raised the ante for the seeds destined to produce feed for livestock. This corporate merger was one of the largest steps toward changing the way animals are raised. The development of feedstocks with bigger yields and lower production costs helped facilitate the growth of factory farms.[17]

By the end of the 1990s' multibillion-dollar spending spree, the company was also very cognizant of its liabilities, and it spun off its chemical operations as a separate company, Solutia. The chemical arm of Monsanto faced tremendous legal liabilities related to the manufacturing of PCBs, and it declared bankruptcy in 2003.

As Monsanto continued to transform itself into a biotech giant, it merged with Pharmacia & Upjohn and went through a confusing reorganization that included name changes, spin-offs, acquisitions, and divestments, including selling its artificial sweetener, NutraSweet, to a Boston investment group. Another game changer for Monsanto was coming up because Roundup's main ingredient, glyphosate, lost its patent protection in 2000. This opened the floodgates for other companies to manufacture similar herbicides and, along with the large amount of money that Monsanto had spent in the race to acquire companies, meant that the company was temporarily short of cash.

In the meantime, the race for humankind's genetic material was charging forward, and DuPont was eager to take on Monsanto. In the late 1990s rumors flew around Wall Street about a possible merger between the two giants. Longtime agricultural journalist Al Krebs noted the companies were competing to be transformed into the biotech equivalent of Coke and Pepsi. He went on to say: "In the three years since the first transgenic seeds were introduced, crop biotechnology has grown from a young science to a hot business. . . . Now in a stunningly swift concentration of power, much of the design, harvest and processing of genetically engineered crops is coming under these two companies' influence." [18]

Long before John Queeny established Monsanto, the French Huguenot E.I. du Pont had established DuPont, the second-largest holder of agricultural patents, with 565. A company that has benefited enormously from war, it was founded in Delaware by an immigrant who had left France to escape the French Revolution. The DuPonts went on to become American aristocracy—one of our country's richest and most prominent families. As a wealthy man trained as a chemist, DuPont saw the opportunity for establishing a company in America that made high-quality gunpowder. The company expanded quickly, and by the time of the Civil War it was able to provide half of the gunpowder used by the Union armies and moved into the production of dynamite and newer types of gunpowder. In the early years of the twentieth century, under the leadership of the founder's great-grandsons, the company bought several smaller chemical companies and reinvented itself as a chemical company; it went on to develop a range of products, including nylon, Teflon, and Lycra. DuPont emerged as a leader in biotechnology a century and a half after it was founded.

In 1997 DuPont acquired an interest in Pioneer Hi-Bred, the world's largest seed company. Although it was first established in 1926 as Hi-Bred Corn Company by Henry Wallace and a group of Iowa businessmen as a way to commercialize the hybridization of corn, the company added the word Pioneer in 1935 to distinguish it from other corn seed companies. Wallace probably would be shocked and appalled that the company he founded was purchased for $10 billion by DuPont in 1998. Wallace, who as a young man helped develop hybrid corn varieties with high yields, was secretary of agriculture during Franklin Roosevelt's first two terms and vice president during his third. A strong and vocal advocate for farmers, he lost his bid for the presidency on the Progressive ticket in 1948.

The company Wallace founded went on to be a giant in the seed business that invested in research. It went public in 1973 and merged with a number of companies. By 1981 it dominated sales of corn seed, and by 1991was the number one brand of soy seeds. It partnered with Mycogen Seeds to develop Bt insect resistance in corn, soy, canola, sunflower, and other seeds. Bt is an abbreviation for the name of a bacterium that is toxic to insects but nontoxic to humans. Widely used by organic farmers, there is significant concern that the use of this genetic material will lead to the emergence of pests that are resistant to the Bt toxin, making it worthless for organic cultivation.

Pioneer went on to purchase the rights to Roundup soybean seeds from Monsanto in 1992 and the rights to Bt corn that is resistant to European corn borers in 1993. With its full acquisition of Pioneer in 1999, DuPont had entered biotech in a really big way. But the struggle for control of the biotech industry was just beginning.

Continuing the crazy and confusing merry-go-round of mergers and acquisitions was the creation of Syngenta. Up until 2000, Monsanto and DuPont were battling for market domination with Novartis, a multinational pharmaceutical company based in Basel, Switzerland. In 2000 Swiss Novartis and AstraZeneca, both the products of mergers and acquisitions, combined their agribusiness companies to create a new company called Syngenta, which specializes in genetically engineered seeds and pesticides. Today the company is the third-largest biotech seed company.

Vandana Shiva saw very clearly how biotechnology and the consolidation of the leading companies would affect farmers and eaters. In 1987 she was at a biotechnology seminar and heard a staff person from Sandoz speak.

> I started to get called into biotechnology seminars because it was the next step. In '87 at one of these seminars, the industry laid out its grand dream of controlling the world. They talked about needing genetic engineering so that there's a technology that they have that peasants can't use, so that they can have a monopoly through technology. Patents. Because without it they cannot consolidate power.
>
> That was said by Sandoz. . . . All of them merged to become Syngenta. What they had said at that time was, "By the turn of the century we will be five," in '87. I said, "I don't want to live in a world where five giant companies control our health and our food." [19]

Sadly, this was more than boastful audacity on the part of the biotech industry. The industry has successfully used its political clout to shape public policy, and subsequently it now controls many of the Earth's genetic resources. Genetic engineering requires large amounts of capital to fund scientists, technicians, and laboratories, and so it is inextricably linked to the foodopoly and its model of large-scale, commercialized food production.

Biotechnology is a primary example of how science has been hijacked to remove food from its proper role in society—providing sustenance to the world's population in an ecologically balanced way. Historically, people around the world have used local ingredients and culture to create a healthy cuisine, using variations of these foods. We have decades of research verifying the foundation of a healthy diet—vegetables, whole grains, fruits, and a dash of protein. It is well established that processed food based on ingredients processed from genetically engineered crops is the antithesis of healthy food. But the science establishment has been a partner with the foodopoly in manipulating taste preferences, food processing, and agricultural production to make food a profit center, not the basis for a healthy diet.

We should not be surprised. Many of the companies now making decisions about the future of the gene pool have been exceedingly irresponsible in the past and have poor track records in protecting public health or the environment. Their use of genetic engineering has been marked by a complete lack of concern about the long-term unintended consequences of the technology. The primary interest of the foodopoly leadership is not nourishing people, contributing to public health, or even ensuring the long-term viability of their companies. Their agenda—shaped by the structure of our economic system—is focused on short-term economic gains. The focus is on next quarter's profits, the price on a share of company stock, the bonus generated from a merger or acquisition, stock options, and golden parachutes.

It is time to use science for the benefit of people rather than corporate financial statements. Farmers have used selective breeding of plants and animals for untold centuries, producing the seeds and animals best acclimated for their local environment. We must reestablish this sensible system of food production, based on techniques that have enabled humanity to produce food for millennia. This has never been more critical than today. As we face new challenges from climate change and an increasing population, it is time to regain control of our genetic future and our food system. Today the biotech industry is almost unregulated. We must build the political power to rein in this industry so that future generations can eat.

13

DAVID VERSUS GOLIATH

No science is immune to the infection of politics and the corruption of power.
—Jacob Bronowski (1908–74), British scientist

Just a decade into the twenty-first century and more than 365 million acres of genetically engineered crops are being cultivated in twenty-nine countries; this represents 10 percent of the globe's cropland.[1] The United States has become the world leader in GE crop production, with 165 million acres, or nearly half of the globe's production.[2] In just fifteen years, U.S. cultivation of these crops grew from only 7 percent of soybean acres and 1 percent of corn acres in 1996 to 94 percent of soybean and 88 percent of corn acres in 2011.[3]

How did our regulatory agencies allow biotechnology to escape regulation, almost without exception? The answer is a story of public policy for sale across five presidential administrations—regardless of political affiliation. This failure to adequately regulate the biotech industry has had a long-lasting and negative effect on food and farming. Genetic engineering has many unintended consequences, from causing food allergies to increasing consumers' exposure to carcinogenic artificial hormones. Genetic engineering of crops that are tolerant of specialized co-branded herbicides has been documented to have dramatically increased the use of these polluting and health-threatening agrochemicals. No one really knows what the long-term consequences of manipulating the gene pool of plants and animals will be.

Escaping regulatory oversight for biotechnology was not a scheme that began in the cutting-edge laboratories of the University of California or Stanford, where the biotech revolution began, or in the dozens of start-up companies that employed ambitious young technocrats. The subversion of the regulatory system started at Monsanto, the industrial chemical company that had survived the regulatory wars of the 1970s and the hundreds of legal challenges that were a result of manufacturing deadly PCBs (now banned industrial coolants) and Agent Orange. Monsanto's corporate culture made it a cunning and ruthless giant with the experience and resources to challenge nature at its very core.

In 1979 a scientist with a vision for biotechnology's future took Monsanto by storm. Howard Schneiderman was recruited to revitalize the company's research and development program. A developmental biologist by training, he had been dean of the School of Biological Sciences at the University of California–Irvine before being hired as chief scientist at Monsanto.

Schneiderman had a colorful history. He grew up in New York City and, ironically, attended Ethical Culture Society schools—they were sponsored by a humanist organization founded to encourage respect for humanity and nature, and one his parents were active in. He and his wife were married at the Brooklyn Ethical Society School after he graduated from Harvard, having finished his postdoctoral work there. In 1953, he began his career at Cornell, later moving to Case Western Reserve University, where he obtained a large grant from the Ford Foundation as part of its green revolution research initiative. Schneiderman's work shifted to cell-related research at this time and he became interested in the advances in gene splicing. During his employment at Monsanto, which lasted until 1990, when he died of leukemia, he lectured widely on the ethical dimensions of the private ownership of genes.[4]

Monsanto may never have made the big and expensive jump to biotech without Schneiderman, a man of his times. Biotechnology was making headlines, and capital was rushing in to fund the start-ups. University scientists were breaking the boundaries between academia and corporations. Schneiderman was part of the new generation of scientists who believed academia had to work hand-in-hand with industry, so he left sunny Southern California for Monsanto's headquarters in St. Louis.

When he arrived at Monsanto, he quickly deputized a career scientist there who had dreamed for a long time that biotech was the future. Nicknamed "Ernie the Cork," Ernest Jaworski was the child of Polish immigrants and a graduate of the University of Oregon who had joined Monsanto in 1952. The corporate culture at Monsanto was rough and ready—not an easy place to work. When the most successful products garnered runaway profits, management would devote large sums of money to new programs, but when profits dropped, budgets were slashed, and new programs were dropped overnight. But his colleagues felt that Jaworski always "somehow bobbed to the surface." He was supposed to be working on the next generation of herbicides, but with the gene-splicing frenzy electrifying biologists, he became interested in creating a gene that allowed corn to survive dosages of Roundup. Until Schneiderman arrived on the scene, there was little interest at Monsanto in what seemed like a crazy idea.[5]

Schneiderman believed that biotech was the second industrial revolution, and he joined forces with Jaworski to recruit young scientists with the skills to move Monsanto into the biotech age. He had the credibility to persuade the management to invest in the Life Sciences Research Center that was built in 1984 and housed one thousand scientists. It was known around the company as the "house that Howard built." He used his prominence and his facility with data to gain the support of each influential executive at the company, and he was able to marshal the resources of universities to advance Monsanto's agenda. Washington University and Monsanto signed a $23 million agreement in 1982, and Schneiderman developed close research collaborations with other universities, such as Oxford.[6]

In 1980, soon after Schneiderman joined Monsanto, Ronald Reagan was elected president on a bandwagon of deregulation and free-market capitalism. Monsanto was well positioned to influence the administration, because it was already a powerful corporation in Washington's halls of power. It had become clear to Schneiderman that the biotech industry was going to be controversial, and that if new laws were enacted to scrutinize genetically engineered products, they would be a deterrent to financial success.

Monsanto's chief lobbyist in Washington, D.C., in the early 1980s was Leonard Guarraia. According to Daniel Charles, author of *Lords of the Harvest*, Guarraia was a jovial, profane, and oversize lobbyist who was hired to charm and cajole officials in Washington. He and his internal allies saw that it would be necessary to create the illusion of regulation to forestall new legislation from passing. It was a cynical ploy. Consumers could be lulled into complacency by pseudoregulation and a few easy regulatory hurdles for product approval. A process with predetermined steps that a large and established company like Monsanto would have no trouble completing would also give them a competitive edge over smaller biotech firms with fewer resources.

Guarraia was alarmed by the growing resistance to biotechnology. He wanted to convince his colleagues that something had to be done, so he took a tape showing biotech's "most vociferous and implacable foe" to St. Louis for Monsanto executives to view. The tape showed Jeremy Rifkin calling environmentalists to action.[7]

Rifkin had co-authored *Who Should Play God? The Artificial Creation of Life and What It Means for the Future of the Human Race*, which laid out a frightening future showing "how precariously we are perched on the genetic powder keg." It predicted that corporations would have the right to own and sell the life-forms created in their laboratories, and that they would soon be able to clone endless replicas of a living organism from a single cell of its body.

Rifkin did not have the profile of a late 1960s activist. He had earned a BS in economics at the Wharton School of the University of Pennsylvania, where he was elected president of the graduating class of 1967. As the winner of the General Alumni Association's Award of Merit, he was commended for "selfless charitable work as President of the Campus Chest," for being imbued with the "spirit of Pennsylvania" and "communicating this to football fans as a cheerleader," and for advancing "the cause of Greek letter societies."[8]

But somewhere along the way, Rifkin was politicized, and he became a vocal opponent of the Vietnam War. As a student at Tufts, where he obtained a master's degree in international affairs, Rifkin pursued antiwar activities and became an environmentalist. In 1973, as the country reeled from high gas prices, he organized a protest against the oil companies at the Boston Harbor during the 200th anniversary of the Boston Tea Party. In 1977, he and a colleague founded the Foundation on Economic Trends (FOET), where he still works today promoting a long-term economic sustainability plan to address the triple challenge of the global economic crisis, energy security, and climate change.

But back at the dawn of biotech, Rifkin was focused on genetic engineering. He frightened the industry, and the fact that Monsanto had in their possession a tape of him speaking to environmentalists suggests that it or some other entity was monitoring his activities.

One of the executives present at the screening was Washington, D.C., insider Will Carpenter, who had represented Monsanto in the battles over chemical production that had pitted environmentalists against industry during the disputes over Monsanto's long record of polluting. He wanted to follow a new path with biotechnology.[9]

At strategy meetings during 1983 and 1984, Carpenter argued that the government's stamp of approval was needed to provide assurance to the public that the technology was safe. He proposed asking the government for regulation and help in shaping the process, so that if companies fulfilled a list of requirements, products would automatically be approved. Monsanto certainly did not want any new laws passed to regulate the industry; rather, it was very likely in search of a regulatory fig leaf.

Carpenter ran into trouble when he went to Washington to sell the idea of regulation: the Reagan crowd didn't believe in regulation.[10] Henry Miller, in particular, was an FDA official strongly opposed to the Monsanto plan. An irascible opponent of environmentalism and regulation, he spent fifteen years at the FDA serving as an important advocate for the biotech industry. During his FDA tenure, he was the founding director of the biotechnology office and a medical reviewer for the first GE drugs who saw to their rapid licensing. Today, the stridently antiregulatory Miller is still speaking out against the regulation of biotechnology as a fellow at the right-wing Hoover Institute.

Miller charged Monsanto with trying to squeeze out the smaller biotech firms such as Calgene, which could not afford to test new products. According to Charles, Miller called Monsanto "traitors" who caused more harm to the industry than Rifkin. Amid this extremely charged atmosphere, Monsanto decided that a detailed strategy was necessary.[11]

Monsanto called upon King & Spalding, one of its hired guns in Washington, to develop a strategy for a regulatory process that would help the company bring its products to market quickly and smoothly. The 125-year-old international law firm still represents big pharma-biotech companies today under the umbrella of its "FDA and Life Sciences" practice, which was established by Michael Taylor. As discussed in chapter 6, Taylor had been a staff attorney at the FDA, and he was hired to head the firm's FDA lobby shop in 1981.

Monsanto's lobbying power was unmatched. The industry-friendly regulation that Taylor proposed was adopted almost verbatim in the policy developed by the White House Office of Science and Technology Policy and published in the *Federal Register* on December 31, 1984. Called the Coordinated Framework for the Regulation of Biotechnology, it remains the basis of regulation today. Monsanto had won; no new law was necessary to regulate biotech.

The Reagan era brought another gift to Monsanto and the biotech industry. It settled the question of patenting life. This story began in 1906, when the plant nursery industry had failed to muscle enough political power to pass legislation amending the Trademark Act in order to prohibit competitors from selling or giving away grafts or cuttings of plants. Failure to achieve its goal led to the formation of the National Committee on Plant Patents, a lobbying group for the American Association of Nurserymen. In 1929, Paul Stark, whose company was famous for developing Red Delicious and Golden Delicious apples, became chair of the committee.

The nursery business lobbied successfully for passage of the Plant Patent Act of 1930, which provided a seventeen-year patent for plants that are grown through grafts and cuttings.[12] This protection was later extended to agricultural products, chemicals, and processes under the Patent Act of 1952. Intellectual property rights were established for plants grown from seeds in

1970, when the Plant Variety Protection Act gave breeders exclusive patent rights for eighteen years, but gave farmers the right to save seeds and sell them to other farmers.

In 1980 a watershed Supreme Court decision bestowed patent protection on GE plants, animals, and bacteria. Writing for the 5 to 4 majority, Warren Burger, the Nixon appointee, argued that laboratory-created living things were not "products of nature" under the 1952 Patent Act and were therefore patentable. This decision has had a far-reaching effect, creating the legal basis for establishing a patent today. Before this landmark decision, living things could not to be patented. The decision allowed microorganisms to be patented, triggering a 1985 U.S. Patent and Trademark Office ruling that plants can be patented under utility patent law. This decision paved the way for other patent law decisions that have made seed saving and sharing nearly obsolete.

Pressure also mounted to allow publicly financed research and technology to be shared with private corporations. Indicative of the rush to embrace the Reagan revolution, the National Cooperative Research Act was introduced by the conservative Senator Strom Thurmond and sixty-eight bipartisan sponsors from both parties in 1984. It relaxed antitrust law for joint research and development ventures, such as those undertaken by biotech firms. Companies from a range of industries, including Monsanto, lobbied for the Technology Transfer Act of 1986. It mandated federal agencies to make technology transfer part of their mission and dictated that staff be evaluated during their annual performance appraisal on their success in doing so.

All of these changes heightened the debate and controversy around biotechnology. As a result, in the early 1980s, two industry trade associations were formed to lobby and counteract opponents. The Industrial Biotechnology Association represented the large established companies, and the Association of Biotechnology Companies acted on behalf of the start-ups. A decade later, as the industry matured, the two groups joined forces to become the Washington-based lobby shop Biotechnology Industry Organization, known as BIO today.

Meanwhile, as biotechnology gathered political steam, the forces opposing genetic engineering were coalescing. The Biotechnology Working Group, an informal network of disparate organizations with a variety of concerns about the technology, began meeting in the early 1980s. Members included the Foundation on Economic Trends, Friends of the Earth, Consumers Union (CU), and a range of other organizations, including religious and labor groups. The coalition called for new regulation to oversee the technology, and their biggest congressional ally was a young congressman from Tennessee named Al Gore. Although an amateur futurist, he was skeptical about the rush of private capital to the unregulated industry.[13]

While a proponent of new technologies, Gore argued that the government needed to become involved in regulation before damage was done to the environment, rather than waiting until after it occurred. He advocated for legislation that would establish new regulations for the industry and protect the public interest. Michael Hansen, who represented CU at the Biotech Working Group and was an early opponent of genetic engineering, says that Gore really knew the issue and was sincere in his concern.

The emerging regulatory framework enshrined the idea that biotech products are the same as those produced conventionally, with no unique or special properties or risks. Because no new legislation was passed to regulate the industry, three agencies would primarily be in charge: the FDA, the EPA, and the USDA. They could apply existing statutes to biotechnology products—a radical idea when compared to the much more protective regulatory framework of the European Union, Japan, Australia, and a host of other countries.

Monsanto was nervous about the growing opposition and the push for regulation. Four executives from the company visited with Vice President George Bush toward the end of 1986. According to a 2001 investigative piece in the *New York Times*:

> [T]he White House complied, working behind the scenes to help Monsanto—long a political power with deep connections in Washington—get the regulations that it wanted. It was an outcome that would be repeated, again and again. . . . What Monsanto wished for from Washington, Monsanto—and, by extension, the biotechnology industry—got. . . . And, when the company abruptly decided that it needed to throw off the regulations and speed its foods to market, the White House quickly ushered through an unusually generous policy of self-policing.[14]

One of Monsanto's hallmarks is the ability to influence regulatory policies even as the faces of the executive branch change. After George H.W. Bush became president, the push for lenient regulation of biotechnology gained momentum. In February 1991, the White House Council on Competitiveness, part of Bush's regulatory relief program and headed by Vice President Dan Quayle, issued recommendations opposing efforts in Congress to more strictly regulate biotechnology. It also recommended extending patent protections for genetically engineered cells and advocated for a shift in federal financing of biotechnology toward applications in agriculture and energy.

Quayle announced the new policy in the Old Executive Office Building, saying, "We will ensure that biotech products will receive the same oversight as
other products, instead of being hampered by unnecessary regulations."

In July 1991 the revolving door turned again when the Bush administration brought Michael Taylor, the King & Spalding strategist, back to the FDA
to the specially created position of deputy commissioner for policy.

With Taylor back at the FDA shepherding the process, in March 1992 the
Bush administration took the recommendation from the Council on Competitiveness and streamlined the process for approving biotech products. The
FDA issued the Biotechnology Policy, which clarified the "agency's interpre-
tation" of federal law regarding genetically engineered food. It was official:
foods derived by bioengineering did not differ from other foods.

MIT's journal *The Tech* gushed about the lenient new policy, saying that
it was likely to stimulate the growth of the industry to genetically alter plant
and agricultural products and as a result "companies might license more MIT
patents." Bush proclaimed, "This $4 billion industry should grow to $50 billion by the end of the decade." [15]

But Monsanto still had not gotten everything that it needed to reap the
economic benefits the company craved from biotechnology. It had been working for several years on the artificial growth hormone, recombinant bovine
growth hormone (rBGH), that increases milk production. In 1981, despite an
ongoing controversy, the company submitted preliminary materials for FDA
approval. Three years later, in an effort to assuage public criticism and concerns about the technology, the FDA took a cursory look at the human health
impacts of drinking milk produced with rBGH. Based on research provided
by Monsanto and other biotech companies, the agency said that there were
no human health risks. And Monsanto pressed ahead despite the continuing
debate about the hormone, and in 1987 it submitted its official application
for approval.

In 1988 it came to light that humans drinking milk produced through the
use of rBGH caused an increased exposure to a protein hormone, insulin-like
growth factor-1 (IGF-1). While humans naturally produce IGF-1, and also
consume it in the milk of untreated cows, much larger amounts of it are created in milk from cows by receiving the rBGH injection. The protein hormone
passes from the cow's blood into the milk and remains after pasteurization.
Regardless of this, in 1989 the FDA completed its health study and gave rBGH
a clean bill of health.

A scientist involved in the review process charged that the agency relied
on data that had been manipulated by Monsanto. These allegations were

serious enough that at the prompting of several members of Congress, the General Accounting Office (GAO, now the Government Accountability Office) conducted an in-depth investigation into the FDA's review of rBGH and recommended that approval be delayed until further studies were conducted. The FDA, however, disagreed and reaffirmed its initial conclusion that the artificial hormone was safe.

Hansen, who was fiercely critical of legalizing the hormone, says that the integrity of the rBGH approval process was in question from the beginning because of the conflicts of interest at the FDA. But despite the appearance of impropriety, in 1993, the Clinton administration's FDA followed in the footsteps of its predecessors and did Monsanto's bidding. Monsanto executives with close Democratic Party ties used their relationship with the administration to obtain approval for the commercial use of the hormone, which is marketed as Posilac.

By 1994, also as a result of lobbying by Monsanto, the FDA had also prohibited dairies from labeling their milk "rBGH-free." Former Monsanto strategist and government insider Michael Taylor wrote the labeling regulation. According to FDA guidance on labeling that was signed by Taylor, "rBGH-free" was misleading, and a lengthy qualifying sentence stating the FDA's conclusions must also be included: "No significant difference has been shown between milk derived from rbST-treated and non rbST-treated cows" (rbST stands for recombinant somatotropin, the industry's preferred name for the drug).

Just days after the FDA released the document, Monsanto filed suit against two dairy farms that had labeled their milk "rBGH-free." King & Spalding, Taylor's former law firm, also sent a number of letters to dairy farmers on behalf of Monsanto, using the guidelines Taylor helped draft at FDA to show that the farmers were out of compliance with FDA regulations.

Michael Taylor became one of the targets of the GAO investigation for conflicts of interest. While Taylor worked for Monsanto in the 1980s, he drafted a memo for the company on the unconstitutionality of state rBGH labeling laws and then subsequently wrote the guidance rules on this very issue at the FDA. The GAO found that the guidance was not a "binding agency decision," and it was thus not covered by the impartiality standards. So Taylor's work on the guidance was completely excluded from the GAO investigation, and he continued with his revolving door career.

Taylor left the FDA in 1994, when he was appointed to the USDA by President Clinton. He left the USDA to return briefly to King & Spalding, until becoming vice president for public policy at Monsanto in 1998. He was reappointed to the FDA as a "special adviser" to the commissioner by President Obama in 2009.

Unfortunately, because of the lack of labeling that was advocated by Taylor, millions of people have been exposed to elevated levels of IGF-1. Recent research has proven that when it is present in the human body at elevated levels, IGF-1 increases the risk of breast, colon, prostate, and other cancers, although scientists still do not fully understand why. Numerous studies have also proven that rBGH (Posilac) causes lower birth rates and birth weights of calves and physical ailments such as mastitis and hoof and leg problems, cystic ovaries, and other diseases. Today the European Union, Japan, Australia, and Canada have all banned the use of rBGH due to animal and human health concerns.

The hormone has remained controversial, and it is no surprise that Monsanto decided to divest its artificial hormone business. In August 2008 the pharmaceutical giant Eli Lilly announced that it was acquiring Monsanto's rBGH operations for $300 million. Posilac became part of Eli Lilly's Elanco division, which was already the exclusive international seller of Posilac in the decade preceding the acquisition. As a maker of veterinary antibiotics, Eli Lilly can benefit on both ends—selling the hormone as well as the antibiotics that are needed to treat the painful and debilitating mastitis that cows injected with Posilac suffer. Sadly, Eli Lilly is focusing its sales on the developing world, where there is a push to increase industrialized animal production.

In the end, here in the United States opponents of the artificial hormone have been successful in dramatically reducing rBGH's use. Because of public interest advocacy campaigns, large national retailers and food companies such as Kroger and Starbucks refuse to use milk produced with the hormone. Dean Foods, the largest milk distributor in the United States, has stated that virtually all of its milk is free of artificial hormones. Yoplait and Dannon announced they are dropping rBGH from their products, and even Walmart's private-label milk is "rBGH free."

While the struggle over bovine growth hormone raged in Washington, D.C., a new fight over seeds and crops had emerged as a new battle in the food wars. Washington faced tremendous pressure from Monsanto and other seed companies that were consolidating. Pressure was building on the Clinton administration and members of Congress to take action on behalf of the industry. In 1994 the United States ratified the International Convention for the Protection of New Varieties of Plants, a Machiavellian move that extended plant patents to twenty years for most crops and prohibited farmers from selling saved patented seeds without the patent owner's permission.

Jeremy Rifkin recognized early on the danger of the giant corporate seed merchants controlling seeds. In *The Biotech Century*, he baptized the biotech industries' release of engineered life-forms into the natural environment "a laboratory-conceived second Genesis." Rifkin hired Andrew Kimbrell, who had formed the Center for Food Safety (CFS) in 1997, after working on biotech issues at the FOET; CFS has filed several lawsuits against the biotech industry.

One of the many early battles that CFS was involved in was over the creation of seeds that were genetically engineered to be sterile—called Terminators, after the Arnold Schwarzenegger film character. Monsanto was again in the eye of the storm, following its efforts to acquire Delta & Pine Land, a seed company that had worked with the USDA to create and patent the sterile seed. In the late 1990s a massive public outcry took place about the biotech industry's development of a seed that could not be saved and that has the potential to cause starvation in places where farmers rely on seed saving.

Monsanto became the target for criticism because it announced that it intended to purchase Delta in 1998. And CEO Robert B. Shapiro was especially embarrassed, because he had predicted that biotech seeds would be a key tool to combat global hunger and poverty. He was so embarrassed that in October 1999 he wrote a letter to the Rockefeller Foundation, whose president had criticized the Terminator technology. Although he confirmed that Monsanto was tabling the technology for now, he left the door open for the future: "The need for companies to protect and gain a return on their investments in agricultural innovation is real. Without this return, we would no longer be able to continue developing new products growers have said they want. . . . We are not currently investing resources to develop these technologies, but we do not rule out their future development and use for gene protection or their possible agronomic benefits." [16]

Although Monsanto was prohibited by federal regulators from purchasing Delta in 1998, it was successful in acquiring the company and its Terminator patents in 2006. Rebecca Spector, CFS's West Coast director, says that the huge public outcry against the terminator seed changed the equation in the debate over biotech: "Farmers have been saving seeds for ten thousand years. People were shocked that the industry would actually breed seeds for planned obsolescence. The whole idea of seeds only being usable for one season outraged a broad range of organizations, including farm groups. It brought enormous new energy to the movement."

Spector is one of the leading voices advocating against genetic engineering. She grew up in suburban Connecticut, where her neighborhood bordered on a dairy farm with its own ice cream shop. Her father would take her walking in and around the neighboring farms that are now covered with condominiums—imbuing her with a keen appreciation of the natural world and an environmental ethic. While in college, she found her interest in the environment intersecting with a desire to create a new vision for agriculture. After graduating from the University of Michigan with an MS in environmental policy, she moved to California, where she helped run an organic farm in Half Moon Bay, California, for nine years. It was because of the implications of GE for organic farming that she became alarmed about the technology. Spector began working with the grassroots organization Mothers & Others—an early voice against the biotech industry—before moving to CFS.

The late 1990s saw a raging debate over organic standards, a debate that galvanized the movement and brought many new actors into play. In 1997 the USDA released the National Organic Standards for public review, and the proposal allowed genetic engineering to be used in organic production. Almost overnight, a mass grassroots movement emerged opposing inclusion of the technology. More than two hundred thousand people wrote in their comments to the agency that genetic engineering should be excluded from the organic standard. The activism around this federal regulatory process galvanized people from around the country into taking action on food issues, and this movement has increased dramatically in size since that time.

Spector says that she was really inspired by the effectiveness of grassroots action after the huge victory in keeping these technologies and practices out of organic food. She joined CFS in 2000 because of the need to gear up grassroots activity and legal action on GE issues in response to the industry push to approve new GE products. The organization decided that there was a deep need for a visual portrayal of how biotechnology was spurring even more industrialization in agriculture.

Spector was one of the editors, along with Andrew Kimbrell, of *Fatal Harvest: The Tragedy of Industrialized Agriculture*, a bible of the good-food movement. The large coffee table–size book has 250 pictures and essays by a range of people, from Vandana Shiva to Wendell Berry. Besides addressing the lack of viability for monoculture agriculture produced with petrochemicals, the book dispels the myths about the efficiency and increased productivity of biotechnology. It created a surge of interest in the issue and, along with its smaller companion "reader," is still an important tool a decade later. Spector also wrote *Your Right to Know: Genetic Engineering and the Secret Changes in Your Food* and the *California Food and Agriculture Report Card on Genetic Engineering*.

As someone who has advocated for effective regulation of the industry,

Spector says, "The Clinton administration was a big booster to the biotechnology industry. Their refusal to take any action during the eight years in office to effectively regulate biotech is an unfortunate legacy. We have no real way to assess biotechnology, because we are left using old and outdated statutes to regulate completely new and dangerous technologies."[17]

The potential long-term risks from eating GE food are unknown. The FDA contends that there is not sufficient scientific evidence demonstrating that ingesting these foods leads to chronic harm.[18] But GE varieties became the majority of the U.S. corn crop only in 2005 and the majority of the U.S. soybean crop only in 2000.[19] The potential cumulative, long-term risks have not been studied. These considerations should be critical in determining the safety of a product prior to approval, and not left to attempt to assess once the product is on the market without labeling.

The FDA allows companies to self-regulate when it comes to the safety of genetically engineered foods. The 1992 FDA guidance that Quayle was so proud of gave the industry responsibility for ensuring that new GE foods are compliant with the Federal Food, Drug, and Cosmetic Act (FD&C Act).[20] In 2001, because of public pressure, the FDA proposed a rule requiring companies to submit data and information on new biotech-derived foods 120 days before commercialization.[21] As of 2012, the decade-old rule still has not been finalized, and industry data submissions remain voluntary.

The FDA regulates genetically engineered animals as veterinary medicines—one of the more bizarre industry-friendly rules. In 2009 the agency decided that the FD&C Act definition of veterinary drugs as substances "intended to affect the structure of any function of the body of man or other animals" includes genetically altered animals. This process allows the biotech company to keep most of the safety data secret because it is of a "proprietary nature." As of spring 2012, only GE salmon and Enviropig™ have been considered for commercial approval, but no genetically engineered animals have been approved to enter the food supply.

For whole foods (foods that are largely unprocessed and unrefined), safety responsibility is on the manufacturer and no FDA premarket approval is necessary.[22] However, for substances added to food, such as biotech traits, the FDA classifies them as "generally recognized as safe" (GRAS) or as food additives.[23] The FDA grants GRAS determinations to GE-derived foods that are considered equivalent to the structure, function, or composition of food that is currently considered safe. The FDA has awarded "generally recognized as safe" status to almost all—95 percent—of the GRAS applications submitted for food since 1998, according to the agency's GRAS Notice Inventory.[24]

By contrast, the FDA must preapprove food additives before they can be sold. However, the FDA trusts biotechnology companies to certify that their new GE foods and traits are the same as foods currently on the market. The company may send information on the source of the genetic traits (i.e., which plants or organisms are being combined) and on the digestibility and nutritional and compositional profile of the food, as well as documentation that demonstrates the similarity of the new GE substance to a comparable conventional food. The FDA evaluates company-submitted data and does not do safety testing of its own.[25] The agency can approve the GE substance, establish certain regulatory conditions (such as setting tolerance levels), or prohibit or discontinue the use of the additive entirely.[26] The FDA evaluates the safety of all additives, but it has evaluated only one GE crop trait as an additive: the first commercialized GE crop, Flavr Savr™ tomatoes.[27]

Once a GE food product has been approved and is on the market (either with the GRAS designation or as a food additive), the FDA is responsible for its safety. The FDA did pressure a company to recall one GE food product: StarLink corn, which was unapproved for human consumption, when it entered the food supply.[28] The FDA's lack of postmarket monitoring can expose the public to unapproved GE traits in the food supply.

GE insect-resistant crops may also contain potential allergens. One "harmless" bean protein that was spliced onto pea crops to deter pests caused allergic lung damage and skin problems in mice.[29] Yet there are no definitive methods for assessing the potential allergenicity of bioengineered proteins in humans.[30] This gap in regulation has failed to ensure that potential allergenic GE crops are kept out of the food supply.

The StarLink corn fiasco is a good example of the problems caused by multiple agencies having conflicting roles in the approval of GE products. Under the bewildering system of multiple agency jurisdictions, the EPA sets pesticide residue limits for food and feed crops, and the agency is required to meet all food and feed safety standards enforced by the FDA. It has jurisdiction over the approval (registration) of crops that are genetically engineered to produce pesticides. These insect-resistant crops contain genes that deter insects but may have other unintended consequences.

In the case of StarLink corn, created by the European pharmaceutical giant Aventis, the EPA approved it in 1998 for livestock feed and ethanol, even though there was evidence that it could cause severe allergic reactions in humans when ingested. Aventis assured the EPA that it would have farmers sign agreements that the corn would not be used for human food, which obviously is a poor method of regulating an item that looks like food but is dangerous when eaten.

In 2000 Friends of the Earth had corn products tested and found that Taco Bell corn tortilla shells had been contaminated by StarLink. Aventis paid out millions in the multiple lawsuits and class actions that resulted from the widespread allergic reactions, as well as the economic impact on farmers and grain elevators caused by the recall.

Despite the FDA's approval of common GE crops, questions about the safety of eating these crops persist. GE corn and soybeans are the building blocks of the industrialized food supply, from livestock feed to hydrogenated vegetable oils to high-fructose corn syrup. Safety studies on GE foods are limited, because biotechnology companies, in their seed-licensing agreements, prohibit cultivation for research purposes.[31]

Some of the independent, peer-reviewed research on biotech crops has revealed some troubling health implications. A 2009 *International Journal of Biological Sciences* study found that rats that consumed GE corn for ninety days developed a deterioration of liver and kidney functioning.[32] Another study found irregularities in the livers of rats, suggesting higher metabolic rates resulting from a GE diet.[33] And a 2007 study found significant liver and kidney impairment of rats that were fed insect-resistant Bt corn, affirming that "with the present data it cannot be concluded that GE corn MON863 is a safe product."[34] Research on mouse embryos showed that mice that were fed GE soybeans had impaired embryonic development.[35] Even GE livestock feed may have some impact on consumers of animal products: Italian researchers found biotech genes in the milk from dairy cows that were fed a GE diet, suggesting the ability of transgenes to survive pasteurization.[36]

Also, evidence suggests that the most common GE-affiliated herbicide, glyphosate, may pose animal and human health risks. A 2010 study published in *Chemical Research in Toxicology* found that glyphosate-based herbicides caused highly abnormal deformities and neurological problems in vertebrates.[37] Another study found that glyphosate caused DNA damage to human cells even at lower exposure levels than those recommended by the herbicide's manufacturer.[38] Nevertheless, glyphosate use on Roundup Ready crops has grown steadily, with application doubling between 2001 and 2007.[39]

In the fifteen years since herbicide-tolerant crops were first introduced, weeds already have become resistant to the herbicides. Ubiquitous application of Roundup has spawned glyphosate-resistant weeds, a problem that is driving farmers to apply more toxic herbicides and to reduce conservation tilling to combat weeds, according to a 2010 National Research Council report.[40]

The EPA regulates pesticides and herbicides, including GE crops that are designed to be insect resistant. A pesticide is defined as a substance that "prevents, destroys, repels or mitigates a pest," and all pesticides that are sold and used in the United States fall under EPA jurisdiction. The EPA also sets allowable levels of pesticide residues in food, including GE insect-resistant crops. Between 1995 and 2008, the EPA registered twenty-nine GE pesticides engineered into corn, cotton, and potatoes.

Bioengineered pesticides are regulated under the Federal Insecticide, Fungicide and Rodenticide Act (FIFRA), enacted in 1947.[41] New pesticides—including those designed for insect-resistant GE crops—must demonstrate that they do not cause "unreasonable adverse effects on the environment," including polluting ecosystems and posing environmental and public health risks.[42] The EPA must approve and register new GE insect-resistant crop traits, just as the agency does with conventional pesticides.[43] Biotech companies must apply to field test new insect-resistant GE crop traits, establish permissible pesticide trait residue levels for food, and register the pesticide trait for commercial production.[44]

The EPA also grants experimental use permits for field tests of unregistered pesticides or of registered pesticides tested for an unregistered use. Applications for permit registration must include management plans that describe any limitation on cultivating the new insect-resistant GE crops. Biotech seed companies are responsible for ensuring that farmers follow these management plans. In 2010 the EPA imposed a $2.5 million fine on Monsanto for selling GE seed between 2002 and 2007 without informing farmers about EPA-mandated planting restrictions.

The third agency involved in the chaotic process is the USDA, which is responsible for protecting crops and the environment from agricultural pests and weeds, including biotech and conventional crops. Companies submit data to the USDA showing that the new GE plant will not harm agriculture, the environment, or nontarget organisms, and the USDA either approves or denies the field-testing application within one month. If the USDA denies the application, the company can reapply under the more involved permit process.

Under this more complicated application process, the USDA determines if the GE field trial poses a significant environmental impact before issuing a permit. The agency reviews scientific submissions for four months before granting or denying the field-test permit request. If approved, the permit imposes restrictions on planting or transportation to prevent the GE plant material from escaping and posing risks to human health or the environment. The USDA approved the vast majority—92 percent—of the applications for biotech field releases between 1987 and 2005. The applying company is required to submit field-trial data to the USDA within six months of the test, demonstrating that the crop poses no harm to plants, nontarget organisms, or the environment.[45] If the applicant violates the permit, the USDA can withdraw it.[46]

After the field tests and before crops are released for commercial use, the USDA must complete an Environmental Assessment (EA) and, in some cases, an Environmental Impact Study (EIS). This is required under the 1970 National Environmental Policy Act and applies to all crops, including biotech ones. The EA determines whether the GE crop will pose significant risks to human health or the environment if cultivated. If the USDA decides there is no significant risk, it issues a "finding of no significant impact." But if the agency finds more significant environmental implications, it must also perform a more thorough EIS.

The process had been further weakened recently. The USDA already relies on company-supplied data for many of its EAs, but a 2011 proposed pilot project threatens to compromise the scientific rigor of the process even more. The two-year pilot project allows consultants that are funded by a cooperative services agreement between the biotech company and USDA to perform EAs, giving firms more influence over the safety designation of their own products.

While Europeans have debated the regulation of genetic engineering, the situation on our side of the Atlantic was settled in the 1990s.[47] The approvals for genetically engineered seeds and other products came fast and furiously, especially as Monsanto gobbled up the seed companies and DuPont followed suit. The controversy continues with each product introduced, especially those introduced for human consumption.

In 2002 Monsanto petitioned the USDA to approve Roundup Ready red spring wheat, the first GE crop designed primarily for human food consumption rather than for livestock feed or for a processed food ingredient. Given that Japan and the EU have different restrictions for GE food crops, the large-scale manufacture of GE wheat could damage options for U.S. wheat exports. A 2004 Iowa State study forecasted that approving GE wheat could lower U.S. wheat exports by 30 percent to 50 percent and depress prices for both GE and conventional wheat. Because of export concerns, Monsanto abandoned GE wheat field trials before obtaining commercial approval, although the company resumed research in 2009.[48]

The USDA approved Monsanto's Roundup Ready sugar beet in 2005 after determining that cultivation poses no risks to other plants, animals, or the environment.[49] In 2008 CFS and the Sierra Club challenged the approval in court on grounds that the USDA's EA ignored important environmental and economic impacts. Even though a U.S. District Court directed the USDA to develop a more in-depth EIS in 2009, the USDA allowed several seed companies to begin cultivation. The court intervened, ordering Monsanto to dig up 256 acres of GE sugar beet plantings pending completion of the environmental review. However, the USDA finalized the EIS in 2012 and approved cultivation of the GE sugar beet.

This will mean that the vast majority of manufacturers that use beet sugar, especially the candy industry, will be using a GE product. The fact that familiar brand names like Hershey's will contain GE sugar is helping to build the momentum and political pressure to label genetically engineered food.

The movement to label is also being strengthened by the ongoing debate over genetically engineered alfalfa, an important animal food crop. The USDA first approved Monsanto's Roundup Ready alfalfa in 2005. In 2007, CFS and a number of organic alfalfa producers challenged the USDA approval on grounds that GE alfalfa could contaminate and wipe out non-GE alfalfa. Because this poses special risks for organic alfalfa and for dairy farms whose crops may be contaminated by GE alfalfa, a California district court ruled for a prohibition on GE alfalfa sales and plantings until the USDA performed an EIS.[50] The USDA's 2010 EIS demonstrated the potential negative economic impacts for organic and conventional alfalfa farmers, including increased costs needed to prevent contamination, reduced demand, and lost markets due to contamination.[51] Nonetheless, in January 2011 the USDA decided to approve GE alfalfa without any planting restrictions.[52]

Spector says of the CFS lawsuit: "Our legal action was successful because it prevented commercialization of GE alfalfa for five years. We are hopeful that the new legal action will be successful, because biological pollution cannot be recalled. Contamination of non-GE crops violates the rights of the vast majority of alfalfa farmers, who do not grow GE alfalfa. We need to protect their rights and the integrity of the food products produced by animals eating alfalfa."

Alfalfa is an open-pollinated crop, meaning that wind or insects can pollinate and contaminate conventional alfalfa fields. It is well documented that a farmer's field can be inadvertently contaminated with GE material through cross-pollination and seed dispersal. A Union of Concerned Scientists study found that 50 percent of non-GE corn and soybean and 83 percent of non-GE canola seeds in the United States were contaminated with low levels of GE residue. Even Monsanto admits that "a certain amount of incidental, trace level pollen movement occurs."[53]

Yet farmers who unintentionally grow GE-patented seeds or who harvest crops that are cross-pollinated with GE traits could face costly lawsuits by biotechnology firms for "seed piracy." Making the situation worse, farmers who grow GE crops are not required to plant non-GE buffer zones to prevent contamination unless this is stipulated in the farm's USDA permit. Yet even the use of buffer zones has proven ineffective because these areas are usually not large enough to prevent contamination.[54]

The USDA prohibits the use of GE crops in any product that carries the agency's "certified organic" label. Certified organic farmers can face significant economic hardship if biotech traits contaminate their organic crops or organic livestock feed. Such contamination can occur either when GE seeds are inadvertently mixed with non-GE seeds during storage or distribution or when GE crops cross-pollinate non-GE crops.[55]

The USDA's approval of Roundup Ready alfalfa in 2010 highlights the significant ramifications that contamination can have for organic producers. Alfalfa is the most important feed crop for dairy cows, yet GE alfalfa can easily cross-pollinate organic alfalfa crops and cause organic farmers to lose their markets if testing reveals contamination. Meanwhile, conventional alfalfa farmers could face seed piracy suits from Monsanto even if their crops are inadvertently pollinated by GE alfalfa. At least one farmer contends that he was sued when his canola fields were contaminated with GE crops from neighboring farms.[56]

Organic dairy farmers already face difficulty securing organic feed, and this challenge will only worsen if GE alfalfa begins to contaminate organic alfalfa more widely.[57] Organic dairy farmers receive a price premium of $6.69 (44 percent) for their milk, but they also face production costs of $5 to $7 more per hundred pounds of milk—38 percent higher than conventional dairies.[58] GE contamination could eliminate this premium that covers the higher costs of producing organic products, making these farms unprofitable.

Organic and non-GE growers bear the financial burden of GE contamination and are fighting instead to make biotech companies liable for these consequences. In 2011 the Public Patent Foundation filed suit against Monsanto on behalf of farmers and organic businesses, asking the court to determine whether Monsanto has the legal authority to sue farmers for patent infringement if their GE traits contaminate a conventional or organic farm. The farmers have lost the case and have appealed.

In 2011 the USDA approved Syngenta's amylase corn, which produces an enzyme that facilitates production of ethanol.[59] Although the corn is intended specifically for ethanol use, the USDA determined that it was also safe for food and animal feed, allowing it to be planted alongside GE corn that is destined

for the human and animal food supply.[60] Contamination of corn crops destined for the food supply is possible, especially in the absence of a buffer zone
to minimize wind pollination. Even the USDA admits that contamination of
high-value organic blue and white corns may produce "undesirable effects"
during cooking, such as darkened color or softened texture.[61]

During the last three administrations, numerous biotech products have
been approved. Under the Clinton administration, fifty approvals took place
in eight years, and the Bush administration legalized twenty during two terms.
In 2010 Bush, Clinton, and Gore gave keynote addresses at the large BIO annual biotech conference held in Chicago and attended by biotech scientists
from around the world.

The Obama administration proved to be an even bigger cheerleader for
biotech. Not only did Obama bring Michael Taylor back to the FDA as deputy
commissioner in a new position at the newly minted USDA Office of Foods,
but his administration granted a giant favor to big seed and agribusiness
companies. The restricted planting of genetically engineered alfalfa was abandoned by the administration as too "burdensome for industry."

Even more alarming was the swiftness with which Obama's appointees
have legalized twelve new products in less than four years. This has meant
legalizing actual GE food products, including sweet corn seed with names
like Obsession II, Passion II, and Temptation II. The "stacked" Performance
Series™ sweet corn contains three distinct traits, but the stacked combination
of these traits has not been through a safety evaluation of any kind. Research
also indicates that two of the corn's traits display resistance to chemical
applications.

According to the vice president of Monsanto's vegetable seeds division, the
corn will be aimed at the 250,000-acre fresh market. If the number of acres
that Monsanto is targeting is reached, it would constitute nearly 40 percent
of all sweet corn.[62] Monsanto told a news source that it is in discussions with
vegetable canning and freezing companies as well.[63] The new sweet corn could
quickly dominate the market.

It is not a surprise that the biotech industry is able to dictate policy to sitting
presidents, members of Congress, and the regulatory agencies. Since 1999, the
fifty largest agricultural and food patent-holding companies and two of the
largest biotechnology and agrochemical trade associations have spent more
than $572 million in campaign contributions and lobbying expenditures.
The companies and trade associations have hired a bevy of well-connected
lobbying shops—where at least thirteen former members of Congress and

over three hundred former congressional and White House staffers work—to promote genetically modified food and agricultural products.

The battle over GE food has moved from an obscure issue in the early1990s to one of the most highly charged issues of our time. The buying of public policy by the biotech industry and the revolving door between corporations and regulatory agencies have made a mockery out of our democratic process. But growing numbers of people representing a broad base of constituencies are organizing to label and regulate GE food.

The industry's overreach is beginning to backfire—people have had enough.

14

THE FUTURE OF FOOD:
SCIENCE FICTION OR NATURE?

No one should approach the temple of science with the soul of a money-changer.
—Thomas Browne (1605–82), English author

Some call J. Craig Venter the Bill Gates of artificial life.[1] He refers to himself as an accidental millionaire who's been gifted large sums of money to start companies that use advanced genetic sciences to generate a profit. Perhaps developing the first synthetic life gives one an omnipotence complex. An iconoclast with an outsized ego, Venter travels around the world in his large research vessel collecting genetic materials from the sea (in what sounds more like piracy to me), plundering our collective genetic resources for private gain.[2]

Venter is a self-made man. After serving a tour of duty in Vietnam, he went on to the University of California to get a PhD in physiology and pharmacology. Afterward, while working at the National Institutes of Health, he developed a new technique for decoding genes. In 1992, Venter founded the Institute for Genomic Research where in the mid-1990s he and a team of scientists decoded the genetic material of the first living organism. He went on to fund a for-profit company, Celera Genomics, to sequence the human genome, which culminated with the February 2001 publication of the human genome in the journal *Science*.[3]

Sometimes reality really is stranger than science fiction—especially when pursued by extremely ambitious men hidden in the labs of private enterprise, scientific institutes, and the business plans of major corporations. And more and more, there is no escaping the impacts upon our food and ecosystems of new technologies that are pursued for profit.

Very quietly—with little public fanfare or regulatory scrutiny—a new field of biotechnology is emerging that is based on the belief that synthetic life can be engineered through chemistry, and that mankind has the right to create new life-forms.

This new breed of scientist believes so deeply in his own abilities that he is willing to make humankind a guinea pig as he seeks to outdo nature in producing synthetic organisms for food. Craig Venter is just one prominent example of how far our food system has gone awry, at the expense of farmers, consumers, and the environment. Corporate control of science and the food system has taken us far afield from the basic ingredients of a healthy food supply—all in the relentless pursuit of profit.

The June 2010 issue of *Scientific American* calls it genetic engineering on steroids. Oil, chemical, and agriculture multinationals joining forces to create artificial life may sound like a low-budget Hollywood movie, but many of the largest corporations in the world are banking on its viability. Among the companies joining the fray are Cargill, Unilever, Bunge, ADM, Dow Chemical, and BASF. Venter's new venture, Synthetic Genomics, funded in part by BP, is working with ExxonMobil to develop biofuels. It is also creating synthetic microbes that will be used in developing high-yielding, disease resistant plants for use as feedstocks.[4]

Chevron and Procter & Gamble are gambling that microbes can ferment plant cellulose fuels and plastics. Unilever is seeking a palm oil substitute. General Motors and Marathon Oil are looking to turn woodchips into ethanol. DuPont is using synthetic yeast to turn forty thousand acres of corn a year into plastic.[5]

The academic proponent Vítor Martins dos Santo recently opined about synthetic biology, "Technological potential is vast, societal impact immense and growing market opportunities substantial and diverse."[6]

In June 2010 the Synthetic Biology Project housed by the prestigious Woodrow Wilson International Center for Scholars released a paper applauding the enormous amount of funding devoted to synthetic biology. This emerging field of study combines science and engineering to redesign living organisms for food or industrial uses. Its proponents believe that man-made genetic code is the next step beyond selectively breeding animals to increase their productivity for human needs. To some it may sound more like a misapplication of the "better living through chemistry" philosophy that has had so many unintended consequences.

Without most people even knowing that synthetic biology exists, the future of the natural world is being decided. More than 230 different entities in the United States and Europe, including multiple government agencies, are funding the development and commercialization of synthetic biology and the sum is predicted to exceed $3.5 billion over the next decade. Currently, only 4 percent of the funds are devoted to the ethical, legal, or social implications of playing god. No funding is currently available for assessing the risks associated from creating synthetic organisms.[7]

Drew Endy, another mover and shaker in the mysterious world of synthetic biology, is a leading member of a "hipster geek"[8] generation of scientists who believe man can vastly improve upon nature. Educated at Lehigh and Dartmouth, he studied genetics and microbiology as a postdoc at University of Texas at Austin and University of Wisconsin–Madison. Since 1998 he has started the Molecular Sciences Institute in Berkeley and co-founded the Massachusetts Institute of Technology Synthetic Working Group, the Registry of Standard Biological Parts, and the BioBricks Foundation. BioBrick is a trademarked name for a man-made biological part that meets the technical and legal standards set forth by the BioBricks Foundation.[9]

According to journalist Rebecca Cathcart, "When Endy envisions the future, he sees giant gourds engineered to grow into four-bedroom, two-bathroom houses. He sees people alerted to nascent tumors in their bodies by internal biological sensors, and cars fueled by bacteria-produced gasoline."[10]

In mid-June 2011 the future of nature and who would profit from it was being decided at a BioBricks Foundation–sponsored conference. (The foundation is so hip that its annual conference is called SB5.0.) It is the Fifth International Conference on Synthetic Biology, which took place at Stanford University with seven hundred people in attendance. The stated goal was to jump-start the widespread academic and commercial development of synthetic biology. In his letter to participants Endy says, "global community will share, consider, debate, and plan efforts to understand life via building, to make biology easier to engineer."[11]

Besides the dozens of scientists on panels discussing the minutiae of creating artificial life, the conference agenda was lightened by an appearance by Darlene Cavalier, the Science Cheerleader. Cavalier was a professional cheerleader for the Philadelphia 76ers basketball team. She is the founder of "Science Cheerleader.com, a site that promotes the involvement of citizens in science and science-related policy featuring the Science Cheerleaders, current and former NFL and NBA cheerleaders-turned-scientists and engineers." Perhaps the conference planners sought to lighten the atmosphere of a scientific gathering that could affect the future of the planet.[12]

Synthetic biology is unregulated and self-governing. No one knows what could occur when synthetic organisms are intentionally and unintentionally let loose into the environment. Creating genes that don't exist in nature is a dangerous business and there is no way to predict how they will behave in living systems. Meddling with genetic material can have long-term unintended consequences in the food system and the natural environment. No one knows what the final outcome will be, especially since synthetic biology is practiced at a nano scale. Working at this scale means that scientists may not even be aware of important biological processes they are altering.

Dr. Allison Snow of the Department of Evolution, Ecology, and Organismal Biology at Ohio State University recently testified that these artificial life forms may seem innocuous or weak but they could evolve to become more successful when they start reproducing. Once they are released into the environment there will be no way to take them back, because they can reproduce indefinitely. Snow questions the ability of regulatory agencies to evaluate or monitor synthetic organisms that are proposed for release.[13]

There is no doubt that synthetic biology is being hyped, and it's hard to tell how quickly it will advance or if it can achieve even a fraction of what its cheerleaders say it can do. Like genetic engineering and nanotechnology, synthetic biology is being touted as the savior of humanity for a hungry population estimated to reach 9 billion people later in the century. Of course, you do not need to be too much of a cynic to realize that these advanced technologies are more about making money than providing food to the world's poor. And in fact the opposite is likely to occur since the energy applications of synthetic biology require immense acreage of feedstocks to be grown that can then be turned into replacements for fossil fuels and other magically produced commodities. Corporations are acquiring large areas of land around the world for the purpose of growing "biomass" that can be turned into energy. From displacing peasant farmers in Brazil and encroaching on the Amazon to massive land grabs on the African continent, synthetic biology could result in even more control in the hands of a few corporations of energy, food, cropland, and resources.

And we've heard the hype about technology feeding the world before. In fact, it's trotted out every time scientists, corporations, and government join together to promote a new and usually dangerous technology. Certainly the same proclamations have been made about nanotechnology feeding the world. Professor John Beddington, the chief scientific adviser to the UK government and professor of applied population biology at Imperial College London, told an Oxford Farming Conference that the world will have to

produce 50 percent more food by 2030 in order to feed the growing population. He named nanotechnology and genetic engineering as the technologies
that would make this possible.[14]

Synthetic biology is a subcategory of nanotechnology—the manipulation
of matter at near atomic scale to create new substances. The genetic manipulation that is the basis for creating new life happens at a nano level. Because
nano-size particles can go places that other particles cannot go, it also offers
a more sophisticated way to manipulate genetic material or to move foreign
DNA into cells. Research suggests that atomic-sized particles are so small that
they can pass through your skin and even through the tight, protective web
of cells that create the blood-brain barrier. (To put these tiny particles in perspective, they compare in size to about 1/50,000th of a strand of hair.)

At the 2010 American Association for the Advancement of Science meeting, the largest annual gathering of scientists in the world, two University of
Idaho academics organized a symposium on how biotechnology, including
genetic engineering, synthetic biology, and nanotechnology, will be used to
produce animal-based foods. The organizers, animal scientist Rod Hill and
food scientist Larry Branen, are representative of the true believers in not only
the human ability to improve upon nature, but also the right of a handful
of scientists to make choices that will have profound effects on the natural
world.[15]

While staying mostly under the radar, nanotechnology is being used for
many purposes related to food, including the creation of more potent food
colorings, flavorings, and nutritional additives; antibacterial ingredients for
food packaging; and more powerful agrochemicals and fertilizers.[16]

Nano particles are submicroscopic and have unique properties that are
different from even slightly larger particles. For instance, an aluminum soda
can does not burn, but aluminum nano particles explode when used as rocket
fuel catalysts. Carbon is a soft element, but a nano-scale cylinder made of
carbon atoms is a hundred times stronger than steel. A multibillion-dollar
industry has grown up around the creation of nano-size particles for use in
food and food packaging.[17]

Nano silver is used as an antibacterial agent in at least 260 consumer products, including food packaging, cutting boards, kitchenware,[18] and fertilizer.[19]
It does indeed have antibacterial properties. A peer-reviewed study found that
mutations occurred more frequently in laboratory tests and in live animals
when exposed to silver nanoparticles. The authors of the study called for a
review of the long-term biohazard issues.[20]

The use of these technologies for food production is the reverse of sustainable, regionally appropriate agriculture and food production. It further promotes industrial-scale, chemically intensive farming and an overabundance of processed foods. It's no accident that the largest and most powerful food companies in the world are researching and developing nanotechnology-related food products. Campbell's Soup, Cargill, DuPont, General Mills, Heinz Company, Hershey Foods, Kraft Foods, Mars, Nestlé, PepsiCo, Sara Lee, Syngenta, Unilever, and United Foods are among the companies that are further industrializing our food system through the use of nanotechnologies.[21]

The federal government is also promoting nanotechnology through a government program called the National Nanotechnology Initiative (NNI), which coordinates the billions of dollars of taxpayer money going into research and development. The NNI is "aimed at accelerating the discovery, development and deployment of this technology," and its budget reflects this.[22]

And it seems that their strategy is having some success, if you believe that getting an unregulated and mostly untested technology to market is an achievement. A recent survey found more than eight hundred consumer products contained nanotechnology,[23] with as many as twenty new items entering the market every month.[24] The environmental group Friends of the Earth's recent investigation found one hundred food and agricultural products containing nanoparticles.[25]

For a snapshot of nanotechnology in the food system today, consider the following: at the farm, fertilizers and pesticides containing nanoparticles of clay and other materials are touted for their slow-release mechanisms and potency.[26]

Food itself can contain nanoparticles, such as cured meats and sausages,[27] nano tea,[28] and the wide variety of nutritional supplements containing nanosilver. Research and development is under way to use nanotechnology in myriad aspects of food processing.[29]

In the kitchen, we prepare food using kitchenware and cutting boards that employ antimicrobial nanosilver technology[30] and store food in refrigerators also coated with nanosilver.[31] When we store food in containers or use wrappings, we can be exposed to the migrating particles used in their manufacturing.[32]

Plans for the future include using nanotechnology to produce ice cream, sodas, and chips that are low in calories and fat. It is being touted as having the ability to "change the color, taste and texture at the press of a microwave button, and products customized to respond to an individual's health requirements."[33]

While proponents extoll the enormous potential of nanotechnology to quell the world's problems, they largely ignore the risks to human health and the environment. In comparison to the money researching its potential, very little money is available to examine the potential unintended consequences. Nevertheless, the field of nanotoxicity has begun documenting the threats of this unregulated industry, such as:

- Damage to DNA[34]
- Disruption of cellular function[35]
- Asbestos-like disease[36]
- Neurological problems like seizures[37]
- Organ damage, including lesions on the liver and kidneys[38]
- Destruction of beneficial bacteria in wastewater treatment systems[39]
- Stunted root growth in corn, soybeans, carrots, cucumbers, and cabbage[40]
- Gill damage and respiratory problems in fish[41]

Regulators seem to acknowledge the unique properties that nanoparticles exhibit; still, they have demurred from confronting head-on the danger that nanotechnology poses. Neither the Food and Drug Administration (FDA) nor the Environmental Protection Agency (EPA) has stepped up to the plate. We are essentially asked to trust a company's own assessment of its product safety, a dangerous prospect.

Unfortunately, this lack of regulatory vigilance by our regulatory agencies is now business as usual, contrary to how new technologies should be introduced into the environment. When a new technology has the potential to harm public health or ecological systems, the burden of proof that it is *not* harmful should be established before it is released in the environment. This prudent course of action, however, is not taken in the United States, where companies put pressure on Congress and the executive branch to prevent the regulatory agencies from following a precautionary path. This is especially true in the area of food, as we have seen with nanotechnology and bioengineering. Next on the agenda is bowing to the pressure of the biotech industry to allow another reckless experiment—genetically engineering animals.

Recently, the FDA moved forward to do the bidding of the economic interests dead set on commercializing genetically engineered (GE) animals for food, beginning with GE salmon. The agency had hoped to slip this one by the public, but public outcry has been so great that as this book goes to press GE salmon has not been legalized.

The biotech industry has a lot riding on the frankenfish. It also has enormous power to influence the political process in Congress and the regulatory agencies through political appointees. Between 1999 and 2010, the biotech industry spent $572 million in campaign contributions and lobbying expenditures, according to an analysis by Food & Water Watch. A bevy of well-connected lobbying shops is busily promoting GE food and products.[42]

The industry is foaming at the mouth to commercialize a whole zoo of genetically engineered animals, with Enviropig—designed for factory farms—right around the corner. But for now the focus is on GE salmon. Public opinion is squarely against legalizing this fish, yet consumer and environmental advocates fear the FDA will approve the GE salmon for human consumption because of the influence pedaling.

AquaBounty, a Canadian company, is a proxy for the rest of the biotech-industrialized food machine that wants this technology to move forward. The company intends to sell the eggs of the genetically engineered fish to aquaculture facilities. Created by inserting the genes of an ocean pout into the genes of a salmon, the fish gets to market size in half the time. At this point it is unclear if the GE salmon could live up to the rapid-growth hype in large-scale commercial production (many other GE products that were promoted as having dramatically increased crop yields have failed to live up to these claims when used on farms[43]). What is clear is that GE salmon poses risks to the environment and consumers that are not being given serious consideration by the FDA.

One of the most outrageous aspects of the GE animal fiasco is that the FDA has no regulatory process for legalizing GE animals. The process the agency is using for GE salmon was actually designed for veterinary drugs, which guarantees a high level of secrecy during the process because company data and research are considered trade secrets. The FDA used the Veterinary Medicine Advisory Committee (VMAC), a group generally in favor of biotechnology, to consider and recommend legalizing the frankenfish. But even this pro-GE panel raised serious concerns around the company's science, citing the poor methodology and construction of the studies, to the disappointment of the biotech industry.

The FDA disclosed four studies that were used in their approval process. One was nearly twenty years old, and the other three were conducted by AquaBounty or its contractors and were not peer-reviewed. The nutrition and allergenicity studies were of special note because they used an extremely small data set: six GE salmon. The sparse data, the poor design, and the fact that the company killed off salmon that were deformed prior to doing a physical analysis for comparison with non-GE fish are scandalous.

Recently, at a long and tedious VMAC meeting, where Food & Water

Watch physically delivered thousands of letters to the committee opposing the legalization of the GE salmon, the committee showed an appalling lack of scrutiny about the risk to human health. The FDA even said in the analysis provided to the public, "Primary deference was given to controlled studies submitted by ABT [AquaBounty Technologies]."[44] This is not good enough when the stakes are so high, especially since the studies did show that GE salmon displayed some statistically significant differences in its composition and nutrition.[45]

In fact, there have been no long-term studies on the safety of eating GE organisms, though scientists recognize and have already documented their ability to harm human and animal health. In the case of salmon, a high-protein food, the risk for allergic reactions is high because protein is more likely to cause reactions than other components of food. A *New England Journal of Medicine* study found that soybeans engineered with Brazil nut proteins caused allergic reactions for consumers with Brazil nut allergies.[46] In another case, a harmless protein found in certain beans that acts as a pest deterrent became dangerous once it was transferred to a pea, causing allergy-related lung damage and skin problems in mice. One study showed high rat pup mortality in litters from mothers fed GE soy flour.[47] Another found irregularities in the livers of rats, suggesting higher metabolic rates resulting from a GE diet.[48]

A 2007 study found significant liver and kidney impairment in rats fed GE corn with the insecticidal Bt gene and "with the present data it cannot be concluded that GE corn MON863 is a safe product."[49] Even GE livestock feed may have some impact on consumers of animal products. Italian researchers found biotech genes in the milk from dairy cows fed a GE diet, suggesting the ability to survive pasteurization.[50]

Another concern is the possible escape of GE salmon into the wild, and it turns out that not all the fish are sterile when you read the small print in AquaBounty's materials. Even a small number of GE fish on the lam could cause extinction of wild populations in as little as forty generations.[51] Because of their competitive advantage as big, voracious fish, GE salmon could out-compete other wild fish for food and habitat. An additional concern about escaping GE salmon is the disease they could spread to wild populations. Farmed salmon, which are raised in stressful, densely crowded environments, have already been linked to the spread of disease in wild populations.

AquaBounty's promises to prevent escapes seem especially weak given the widespread problem of regular farmed salmon escapes from existing farms. In March 2010 nearly one hundred thousand farmed Atlantic salmon escaped into the wild through one hole in a net at a UK fish farm.[52] Globally, the numbers of escapees are much higher, with an estimated 2 million farmed salmon escaping into North Atlantic waters every year,[53] while millions of others escape into the Pacific.[54] One biotech corporation doing experimental GE breeding in New Zealand is even suspected of accidentally releasing genetically modified salmon eggs into the wild,[55] demonstrating the logistical difficulties of preventing escapes, even in tightly controlled experimental settings.

The FDA's sloppy analysis was further exposed in the fall of 2010 when Food & Water Watch received numerous recent internal documents and e-mails from the U.S. Department of Interior's Fish and Wildlife Service (FWS) from a Freedom of Information Act (FOIA) request exposing startling concerns with AquaBounty.

The documents revealed that the FDA had not adequately fulfilled a requirement under the Endangered Species Act to consult with both FWS and another federal agency, the National Marine Fisheries Service (NMFS), to determine whether approval of AquaBounty's salmon might impact endangered wild Atlantic salmon.

"Nice work, Greg," FWS regional geneticist Denise Hawkins, PhD, wrote to a coworker in September. "Especially pointing out that there is no data to support the claims of low survival in the event of escape, which I agree with you all is a big concern. I also agree . . . that using triploid fish [which AquaBounty claims have undergone a sterilization process] is not foolproof. Maybe they [the FDA] should watch Jurassic Park."

Despite AquaBounty's claim to produce only sterile salmon, the company admitted that up to 5 percent of their GE salmon eggs could be fertile, prompting the FDA to label the company's claims "potentially misleading."

According to internal FWS e-mails, contrary to AquaBounty's claims that GE salmon would be grown in closed systems (and therefore unable to escape), FWS employees received news of a proposal to grow the fish in a facility that would discharge into the ocean off the coast of Maine. It appears that the proposal is from Joe McGonigle, the former (as of February 2008) vice president of AquaBounty. "No matter what precautions you take, fish escape and once they do, there is no closing that door. So, that being said, I think it is very bad precedent to set," said one FWS program supervisor.

In the documents, high-ranking FWS employees, including branch chief Jeff Adams, complain of the FDA's failure to consult with FWS, as required by law. "The proposal [to approve AquaBounty salmon] also presents a situation where FDA, whose jurisdiction is not focused on natural resources, is entrusted with the authority to approve an application which poses such a threat to the country's natural resources," Adams said.

Regarding the FDA's level of consultation with FWS, one assistant regional director was quoted as saying, "It's a little hazy to me how we are supposed to be engaged."

"Hazy to me, too," replied Gary Frazer, FWS assistant director for endangered species.

Meanwhile, GE salmon is not the only place that the "god complex" rages unabated in the scientific-corporate realm of research and technology. Many other bizarre and ethically challenged food technologies and products are under development. In early 2008 the FDA announced that it considered meat and milk from cloned animals to be safe to eat, despite years of controversy and a long list of unresolved ethical, health, and animal welfare concerns. The agency asked the livestock industry to continue a voluntary moratorium on putting meat and milk from cloned animals in the food supply but did not ask for the same moratorium on products from the offspring of cloned animals.

And to add insult to injury, the agency will not require any of these foods to be labeled. The appeal of cloned animals to the livestock industry lies in their role as breeders or milk producers. Already, cloned bulls' sperm is shipped all over the country to sire offspring with desirable traits like high milk production.

Cows were successfully cloned for the first time in Japan, and since then hundreds of cows have been cloned in the United States.[56] A Texas-based company began cloning champion horses in March 2006, which can sell for as much as $150,000 per horse.[57] For a mere $32,000, a for-profit company can clone your pet cat.[58]

Researchers are pushing forward with cloning animals for food and have successfully cloned pigs with higher levels of omega-3 fatty acids by blending a gene from earthworms with pig's genetic material.[59] Bet that fact will not be included in the ad campaign.

But the dirty little secret about cloning is that the animals are more likely to have birth defects and health problems, and only about 5 percent of cloned animals survive.[60] A heartbreaking list of maladies has been documented in sheep, cows, and mice, including malformed brains, livers, spleens, lymph nodes, and urogenital tracts.[61] The most common causes of death among cloned animals in the first week of life are internal hemorrhaging, digestive problems, hydrocephalus, and multiple organ failure.[62] As MIT biology professor Rudolf Jaenisch says, "You can't tell me that 95 percent die before birth and the other 5 percent are normal."[63]

And if that doesn't ruin your dinner, consider the newest headline-grabbing idea for feeding the planet. Eighty-seven-year-old Willem van Eelen wants to grow meat in petri dishes quaintly referred to as bioreactors. Van Eelen is someone you might refer to as a character: he was born in Indonesia when it was ruled by the Dutch, and his father was the doctor who ran the leper colony. During World War II he spent several years as a Japanese prisoner of war. He told Jeffrey Bartholet of *Scientific American* that he became obsessed with food when he was starving in a war camp, where prisoners would eat raw dog if a stray one ventured over the wire. Van Eelen studied medicine at the University of Amsterdam, where a professor showed the students how muscle tissue could grow in the lab.[64]

This was of interest because most of the flesh that humans like to eat comes from muscle tissue. Most news reports on the subject of test-tube meat use Winston Churchill as a celebrity endorser. He wrote in a 1932 book that the most popular parts of meat would eventually be man-made using a suitable medium. Van Eelen kept the dream alive, acquiring patents, using his own money, and eventually getting a Dutch grant for research that funded a consortium to work on the project.[65] Meanwhile, across the Atlantic a NASA grant had provided funding for Morris Benjaminson, who used strips of muscle from a goldfish kept in a liquid bath of blood from unborn calves to produce meat.[66]

Van Eelen no longer has a kooky identity, as a range of scientists and animal rights groups pursue his passion.[67] Multiple teams are pursuing the idea of culturing meat, including the University of Technology and the University of Amsterdam, both in the Netherlands. In the United States, the Medical University of South Carolina is trying to find a cost-effective way to grow meat. Vladimir Mironov, who was trained in Russia, is a well-known tissue researcher and has been attracted to the idea for many years. The scientists working on test-tube meat seem to have an idealistic view that it can replace factory farms.

But even if synthetic meat could be grown in a test tube, there is no assurance that the process provides a food that metabolizes in a way that is safe and provides a nutritional benefit. Even more suspect for public health the potential that the processing used and the chemicals necessary for flavor create toxic properties to the test-tube meat. Rather than moving us toward a local and more sustainable type of food production, it puts food back in the laboratory.

Only the largest corporations would have the capital necessary to acquire the patents and to build the processing plants that would be essential to mass-produce test-tube meat. And like all processed food, it provides one more opportunity for large corporations to control the food system and to put profits before people's health.

Jim Thomas, who has written extensively on technology and food for the environmental group ETC, sums it up: "If test-tube meat hits the big time, we will likely to know by its appearance in a Big Mac or when agribusiness buys the patent-holder." [68]

These technologies are about making money, not feeding people or providing an economic base for the vast rural areas of our country, or for that matter the world. Factories and laboratories can never replace farms. This type of industrialized food production relies on a host of chemical components used to mimic the taste of real ingredients. The genuineness of real food grown on appropriately scaled family farms can never be replaced by concoctions dreamed up in corporate boardrooms and created in laboratories. Our food system is already too far out of whack, as the profit motive has fused with science to rely more on chemistry than nature to process and manufacture food products.

Imagine the outcome if the science fiction food that corporations, the science establishment, and governments are funding is commercialized. If their strategy is unchecked, biotechnology of the future—artificial life, nano-scale particles, cloning, and test-tube meat—could make our current unhealthily stocked grocery stores seem old-fashioned and low-tech. Is the antidote for overweight Americans really no-calorie junk food with artificial nutrients—all processed, flavored, texturized, colored, and stored in containers produced through the use of nano particles that can cross the blood-brain barrier and are unregulated and untested? Almost all scientific advancements have had unforeseen consequences. It is not too hard to imagine the hazards and dangers of these newest schemes.

PART VII

Building the Political Power to Challenge the Foodopoly

Our nation has a history of large movements coming together to bring social change. Today a movement for healthy, sustainable, and local food production is inspiring large numbers of people, who are voting with their food dollars, organizing direct-marketing arrangements between farmers and eaters, and engaging in many exciting initiatives to educate and excite people about changing the food system. The next step must be politicizing food activists to engage in changing the federal farm and food policies that have resulted in the dysfunctional food system. Re-creating a food system to benefit all Americans will require a range of policy changes from enforcing antitrust law to regulating the marketing of junk food to children. Ultimately, we must recognize that we share a global commons—and we must build the political power to protect it for future generations.

15

EAT *AND* ACT YOUR POLITICS

Every great movement must experience three stages: ridicule, discussion, adoption.

—John Stuart Mill (1806–73), British philosopher

Throughout history, social movements have played a powerful role in introducing new ideas, values, and beliefs that lead to political change. Over the past fifty years, the civil rights, women's rights, and environmental movements have transformed American society, ultimately resulting in important regulatory and legislative changes. The food movement—a decentralized and diverse phenomenon—is creating an important cultural shift by changing the way many Americans both eat and think about food. Awakening people to the problems and dangers of the food system is the first step to political action.

One of the promising outcomes of this cultural shift is a decline in meat consumption. U.S. demand for meat, particularly red meat, has declined for four years in a row and in five out of the last six years, according to the trade newspaper the *Daily Livestock Report*. The Values Institute at DGWG, a social science research company, recently predicted that the "flexitarian diet"—reducing the amount of meat eaten without becoming a full vegetarian—is a major health trend. It ascribes this change in eating patterns to the Meatless Monday initiative.

FGI Research confirmed in a 2011 study that 50 percent of Americans are aware of Meatless Monday and that more than a quarter of those individuals have been influenced to cut back. What is stunning about this is that the campaign has been a grassroots, word-of-mouth effort with no paid ads, media flacks, or public service announcements.

Meatless Monday is the brainchild of retired advertising executive Sid Lerner, the developer of the "Don't squeeze the Charmin" commercials. The advertising guru, with fifty years of advertising experience under his belt, wanted to do something meaningful to reduce the health impacts of eating too much saturated fat, a leading cause of heart disease. Lerner remembered the idea of Meatless Monday from his childhood, when it was introduced by the federal government during World War II as a patriotic voluntary rationing effort. Monday, the beginning of the new week, represents a common cultural experience.

Lerner, a charismatic and energetic eighty-year-old, teamed up with twenty public health schools to promote the idea of using Monday as a prompt to promote healthy habits. He knew intuitively what research has now proven: that a weekly health routine beginning on the first day of the workweek reinforces good behavior fifty-two times a year. Meatless Monday has taken off across the globe. And it's proven to be a first step for many people to move on to political engagement.

Film, social media, and other forms of cultural expression are also playing an important role in engaging people on food issues. *Food, Inc.*, the Oscar-nominated and Emmy-winning film exposing the devastating impacts of agribusiness, is one of the top twenty documentaries of all time, ranked as one of Amazon's top-selling DVDs of a documentary film, and has nearly half a million Facebook fans. The Norman Lear Center at the University of Southern California conducted a study on the effectiveness of *Food, Inc.* as an activist tool. Seventy-six percent of the fifteen thousand respondents in the study said that, after seeing the movie, they wanted to be part of a social movement to reform agribusiness, and 81 percent said the film changed their life. The entertainment trade journal *Variety* called it "a civil horror movie for the socially conscious, the nutritionally curious and the hungry. . . . 'Food Inc.' does for the supermarket what 'Jaws' did for the beach."

Food, Inc.'s director, Robby Kenner, who is humble about the film's success, calls himself an accidental activist once he realized how passionate people are about food. He found himself speaking in communities and colleges and was surprised to find hundreds, often thousands, of people show up to hear him talk. He notes, "I speak on many college campuses, and the students always ask me what they can do next. These issues inspire so much passion that many people want to take political action."

Kenner is now busy at work on his latest venture, FixFood, a cross-media social action project using videos, an interactive Web site, and community engagement to activate a mainstream audience to help transform the food system. Targeting the key food and health issues of our time, FixFood will identify immediately available solutions, support groups already redefining the food system, and grow the base of consumers eager to take action. Kenner hopes that these efforts can lead to broader social change. He says, "The goal is to transform people from passive eaters to informed shoppers who step into the political fray, either by legislation, through retail and marketplace pressure, and/or by holding their elected officials accountable."

Destin Layne, director of food programs at GRACE Communications Foundation, an organization that uses Web-based initiatives to educate and activate people, says the food movement is becoming more political. Among the programs she oversees are the *Eat Well Guide,* a carefully curated online directory of over 25,000 locally grown and sustainably produced food listings, and Sustainable Table, a consumer-friendly resource center that encourages people to become food activists. She says, "Viewing *Food, Inc.* and reading books like Eric Schlosser's *Fast Food Nation* expose large numbers of people to the industrial food system in a very visceral way. Some people just react by wanting to eat more responsibly, but many others are driven to take political action."

New media and other forms of cultural expression that are geared toward digital natives—people who are well versed in digital technology—are critical for mobilizing youth into political activism. *The Meatrix,* a flash animation featuring three superhero farm animals, is one of the online tools that Layne's organization developed and that she's actively promoting. Using pop culture and satire, the viral film broke new ground as a Web-based grassroots advocacy tool for motivating people to get involved in stopping factory farming. *The Meatrix* films, now a series, have been translated into more than thirty languages and are one of the most successful online advocacy campaigns ever, with well over 30 million viewers worldwide.

Diane Hatz, a creative marketing expert who was involved in producing *The Meatrix,* believes that art and popular culture can be used not only to educate people but to move them from political apathy to political action. As a co-founder and the director of the Glynwood Institute for Sustainable Food and Farming, Hatz organizes TEDxManhattan: Changing the Way We Eat, an annual event that combines music, good food and wine, and dynamic presentations to mobilize people to action. Speakers are given twelve minutes each to "charge people up" on a food issue. Besides the 350 people attending, dozens of viewing parties—sometimes with as many as 450 participants—are organized around the country to encourage people to become involved. The taped presentations then become twelve-minute video tools used to mobilize people on the different issues. Hatz says, "The independently organized event is based on the international TED events that are about 'ideas worth spreading,' but we also try to take it a step further and put our ideas into action."

While many good-food organizations like the ones above are using traditional and online organizing to urge political intervention, others are focused primarily on developing local and alternative food systems. These local food initiatives are extremely valuable for farmers and eaters to "build community" and to educate a broad range of people about food issues. However, it is doubtful that a wide-reaching alternative food system that serves a broad segment of the population and replaces the dominant model can be achieved through local initiatives alone. Despite the value and worth of these efforts, addressing the needs of a large share of the American population will take fundamental structural changes.

The USDA has documented that although direct marketing through CSAs and farmers' markets is growing, it tends to be isolated to urban markets. While the $4.8 billion in sales is impressive, it is very small in comparison to the $1,229 trillion in sales of conventional foods. Furthermore, only farmers who live in close proximity to a population center really have a way to participate in a direct sales market. Half of all farms that sell directly to consumers are located near metropolitan areas, whereas two thirds of all farms are rural with no nearby urban population.

This speaks to the difficulty that farmers in sparsely populated regions face in trying to shift from growing commodity crops to vegetables or fruit. Rural commodity farmers do not have access to a distribution network for their products in a food system where distribution is controlled by the consolidated grocery industry. In most cases, farmers do not have a climate or growing conditions that allow them to compete with California or the globalized produce market. They do not even own or have access to the equipment available to seriously enter into the produce growing business.

Reforming the dysfunctional food system must go further than offering direct sales opportunities or building food hubs. It requires food activists to involve themselves in organizing for political change. We must build the political power to take back our democracy and our food system.

Michele Knaus, president of Friends of Family Farms (FoFF), says that her organization is using multiple approaches in Oregon to change the food system, including education, advocacy, and grassroots organizing. FoFF is working on the problems farmers have today, with an eye for making major changes in the future. As an organization that has built a strong and united voice for independent family farmers and eaters, FoFF has already been able to pass legislation in the Oregon legislature that makes it easier for producers to sell and process local food in the state.

Knaus says, "The legislative victories we had this year will make it possible for small farmers to process chicken on their farm and to sell jams, jellies, and other food products made from the fruits and vegetables they grow. This is just a first step in a much larger legislative agenda to create a viable food system. Eaters are realizing that they are stakeholders in agriculture policy. When eaters join farmers, their collective voice gets the attention of legislators from urban and rural areas alike and demonstrates how integral agriculture is to our local economies."

Rural sociologist Mary Hendrickson believes that creating alternative food systems that benefit the community and urging political action must both be pursued. Hendrickson, who along with her colleague Bill Heffernan was among the first academics to research and disseminate information about the extreme concentration of the food industry, says that reforming antitrust law is key, but there is an important place for creating local food systems, too. She notes, "We have used our special 'standpoint' as land-grant university researchers and extension educators working out of the University of Missouri to illustrate the size and scope of the corporations involved in this global food system, and to help farmers understand these companies' strategies."

Hendrickson says that simply reporting on the data without providing a framework for understanding how change can come about disempowers people. With this in mind, she began working with the Kansas City Food Circle (KCFC) in 1994, using a pragmatic approach as an extension agent to help create an alternative community food system. KCFC's roots are in an initiative put together by the Greater Kansas City Greens in the 1980s out of concern over agriculture's dependence on fossil fuels. The group created a vision of an alternative food system that was based on solid relationships between farmers and eaters. Hendrickson notes that while it's easy to dismiss a group like this, they "challenged the existing economic and political structures" in an important way for the community.

In the late 1990s Hendrickson and her colleagues at the University of Missouri Extension received federal and state grants to develop the Food Circles Networking Project, which embraced KCFC's original vision. The project drew in other academic and public interest organizations to help locate farmers and redevelop the local infrastructure and distribution system. Hendrickson says, "Chefs and grocery stores still want local and seasonal food on their own terms—at the price and with the packaging and the delivery options that work for their own businesses. This is where 'scaling up' may conflict with the vision and needs of farmers." She explains this is always an ongoing challenge for communities trying to establish "food hubs."

In Kansas City, with the help of government and foundation money, Hendrickson and the food activists were able to move the food circle concept by teaming up Balls Food Stores, a regional chain with thirteen Hen House Markets and fifteen Price Choppers, with Good Natured Family Farms (GNFF), an alliance of farms in eastern Kansas and western Missouri. The farmers needed a market, and Balls Foods wanted to attract customers and differentiate its business from the large consolidated national chains by offering local food. Importantly, Balls was willing to invest in a warehouse as a distribution center for receiving farm products and providing cooling before items were sent to the chain's stores. The GNFF alliance's success is due in large part to the organizing capacity, hard work, and vision of Diane Endicott.

Endicott, an energetic woman who is constantly on the go, had moved back to the family farm from Texas and wanted to make a living on the farm. She remarks about her initial experience: "First of all, I didn't realize what small, family farms were up against until I set out to sell our tomatoes. You have to initially find a market for your product. Then, if you are fortunate enough to locate a store willing to carry the products, they have to let people know that the product is local and better tasting and healthier. I guess what really surprised us all was how much money it takes to do that."

Initially, Hendrickson and her colleagues worked with Balls's management to identify more local producers, and they shared research on consolidation in the grocery industry and held seminars on food safety and quality issues. Over the past ten years, GNFF and Balls Foods have worked to successfully create personal relationships between farmers and eaters. Hendrickson notes, "Farmers who market to the grocery chain are required to attend at least one summer Saturday event that showcases the 'Buy Fresh, Buy Local' products, and to participate in training in food safety or marketing." All food sold at the Balls stores is sourced within two hundred miles. More than one hundred farmers are part of the GNFF, and Balls has relationships with twenty-five additional growers.

Balls wanted to be able to offer "local" food all year long, and one way to achieve this is by providing locally grown meat. Endicott helped organize the All-Natural Beef Co-op, which comprises eighteen family farms. This endeavor has been made possible by the Endicott's purchase of a meat-processing plant that is USDA certified. The co-op currently raises beef and chickens.

Good Natured Family Farms is now the brand name for all of the local products sold through Balls stores, including meat, eggs, milk, honey, and a wide range of produce during the growing season. According to a report on "Innovative Models" by the Wallace Center at Winrock International, "BFS' [Balls Food Stores] partnership with the GNFF alliance is the role it plays in helping farmers to remain financially viable. For example, the grocery chain works with farmers on negotiating price."[1] Pricing is based on a communications process between the farmers and the Balls buyers that allows both the farmers and Balls to achieve profitability. Balls acknowledges in the report that promoting locally grown food throughout its stores attracts customers and has increased overall sales, including conventional food and nonfood items.

Hendrickson is very enthusiastic about the Kansas City project but says that, while the local community benefits cannot be overstated, legislative and regulatory changes are necessary to create a long-lasting alternative food system that serves everyone. She notes that her work with this group and others demonstrates the tensions between creating a local food system and operating in the existing economic structures. For instance, the expansion of the grocery market described above required significant investments both from GNFF, which invested in infrastructure such as meat processing, distribution, and marketing, and from Balls Foods, which not only invested in a central warehouse to store and distribute local products, but developed significant employee education to thoroughly implement the Buy Fresh, Buy Local marketing plan.

Hendrickson notes: "If we don't pay attention to how our economic structure currently works, without making it more democratic and competitive, local food systems will end up the same way as our current food system. Without thinking about how this could happen, farmers could again be at the mercy of those with economic power. I've already heard from a small distributor who is working to integrate more local foods into her business that the big companies 'get' that local is not just a fad. But they are going to adapt local—or green—to *their* way of doing business, like buying farmland and hiring farmers to produce. Now, how is that going to be fair? It might be affordable, but will it be fair? And will those who produce it get to eat it? Geography alone doesn't make a fair, affordable, democratic and green food system."

Scott Cullen, executive director of Grace Communications Foundation, who works extensively with organizations and foundations on food-related issues, sums it up. He says that food is "one of the defining issues of our age because of the impact its production has on our air, water and health." He observes, "Good food is starting to percolate in all sorts of surprising places and is such a powerful social-change vehicle because everyone eats. Food really does bring people together."

However, Cullen cautions: "Regionally produced food must be more than just a luxury; learning about it and producing it must be part of our lives, our communities, and our schools. The elephant in the room is always the perverse federal policies that set up such an uneven playing field and incentivize mostly the wrong things. In my mind, we need to build the political power to take the Farm Bill apart and completely rethink federal food policy. The food movement needs to get much bolder, more ambitious and to really become a movement that builds political power for change. Eating our politics is a first step; we need to take the second step and take political action, too."

16

THE WAY FORWARD

To embrace the worldview of the commons and act upon it with integrity is a pathway to generous sanity, which is the antithesis of the horizon toward which we're going.

—Harriet Barlow, co-founder of On the Commons,
Wisdom Voices, January 16, 2012

Breaking the foodopoly and fixing the dysfunctional food system require far-reaching legislative and regulatory changes that are part of a larger strategy for restoring our democracy. Food activists must engage with other progressives in building the political power to reform and restructure public policy to serve the interests of all Americans. To do so means overcoming the deep disenchantment with civic affairs that plagues our nation. Too many well-intentioned people have come to believe that the political system is intractable and that political engagement is hopeless.

The real beneficiaries of this apathy and cynicism are the economic interests perverting our political system. We cannot afford to be discouraged from challenging the corporate control of our food system, our genetic commons, our shared resources, or our democracy. The history of social change in our nation shows that the political system can be reformed, even if the road is long and zigzag. It requires political engagement, a commitment to organizing, and the patience for taking the long view. It demands a clear vision of what we are fighting to achieve.

Creating a just society where everyone can enjoy healthy food produced by thriving family farmers using organic practices can only be realized by making fundamental structural changes to society and to farm and food policy. A robust regional food system that benefits eaters and farmers cannot be achieved in a marketplace that is controlled, top to bottom, by a few firms and that rewards only scale, not innovation, quality, or sustainability. Re-creating the food system means taking back the control from the tiny cabal of agribusinesses, food manufacturing conglomerates, and bankers that has a stranglehold on every link of the food chain.

Rejecting the status quo—the domination of the food industry by a handful of giant companies, the offshoring of food production, the reliance on chemical agriculture, and dangerous technologies—is key in the long term to creating a new food system. We must challenge the dominant paradigm: the idea that the existing, noncompetitive market dominated by a few players can fix the problem. It is fairly easy to have access to the media or to be funded for a project if you say that Walmart can re-regionalize food production or that the market alone will eventually dictate a better food system if people just vote with their dollars. False promises will not create a better food system.

We Cannot Shop Our Way into a Sustainable Food Future

The market alone cannot solve the problem. Many in the food movement have become so discouraged by the undue influence that corporations have on public policy making from campaign contributions, lobbying, and sheer political power that they have turned to the market for solutions. Market-based solutions are popular because they fit well into a society that has been imbued with the libertarian philosophy: government bad, regulation bad, individual liberty and choice good. Many influential food activists believe that if enough people want healthy, organically grown food, the market will respond by providing it. A careful examination of this strategy shows that we cannot shop our way to a better food future.

The organic industry demonstrates how "shopping well" cannot by itself fix the broken model. Failure to enforce antitrust laws and to address the monopoly power of the retail and distribution network is causing the structure of the organic industry to mimic that of conventional foods. Although people pay more for organic food, their dollars have not reshaped the food system. To the contrary, the large corporations that have entered the organics market view it as a profitable niche and have gobbled up many of the small organic companies that were founded by people with vision and good intentions. The failure of effective antitrust enforcement has resulted in few outlets for organic food, and the monopoly control has raised the price. Corporate control of the industry has also undermined its integrity as the large corporations have successfully lobbied to use synthetic ingredients in organic food.

As this demonstrates, it is not the market that will reform the food system, it is regulation of the market—from antitrust enforcement to commodity trading—that will begin to solve the root causes of the problems. Unfortunately, over the past three decades our movement has fallen prey to allowing large and conservative foundations to dictate policy prescriptions. Rather than funding the type of organizing that is necessary to build the political power needed to rein in corporate greed and to hold elected officials accountable, market-based solutions have been promoted.

This has been a failed strategy. It is time to begin the hard work of organizing to achieve the victories we need to protect people and the planet.

Demanding a Functional Market

A sustainable food system cannot exist without fair and functioning markets. The century-old U.S. antitrust laws were not designed to address the scale, shape, or structure of today's agricultural marketplace, and federal enforcement has failed to effectively moderate the impact of consolidated power on consumers, farmers, or the marketplace. While the problem infects every industrial sector, consolidation in the food industry is stunning because it affects every link in the food chain.

An investigation by Congress into the state of competition in agriculture markets is long overdue, and it is time for food activists to add antitrust issues to their agenda for a better food system. Unfortunately, the technical and arcane nature of the problem, and the lack of funding for staffed food organizations to work on antitrust reform, means that few in the food movement have engaged in this important set of issues.

Leaders on food issues cannot afford to ignore the failure to enforce and strengthen antitrust laws, because it is at the very root of our dysfunctional food system. The companies that want to sell the foods consumers are demanding—healthier, more local, more organic—cannot get onto grocery store shelves because of consolidation in the food industry. It is time to start demanding that the federal government embark upon a program of enforcement and regulation that restores competition in the marketplace for the benefit of consumers and producers.

Currently, lackluster enforcement is divided between three agencies—the Department of Justice (DOJ), the Federal Trade Commission (FTC), and the U.S. Department of Agriculture (USDA)—preventing a real and coordinated examination of market power in the food chain. Congress should reexamine this artificial division of labor and jurisdiction. The convoluted enforcement is not only confusing, it has resulted in little oversight of the food industry, even though the anticompetitive impacts of consolidated retail market power span the entire food chain.

The DOJ and FTC have joint or overlapping antitrust authority over mergers, monopolies, and collusion like price fixing. The FTC has authority over industries with a significant impact on consumers, including supermarkets, food manufacturers (anything in a bag, box, or can), and farm inputs like seeds and fertilizer. The DOJ oversees most other farm-related items such as the dairy sector and the manufacturing of farm implements. The USDA has jurisdiction over the meat and poultry industries under the Packers & Stockyards Act.

Although the FTC has authority over supermarkets, over the past two decades, as a wave of mergers has swept the grocery store industry and Walmart has emerged as the largest food retailer in the country, the FTC has largely ignored this consolidation. The FTC should reenergize its enforcement of anticompetitive behavior in the retail grocery industry, which has been largely dormant since the 1970s. The FTC should investigate the impact of consolidated retailers on the entire food supply chain. This should include looking at the coercive marketing arrangements that the big retailers impose on suppliers, and it should examine the effect on consumers in terms of price, quality, and product choice.

The confusion and lack of action on antitrust is demonstrated by the DOJ's failure to finish its inquiries into the consolidation and market power of Monsanto and the dairy industry. These investigations have languished at the agency for years; some even predate the Obama administration. They should be vigorously pursued and finalized as soon as possible, along with any others that are in the pipeline.

However, in 2011, we saw how afraid our policy makers are of making the food industry operate fairly. When President Obama ran for office in 2008, he heard firsthand about the meat industry's noncompetitive practices and how its consolidated market power reduced the earnings of livestock producers, forced them to become significantly larger, and encouraged them to adopt the more-intensive practices used on factory farms. Based on the first-ever livestock title in the 2008 Farm Bill, which was fought for by family farm advocates, the USDA was directed to develop new rules to ensure that livestock producers are treated fairly by meatpackers and poultry companies. The Obama administration lacked the courage to finalize these commonsense rules and capitulated to industry pressure, leaving farmers vulnerable to the market power of the consolidated meat industry.

The Western Organization of Resource Councils, a regional network of seven state-based grassroots organizations that has worked tirelessly to save family farms and ranches, chided the administration.

> WORC members had high hopes for these rules based on the content of the original proposal and have been waiting for their final publication for over 18 months with great anticipation. With this final rule, USDA and the Obama Administration have let down the independent farmers and ranchers of this country. In his campaign, President Obama said he would fight to ensure family and independent farmers have fair access to markets, control over their production decisions, and transparency in prices. Instead of taking this opportunity to keep that promise, with these rules the Administration has caved in to pressure from big meatpackers, and is allowing unfair and deceptive practices to continue.[1]

Unfortunately, the food movement was largely silent in this important battle—a lost opportunity for challenging the market power of the meat industry and all of its associated abuses of consumers, farmers, and the environment. Implementing the needed reforms would have had a direct effect on curtailing the unfair and artificial advantages and profits from factory farming and providing family livestock producers the ability to make a living.

As Niel Ritchie, executive director of the League of Rural Voters, says, "By its very nature, industrial agriculture has done tremendous damage to our land and water resources. It has also drained the economic life out of once vibrant rural communities. But it doesn't have to be this way. We can rebuild our food system by supporting small and medium-size farms that serve local and regional markets."

We will have another opening to move in this direction in future Farm Bills, where the process of developing a new set of rules to make the meat industry more competitive can be initiated. The beginning elements of competition policy included in the 2008 Farm Bill livestock title need to be strengthened and expanded into a competition title that would address the consolidated market power across the entire food chain, including dairy, livestock, and poultry processors; seed companies; fruit and vegetable buyers; and grocery stores.

Advancing an environmentally and economically viable food system that serves all Americans means busting apart the agribusiness monopolies that control every step of our food chain and rebuilding the economic and physical infrastructure to link farmers to eaters in their region. We need the food movement's full engagement in this battle. And central to the food fight is changing the Farm Bill.

Tackling Future Farm Bills

The broken food system did not come to be by accident, nor is it the natural result of economic efficiencies. Agribusiness, the food industry, and other corporate interests lobbied Congress for the policies and programs that have shaped it. The policies dictate what food is available to the public, how it is produced, where it is sold, and who can afford to buy it. Many of these policies are contained within Farm Bills, major pieces of legislation that are revised and renewed about every five years.

Instead of catering to agribusiness's desire for cheap raw materials, our next Farm Bills should ensure functional, fair markets so that farmers and farmworkers who grow our food can earn a decent living, promote environmental stewardship, and rebuild the infrastructure we need for consumers to access sustainably grown, regionally produced food. Farm Bills are the vehicle for dealing with many of the issues that adversely impact farmers and consumers, including issues related to consolidation. This large piece of legislation, with its multiple titles, frames and establishes the policies and government support for U.S. agriculture, nutrition programs such as food stamps, rural economic development programs, agricultural research, and much more.

Most of the spending in recent Farm Bills—about two thirds—has gone to the nutrition title, which establishes government programs that provide food assistance such as the Supplemental Nutrition Assistance Program (SNAP, formerly known as food stamps). However, the most controversial section is the commodity that deals with the crops that serve as the raw materials of our industrial food system—namely, corn, wheat, and soybeans, along with other grain and oilseed crops.

Although the current structure of commodity programs contributes to many of the excesses of industrialized food production, simply eliminating commodity program payments to corn, soy, or wheat farmers will not restore the balance to the farm system that most consumers and farmers want. The agriculture policy must be reformed to restore a safety net for farmers; simply eliminating the agricultural policies that cover commodity crops is not a road map to a fair food system. It would lead to an even larger loss of family farms, making it even more difficult to transition to a sustainable food system.

As we have seen, farm programs encourage overproduction of commodities and allow their price to fall below the cost of producing the crops; when prices fall enough, the government reimburses farmers to cover only some of their losses. Taxpayer dollars end up subsidizing meatpackers, factory farms, and food processors, effectively laundering the money through farmers. The beneficiaries of the farm payments are not farmers but the buyers of the cheap crops, because government payments to farmers allow these buyers to pay less for the crops that are their raw materials. Ending government payments to farmers will not fix the problems in our food supply because the payments are the result, not the cause, of the low prices farmers receive for their crops.

Timothy A. Wise, director of the Research and Policy Program at the Global Development and Environment Institute at Tufts University, who has researched the role of subsidies in farm policy confirms: "Government subsidy checks are written to farmers, but they aren't the true beneficiaries of U.S. farm programs. Agribusiness is. Input suppliers like Monsanto and John Deere benefit from the increased production—and demand—the subsidies encourage, and agribusiness buyers like Smithfield and Cargill reap the benefits of the lowered prices that result. Farmers remain squeezed between these oligopolies, paying dearly for inputs and getting lower prices for what they produce."

Farmers lose when crop prices collapse, but buyers of those crops win. Cheap grain prices are exactly what big agribusiness processors want and what they have used to design their business models. With lower-cost inputs of corn and soybeans, they can produce their processed junk foods and high-fructose corn syrup much more cheaply. And instead of raising livestock on pasture, animals can be crammed into factory farms and fed artificially cheap corn- and soybean-based animal feed.

Instead of encouraging overproduction and maintaining farm programs that benefit the big agribusiness companies, it is time to restore commonsense supply management policies and price safety nets that make agribusiness, not taxpayers, pay farmers fairly for the food they grow and we eat. This means bringing back strategic grain reserves, requiring that farmers leave a portion of land fallow, and maintaining minimum price floors for crops to ensure that at the very least farmers are paid for the cost of producing their crops.

Jim Goodman, a Wisconsin family farmer and a Family Farm Defenders board member, adds, "Farmers need access to farm credit, a fair mortgage on their land, fair prices for the food they produce, and seeds that aren't patented by Monsanto or other big corporations. Consumers need to be able to purchase healthy and local food, and to earn a living wage."[2]

If we are to really change the politics-as-usual debate on the Farm Bill every five years and create a sustainable food system for the long term, it is critical for progressives in the environmental and conservation movements to stop blaming farmers for the food policies that agribusiness and the food industry have lobbied for and achieved incrementally over the past eighty years. Scapegoating farmers is not only unfair, it also will not help us build the farmer-eater coalition that we need to advance enlightened agriculture and food policy.

While it is easy to rile up food activists and get media attention by focusing on the subsidies received by "greedy" farmers, this alienates a critical constituency for changing farm policy. The name-calling has provided the shills of corporate agribusiness, food manufacturing, and other regressive economic interests, such as the American Farm Bureau, the ammunition to divide farmers from progressive food and farm advocates.

We must be clear. We will not be able to change farm policy it we do not join forces with farmers so that we can elect and hold accountable the members of Congress from the farm states. Food & Water Watch researched the composition of the agriculture committees of both houses of Congress over the past thirty years. Research director Patrick Woodall explains: "The members of the congressional agriculture committees have always represented more rural states and constituencies. Farm policy is written by legislators from the Midwest, high plains, and across the Southeast, not from the biggest cities. Progressives need to make common cause with the farmers and rural communities and not work at cross purposes by demonizing the farmers and rural voters that live under the boot heel of agribusiness power."

Rural Economic Development

It is in the self-interest of both food advocates and farmers to solve the problems that have devastated farming communities. The rise in corporate consolidation in the food industry has driven independent, locally owned food manufacturers, processors, and distributors out of business. The Main Streets of rural towns have been boarded up as distant companies siphon the profits from rural America. Independent grain elevators; flour and corn mills; small meat processors; fruit and vegetable distribution terminals; and independent suppliers of farm inputs such as seed, feed, and machinery are increasingly hard to find. Today the businesses that do exist are most likely corporations supplying inputs through contracts with large farms. The mills or slaughterhouses that process the grain and livestock for all of the farmers in surrounding towns are very large and much farther away.

Economically viable independent farms are the lifeblood of rural communities. The earnings from locally owned and locally controlled farms generate an economic "multiplier effect" when farmers buy their supplies locally and the money stays within the community. When these businesses disappear and are replaced by just a few larger facilities, rural economies suffer. The earnings and profits from meatpacker-owned cattle feedlots and hog production facilities are shipped to corporate headquarters instead of invested locally. Independent slaughterhouses, milk and meat processing firms, locally owned grain elevators, and local feed and equipment dealers provide employment, investment, and stability to rural communities.

Farm Bills trigger hundreds of millions of dollars of USDA spending on rural development, ranging from grants to local governments and community organizations to government-backed loans to businesses. Unfortunately, many past bills have focused funding only on larger projects like broadband Internet access or businesses that don't help rebuild food systems, like hotels or convenience stores selling processed food.

What has been sorely lacking is investment in agriculture-related industries and infrastructure that would support the vegetable, grain, dairy, and livestock farmers who need distribution, packing, and processing facilities before they can bring their products to market.

Future Farm Bills should focus on leveling the playing field for independent farmers, ranchers, and food processors and redirecting rural development programs to rebuild missing infrastructure that can serve regional food systems, not corporate supply chains. One way to do that is to dedicate rural development funding in future Farm Bills to facilitate the growth of small and medium-size independently owned agriculture and food enterprises that can reinvigorate local economies and increase the role of local food.

While local food sales are increasing, they are still small in comparison to conventional foods. The statistics speak to the challenge: $4.8 billion in all local food sales compared with $1.2 trillion for all conventional food sales. Yet the rising demand for locally grown food will not lead to sufficient availability. When consumers and farmers seek to make more local food available in supermarket aisles, they run up against a few supermarket chains, each with a highly concentrated distribution network that supplies thousands of stores— a model that is inaccessible for small or midsize independent producers or processors. This is especially true in the farm states, where millions of acres of commodities are grown. These farmers have a small "local" market and no way to distribute locally grown fruits, vegetables, or meats.

Judy LaBelle, past president of and now a senior fellow at the Glynwood Institute for Sustainable Food and Farming, has been working to build a local food system in New York's Hudson Valley for more than a decade She articulates the challenges of creating an alternative food system: "After decades of government and corporate policies that encouraged farmers to 'get big or get out,' we need to rebuild virtually the entire system all at once. Smaller, independent farmers need the off-farm infrastructure—the processing, distributing, marketing—that allows them to serve and benefit from growing consumer demand for their product. We need to change federal policy to truly change the food system so that it serves everyone."

This does not mean that we should not be excited by the healthy food projects that are being organized around the country. Food hubs, food circles, farm to school programs, farmers' markets, CSAs, and all of the other inspirational projects around the country are building the case for a new food system. They are creating community by educating and inspiring people to appreciate locally grown food and the farmers that produce it. Combined with advocacy for addressing the root causes of the dysfunctional food system, these grassroots self-help projects can be a powerful tool for good. But rebuilding the local and regional food system that can serve all Americans cannot be done without addressing consolidation and making federal policy changes.

Increasing Access to Healthy Food

Besides setting the rules for how agriculture markets will work, Farm Bills create programs that help create access to fruits and vegetables. One popular prescription for our broken food system is to divert money from the commodity programs to the "specialty crops" title, which covers fruits and vegetables. In past Farm Bills, money for fruits and vegetables has been misspent. For example, the Alliance for Food and Farming received a $180,000 grant to "correct the misconception that some fresh produce items contain excessive amounts of pesticide residues," in an effort to rebut the Environmental Working Group's "Dirty Dozen" list of the twelve most pesticide-contaminated fruits and vegetables.

In future Farm Bills, more guidance is needed in the specialty crops title to ensure that the money and resources go toward supporting small and medium-size fruit and vegetable farmers instead of just the largest players in these industries. Programs that support specialty crops and sustainable, organic, or diversified production of these crops should also be included in other titles, such as research and conservation, which have focused predominantly on commodity crop and livestock production.

However, to really promote more access to fruits and vegetables, the consolidation of the retail grocery industry needs to be addressed. As we have seen elsewhere in this book, four grocery chains dominate the U.S. market. Academic studies have found that higher levels of local retail concentration are associated with higher grocery prices. These retailers are exerting their market power over the produce industry, causing consolidation down the entire supply chain, all the way to the growers.

The retail chains favor the largest suppliers, who can best negotiate with the retailer and who then pass on the cost-cutting pressure to their farm suppliers. Produce packers and shippers are becoming larger and more consolidated to match the size and demands of large chains such as Walmart. An empirical USDA study found that retailer market power enabled supermarkets to push the prices paid to shippers of grapefruit, apples, and lettuce below the prices they might receive in a functioning competitive market, and revealed that consumer retail prices were higher than "purely" competitive prices for apples, oranges, grapefruit, fresh grapes, and lettuce. Enforcing and strengthening antitrust law would go a long way toward creating more affordable access to fruit and vegetables.

Additionally, to date there has been little analysis of anticompetitive behavior by the largest produce packers, from Dole to Earthbound Farm. An investigation should include an examination of the terms of company contracts with growers and their impact on crop production. The majority of processed produce, fresh fruit, fresh vegetables, sugar beet, and tobacco production is done under some form of contract, and a significant percentage (about a fifth) of many commodity crops are grown under contract. The USDA should survey and study contract crop producers, their contracts, and the impact of vertical integration on farmers and the marketplace, with special emphasis on the crops with the highest shares of contract production.

Not only does the shape of the produce industry affect the cost and composition of the fruits and vegetables available in grocery stores, it also dictates the end price paid by the consumer. The concentrated nature of the produce industry makes it impossible for farmers that want to transition into growing more fruits and vegetables to compete with the large concentrated produce packers. These large players can do more than just undersell smaller growers—they can submit to the various demands of large retailers, including paying fees for shelf space.

However, even a reformed food system, with plenty of fruits and vegetables available, cannot ultimately address the problems related to poverty and lack of access to adequate food resources. Fixing the dysfunctional food system must be part of a larger, broad-based strategy for dealing with the economic inequities of our society. A 2011 Congressional Budget Office study documented the stratification in the United States, which has increased dramatically since the 1970s: "As a result of that uneven income growth, the distribution of after-tax household income in the United States was substantially more unequal in 2007 than in 1979. . . . [T]he after-tax income received by the 20 percent of the population with the highest income exceeded the after-tax income of the remaining 80 percent."[3]

Noted author Mark Winne, an expert on food security, observed on his blog, "Food insecurity has cast a dark shadow across our national landscape for decades, primarily because we cannot bring ourselves to confront its root cause, poverty. Our elaborate and not inexpensive network of private and public food programs make a noble effort to mitigate the worst aspects of poverty, namely hunger, but even on their best days, they only succeed in managing poverty, never ending it."[4] In an earlier post on his blog, he wrote:

> Until our public policies once again take on the task of ending poverty, and private industry is forced or shamed into paying a living wage to all its workers, hunger and food insecurity will be business as usual for tens of millions of Americans. The recent flare ups in our stressed food system may remind us how vulnerable we all are to economic and natural forces, but for the poor and those now joining their ranks, it's just another bad day in the food line.[5]

Even Warren Buffett, chairman and chief executive of Berkshire Hathaway, is among the people arguing for reforming economic policy. Buffett is advocating for the superrich to pay their fair share of taxes and to lower the rate for the poor and middle class, an important component in fixing income disparity and having an adequate federal budget. Buffett wrote in a *New York Times* opinion piece: "These and other blessings are showered upon us by legislators in Washington who feel compelled to protect us, much as if we were spotted owls or some other endangered species. It's nice to have friends in high places. . . . My friends and I have been coddled long enough by a billionaire-friendly Congress. It's time for our government to get serious about shared sacrifice."[6]

Organic Agriculture and Environmental Stewardship

Organic agriculture has seen incredible growth in the past several decades, yet farm policy has not kept up with the pace. The Farm Bill needs to include provisions targeted to organic production throughout the legislation, including the research, conservation, credit, and insurance titles. The USDA research agenda should increase the proportion of time and resources it devotes to organic agriculture.

Certified organic farmers should have easier access to conservation programs, since organic operations are already using practices intended to be better for the environment. The 2008 Farm Bill contained a cost-sharing provision that gave money to farmers to offset the cost of becoming organically certified. This helps conventional farmers who want to make the transition to organic farming and should not just be maintained but expanded.

A major portion of spending generated by Farm Bills goes to conservation programs that either encourage farmers to take vulnerable land out of production, require certain environmentally preferable farm practices, or help farmers financially to change their practices. Programs such as the Environmental Quality Incentives Program (EQIP) can help farmers implement cost-effective, environmentally beneficial production methods such as grasslands and wetlands management and better management of manure.

But too often, these funds are used to subsidize short-term, technology-heavy "fixes" to the pollution problems of industrial livestock and crop operations. Under EQIP, the 2008 Farm Bill delivered a subsidy to factory farms by dedicating 60 percent of program funding to livestock operations. Between 2003 and 2007, taxpayers paid $179 million to cover manure management costs for factory dairies and hog operations (not counting chickens or cattle) under EQIP. One way to address this problem is to cap the size of individual grants given under the program, and to limit grants to smaller and medium-size facilities.

Farms receiving benefits from Farm Bill programs should have to comply with good conservation practices, and conservation compliance provisions in the Farm Bill can help ensure that they do. These provisions focus on maintaining proper soil conservation on land that is subject to erosion and protecting wetlands by cutting off access to some Farm Bill funding if minimum environmental standards are not met.

These requirements should be maintained and strengthened in the next Farm Bill. Conservation program funding should be targeted to operations that provide the greatest long-term environmental returns. Funds should facilitate the transition to and maintenance of farm management strategies that improve biodiversity; minimize air and water pollution; and conserve soil, water, and other essential resources. Conservation programs should support the transition to organic farming and help farmers identify crops and techniques appropriate to their region's water resources and climate.

Farm Bill research dollars should be geared toward supporting organic agriculture, helping farmers transition to more sustainable methods, and developing publicly available technologies and plant and animal breeds. In the past, federal dollars have been shifted toward applied research that is conducted jointly with the private sector. Agribusinesses are thus able to develop and market the fruits of USDA-funded research for private gain.

Future Farm Bill research titles should shift toward policies that will benefit smaller-scale farmers instead of agribusinesses, with research being performed for the public, not private interests. It should focus on alternatives to industrialized production, including reviving research into non–genetically engineered crops and livestock breeds and low-input and sustainable production methods such as grazing livestock.

Putting an End to Gene Tinkering

Genetic engineering involves manipulating the genetic makeup of plants or animals to create new organisms, a process with many unintended consequences. Promoted by the biotech industry as an environmentally responsible and profitable way for farmers to feed a growing global population, despite all the hype, genetically engineered plants and animals do not perform better than their traditional counterparts.

Lack of vigilance by the federal government, including a dearth of responsibility, collaboration, or organization from the three U.S. federal agencies responsible—the FDA, the USDA, and the EPA—has put human and environmental health at risk through inadequate review of genetically engineered (GE) foods, a lack of postmarket oversight that has led to various cases of unintentional food contamination and a failure to require labeling of these foods.

The real winners are the handful of biotechnology companies that patent specific genetic crop traits and sell the GM traits in seeds and affiliated agrochemicals to farmers. A few major chemical and pharmaceutical giants now dominate the seed industry, which once relied on universities for most research and development. Between 1996 and 2007, Monsanto acquired more than a dozen smaller companies, and it now controls an estimated 70 percent of all GE corn and 99 percent of all GE soybeans planted in the United States. The few firms that do exist often have cross-licensing agreements for their patents that create partnerships between companies to sell seeds with specific combinations of traits from multiple firms. This high level of concentration is raising seed prices and could threaten access to and availability of food in the future.

In the long term, we need a world where scientists do not tinker with our genetic commons and where a handful of corporations does not control the seeds we depend on for survival. We should not be satisfied until the genie is back in the bottle.

In the shorter term, the United States should enact a moratorium on new approvals of GE plants and animals until the federal government develops a new regulatory framework for biotech foods. Congress should establish regulations specifically suited to GE foods, and the federal agencies responsible should adequately monitor the postmarket status of GE plants, animals, and food. Currently, most GE foods in the United States are generally considered safe for consumption and the environment until proven otherwise. A rigorous evaluation of the potentially harmful effects of GE crops and food should take place before their commercialization.

At the same time, consumer and food organizations must continue to pursue a vigorous labeling campaign at the state and federal levels. Momentum is building for labeling, especially since the Obama administration has legalized so many new GE foods, including beets, which are ubiquitous in the food supply as a form of sugar. An affirmative label should be present on all GE foods, ingredients, and animal products.

Additionally, we must work to prevent the research funding and agenda set by the Farm Bill to benefit the biotechnology industry. Far too much of past farm policy and the USDA's agenda has been skewed toward encouraging and supporting genetically engineered crops. The research agenda, set through the Farm Bill, provides federal money to private companies researching patentable biotechnology, and it provides discounted crop insurance rates to farmers if they plant GE corn under the USDA's Biotechnology Endorsement program.

Future Farm Bills also needs to address the issue of liability that patent holders of GE seeds should bear for the contamination of non-GE crops. Currently, if GE crops contaminate non-GE crops, the GE seed company is not responsible for the damage. Certified organic farmers can face significant economic hardship if biotech traits contaminate their organic crops or organic livestock feed. Many domestic and global markets for non-GE and organic products have zero-tolerance policies, which means that unintentional contamination can cause farmers to lose their markets, and the company that created the GE seed remains unaccountable. Future Farm Bills should include contamination-prevention policies to protect organic and non-GE producers from losing their markets. The financial responsibility of contamination should be on the patent holders of the GE technology, rather than on those who are economically harmed.

The Trade Myth

Another wrinkle in the debate about creating a new food system is the World Trade Organization (WTO), which promotes international trade but also decides if domestic policies created by member countries are serving as barriers to trade. When someone proposes a new idea for farm and trade policy, the question arises, "But is it WTO-compliant?" In 2009 Canada and Mexico challenged U.S. country-of-origin labeling (COOL) for meat as an illegal trade barrier under the WTO. This unaccountable international institution found COOL to be a barrier to trade in 2011. A WTO panel also argued recently that labeling of canned tuna as "dolphin safe," which lets consumers know if dolphins were accidently harmed in fishing for tuna, is trade-restrictive.

The WTO's Agreement on Agriculture actually prohibits member countries from adopting any farm policy that affects the supply or price of crops. Instead, farm programs must be "decoupled" from price or supply considerations. For years, U.S. lawmakers have tried to change U.S. farm policy to fit into the rules set out under global trade agreements. In fact, direct payments to farmers, in the form of subsidies, were adopted largely because they were unconnected to price or supply conditions, and therefore WTO-compatible.

Instead of trying to find ways to make U.S. farm policy compliant with global agricultural trade rules, lawmakers should adopt policies that ensure that farmers receive fair prices for their crops and livestock. Indeed, the WTO Agreement on Agriculture has encouraged a flood of cheap commodities into the developing world (when prices are low), and food import–dependent countries have faced prohibitively expensive food staples (when prices were high, as in 2008 and 2011). The WTO required member countries to reduce tariffs on agricultural goods. As a result, floods of U.S. corn, soybean, and other commodity crops inundated developing countries most years when prices were low. This dumping of commodities pushed many farmers in the developing world off their land.

Meanwhile, U.S. fruit and vegetable producers have been undercut by a sharp increase in imported fruits and vegetables from corporate-owned plantations in developing countries with weaker labor and environmental standards. The globalized market in agricultural products allows the same agribusinesses that squeeze farmers in the United States to take advantage of farmers and farmworkers worldwide in the pursuit of cheap farm goods.

The WTO offers a failed model for agriculture policy that should not be the foundation for agriculture policy in the United States. Instead, U.S. farm programs should ensure that farmers receive fair prices for their crops and livestock, which would stabilize prices for farmers worldwide. When prices are low, agribusiness can export U.S. crops at below the cost of production, which hurts farmers in the developing world. If agribusiness export companies had to purchase their grains from U.S. farmers at a fair price, they could no longer dump these products on the developing world at below the cost of production and destroy local markets in poor countries.

Creating a truly fair food system that rewards sustainable local and regional food production requires taking decision making about agriculture and food away from the jurisdiction of the WTO and other trade pacts. Trade deals like the North American Free Trade Agreement (NAFTA) have swamped Mexico with low-priced corn; when crop prices were especially low in the late 1990s, this drove millions of rural Mexican farm families off the land. At the same time, many agribusinesses relocated to Mexico or sourced their fruits and vegetables from Mexico to take advantage of cheaper labor and weaker environmental standards. Imports of tomatoes and peppers and other produce from Mexico undercut American farmers as well as the workers at canneries and processing plants.

All of these trade deals reward the transnational agribusinesses and punish farmers and workers both at home and abroad. Food is simply too important to be treated like a widget under international trade rules.

Safe and Drug-Free Food

One of many battles before us in the short term is stopping the use of antibiotics in industrialized animal production. Not only is this critical for creating a sustainable food system, but the loss of these important drugs—through their overuse and the ensuing rise in antibiotic resistance—will have a devastating effect on the health of humanity into the future.

The FDA began raising questions about the use of antibiotics in animal feed, beginning with a task force to review the issue, in 1970. But more than forty years later, little has been done, even as the medical and public health communities have continued to sound the alarm about the consequences for human health when antibiotics are no longer effective. The science is clear: overuse of antibiotics in food animal production is driving antibiotic resistance. Yet there has been little movement from regulators or Congress to rein in the use of these critically important drugs by agribusiness. Science is not enough.

We need to build the political power to force our elected officials and regulators to take action instead of dragging their feet with talk of more studies and guidance to industry about what they might do voluntarily. Congress should pass the Preservation of Antibiotics for Medical Treatment Act, and FDA should adopt regulations to end the nontherapeutic use in livestock of medically important antibiotics.

We also need a strong, well-funded grassroots campaign on a range of food safety issues. Ceding food safety to the food industry is not acceptable, despite how popular this proposal has become in an era where all sides of the political spectrum compete to shrink the role of government and regulations. Leaving something as critical as food safety up to the industry to figure out leaves those eating the food without anyone looking out for them. A well-run government food safety program can look out for consumers and make sure that food safety practices are not determined by the largest players in the industry and used as a competitive advantage to push smaller players out of business.

This means reorienting the government food safety research agenda to look at the risks created by different types of food production and considering the best practices for preventing contamination, not just treating it. It means adding the long-term health risks of chemical residues, genetic engineering, and other food technologies to the list of what is considered under the umbrella of food safety. And it means developing regulations that are based on good information about the risks of food safety problems that work for all types and scales of food production. Achieving a safe, smaller-scale food system requires enough funding for government inspection and technical assistance, so access to these programs is not a barrier to new players who want to rebuild regional food systems.

Getting Tough on Advertising

The recent flurry of activity on preventing obesity—from First Lady Michelle Obama's initiative to the USDA's National Institute of Food and Agriculture research agenda—has failed to recognize the role of advertising in the obesity epidemic, which is yet another symptom of our dysfunctional food system. Attempts to regulate advertising have met with strong industry resistance. In 2011 the Interagency Working Group, comprising four federal agencies, attempted to set voluntary nutrition standards for foods allowed to be marketed to kids. Their recommendations met so much industry blowback that the FTC's Bureau of Consumer Protection backed down on the weak voluntary standards agreed upon by the group.

Prominent author and food activist Michele Simon notes, "Industry keeps right on lobbying; it's what they do best. And for Congress, it's just business as usual. But the very real consequence of maintaining the status quo is that children will continue to be exploited for their emotional vulnerability, while getting lured into bad eating habits that can last a lifetime. Cost/benefit analysis? Industry benefits, while children pay the cost."

The role of advertising in affecting children's behavior is well known. In the 1970s Action for Children's Television took on the battle of advertising to children, eventually culminating in the 1990 Children's Television Act, which limits advertising time, but not which products are advertised. A 2007 Kaiser Family Foundation study, the largest of its kind, documented how much food advertising children are exposed to on television. Children ages two to seven see an average of twelve food ads a day—a total of 4,400 food ads annually, or nearly thirty hours of food advertising a year. Children ages eight to twelve see an average of twenty-one food ads a day—7,600 food ads, or over fifty hours of food advertising a year. And teenagers ages thirteen to seventeen see an average of seventeen food ads a day—more than 6,000 food ads, or over forty hours of food advertising a year.[7]

Meanwhile, an abundance of research proves that children eat more junk food after seeing it advertised on TV. Advertising is almost all geared toward unhealthy foods full of artificial coloring and flavoring, as well as sugar, fat, and salt.

The manipulation of fat, sugar, and salt in processed food to make it addictive—and the deployment of mind-controlling ads to drive people, especially children, to eat high-calorie, low-nutrition products—is unconscionable. Marketing techniques should be called out as one of the causes of not only obesity but diabetes, heart disease, and all of the diseases related to overeating junk food.

To stop obesity and the ravages of the diseases it causes, food advocates must begin organizing to regulate the advertising of food to children. Marion Nestle sums it up well: "The intent of the First Amendment was to protect political and religious speech. I cannot believe that the intent of the First

Amendment was to protect the right of food companies to market junk foods to kids. Marketing to children is unethical. It should be stopped. And it's the government's responsibility to do it."[8]

Pursuing Legal Remedies

As a result of successful organizing and advocacy, we have already won many important public and environmental safety laws and regulations. When the industrial food producers seek to erode these protective standards or take production shortcuts that bypass these safeguards, the courts play can play a pivotal role in protecting our natural resources and citizens. Michele Merkel, co-director of Food & Water Watch's justice program notes: "Often, the courts are the last line of defense when special interests and lobbyists bankroll harmful legislation and strong-arm state and federal agencies. From ending segregation to promoting freedom of speech to the recent California case striking down a ban on gay marriage, court victories have set the stage for greater equity and justice. The same holds true in the environmental arena, where courts have the power to stop polluters and safeguard our communities."

Today one of the biggest problems impacting our waterways is the excess loading of nutrients—nitrogen and phosphorus—coming from the mega factory farms. These animal factories have been regulated as sources of pollutants under the Clean Water Act for many years, but both the EPA and state environmental agencies have struggled to properly implement the act in the effective ways they have used against other industrial dischargers. That means that factory farms, though regulated on paper, are largely unregulated by the federal and state regulatory authorities.

Yet while government has looked the other way, citizens continue to play a critical role in monitoring and enforcing the CWA and other laws against CAFOs and holding the regulatory agencies feet to the fire to force better laws and policies.

Legal actions against industrial agriculture can take many forms, from citizens' suits against factory farms that discharge pollutants to our air and water under a host of environmental laws to nuisance actions for the impacts on the surrounding home owners. Challenges can be brought against state agencies for the issuance of permits to these facilities, while rule challenges can be brought against the EPA when unprotective final rules are promulgated under the Clean Water Act, the Clean Air Act, or other environmental statutes. In addition, where no regulations exist, concerned citizens can petition the EPA to fill the gap through a rule making. We can use all of these legal tools in the future to hold industry accountable for their production systems and to remedy harms to our public trust resources and communities when industry acts irresponsibly.

Creating a New Paradigm: The Global Commons

Food—basic to the human experience, culture, and health—provides an opening to redefining how the world is viewed. We need a new paradigm based on perceiving the world as a global commons with collectively shared assets, from air, water, soil, and genetics to taxpayer-funded research, libraries, roads, and all of the other resources we share.

Using the commons as a prism for conceptualizing how we live together in an increasingly crowded world offers an affirmative vision for creating a more equitable and sustainable society. It provides a countervailing force to the selfish, profit-driven mind-set that rules our culture and political system. In all of our work, we need to promote a more humanistic and environmentally friendly future. We need a new and inspirational frame that can be used in our work—from academia to grassroots organizing.

Jay Walljasper, author of *All That We Share: A Field Guide to the Commons*, says:

> The commons is more than just a nice idea; it encompasses a wide set of practical measures that offer fresh hope for a saner, safer, more enjoyable future. At the heart of the commons are four simple principles, which have been practiced by humans for millennia: 1) serving the common good; 2) ensuring equitable use of what belongs to us all; 3) promoting sustainable stewardship so that coming generations are not cheated and imperiled; 4) creating practical ways for people to participate in decisions shaping their future.[9]

Maude Barlow, former UN adviser on water and a leader in the progressive movement, speaks of its importance: "The commons as a concept is pure gold because it bridges the solitudes and silos that divide us. Bringing together food, farm, and justice advocates to understand one another's work and perspective is crucial. All parts of the progressive movement have to dream into being—together—the world they know is possible and not settle for small improvements to the one we have. This means working for a whole different economic, trade, and development model even while fighting the abuses existing in the current one."[10]

Building Political Power

Our movement must deepen and expand the strategy for moving people to political action. It must join with the broader progressive movement to organize across the country in each state and in a majority of congressional districts. There are no shortcuts to building the long-term political power to reclaim our food system, our democracy, our commons.

We must organize and mobilize people around the issues that affect their lives and their families. Because most people come to politics via their interest in an issue, and based on a perceived self-interest, food-related issues could play an important role in changing politics and encouraging civic engagement. As the food movement continues to grow, it can be a catalyst for action and can politicize Americans who have not been active participants before. Food by its very nature can be part of a unifying strategy.

This can even be true in the rural areas of the country where we must build capacity if we are to work on future Farm Bills. It requires a long-term vision of fighting for what we really want, not just the best that can be negotiated. It means breaking our long-term goals for the future down into shorter-term, winnable goals that help us reach the sustainable and equitable future we want.

The issues surrounding food, if framed correctly—from antibiotic resistance to food safety to reestablishing a fair market—can be part of the majority strategy we need to win. This means an action plan based on issues that a large segment of the population cares about and agrees on, when the problem is brought to their attention. Food is already attracting large numbers of people; however, we must provide a step-by-step road map for engaging large numbers of people in the political process, including electing people who agree with us and holding those elected officials accountable.

We have a long history of successfully organizing for social change. All of the movements of the past began from people fighting to seek concrete improvements in their lives, and eventually through mass actions altering the relations of power.

As Steve Max, who has been teaching grassroots organizing skills for the Midwest Academy for forty years, says: "Social change comes from making the personal political. Central to this is understanding people's self-interest, which should not be interpreted in the narrowest sense of material benefits. Self-interest is a much broader concept. People will not only fight to end their own oppression; they can be motivated across generational lines to help their children and grandchildren. Self-interest is what makes people feel good, connected in the sense of being active in the community, being useful, and doing what is morally right. More broadly still, many people can be inspired to take on the responsibilities of citizenship and to play a role in shaping public affairs because they enjoy working for the common good."

Max adds: "Most elected public officials are haunted by the fear that they will be defeated by angry constituents if they serve corporate interests rather than the public interest. Our job is to organize enough of their constituents to hold their feet to the fire. To do so, we need to use a systematic approach to organizing, building and using political power and creating lasting institutions that can achieve and defend progressive wins and that are avenues for citizen participation in public life."

Many people say this is impossible, because they have been too discouraged by the influence of money in politics—a system of legalized bribery—to see a road map for the future. We must get beyond this defeatist attitude if we want to reclaim our country from the selfish forces that have not only ruined our food system but also damaged our democracy.

Many exciting initiatives are bubbling around the country, proving that "we the people" can prevail. For instance, we can all take heart from the growing movement to challenge the undue influence of money on elections and public policy. Momentum is building for passage of a twenty-eighth constitutional amendment to overturn the latest assault on fair elections, the U.S. Supreme Court's 2010 ruling in *Citizens United v. Federal Election Commission*. This abysmal court decision opened the floodgates to unlimited corporate spending on elections by giving corporations the same free speech protections as individuals. Dozens of groups and thousands of citizens are participating in the Democracy Is for People campaign. Because so many people are discouraged by the large amounts of dollars spent on lobbying and campaign contributions, passing this amendment could provide a boost to organizing on a range of issues, including food.

Robert Weissman, president of the consumer group Public Citizen, one of the lead organizations promoting the adoption of a twenty-eighth amendment, says the choice is simple: "Accept the further debasement of our democracy, the hijacking of government by giant corporations; or take action to remove the corporate stranglehold tightened by [the *Citizens United* ruling]. We need a constitutional amendment to overturn Citizens United—the pernicious decision holding that corporations have a constitutional right to spend as much as they choose to influence elections—end corporate spending in elections, and clear the way for adoption of a system of public financing for public elections." [11]

Food activists, who are involved at the community level, can play an important role in this battle over the future of our democracy. Three quarters of the states must vote in favor of ratifying the proposed amendment. Achieving this will necessitate statewide campaigns that are made up of many coordinated local organizing initiatives. If successful, this effort can infuse energy into and unite a number of different movements under a common banner. Success on this campaign could help build the broad-based coalition that is needed to initiate many reforms, including those needed to fix the broken food system and to strengthen our democracy.

The changes made in our society are only possible when people take political action and organize at the grassroots level. Change does not come from others—it comes from each of us being willing to work in our own community to mobilize people to action and to become part of a larger statewide, and eventually national, movement for change. To do so, we must articulate a far-reaching vision and the path by which to attain it.

NOTES

The following abbreviations are used in the notes:

DOJ   U.S. Department of Justice
EPA   U.S. Environmental Protection Agency
FDA   U.S. Food and Drug Administration
FTC   Federal Trade Commission
UFCW   United Food and Commercial Workers International Union
USDA   U.S. Department of Agriculture
USDA ERS U.S. Department of Agriculture Economic Research Service

Introduction

1. Sarah Low and Stephen Vogel, "Direct and Intermediated Marketing of Local Foods in the United States," U.S. Department of Agriculture, Economic Research Report No. 128, November 2011, iii.

2. Ibid.

3. Ibid.

4. Timothy A. Wise, "Still Waiting for the Farm Boom: Family Farmers Worse Off Despite High Prices," Policy Brief No. 11–01, Global Development and Environment Institute, Tufts University, March 2011.

5. Fred Kirschenmann et al., "Why Worry About the Agriculture of the Middle? A White Paper for the Agriculture of the Middle Project," available at www.agofthe middle.org.

1. Get Those Boys Off the Farm!

1. Foster Kamer, "Hedge Farm! The Doomsday Food Price Scenario Turning Hedgies into Survivalists," *New York Observer*, May 18, 2011.

2. Timothy A. Wise, "Still Waiting for the Farm Boom: Family Farmers Worse Off Despite High Prices," Policy Brief No. 11–01, Global Development and Environment Institute, Tufts University, March 2011.

3. Doug Smith and Richard Fausset, "Rural America Gets Even More Sparsely Populated," *Los Angeles Times*, December 15, 2010.

4. I am indebted to the late Al Krebs, author of the monumental history of agriculture *The Corporate Reapers: The Book of Agribusiness* (Washington, DC: Essential Books, 1992). Much of the history of the farm movement in this chapter is drawn from this detailed work.

5. Committee for Economic Development, "An Adaptive Program for Agriculture: A Statement on National Policy by the Research and Policy Committee," New York, 1962.

6. Ibid.

7. Krebs, *Corporate Reapers*.

8. Peter Carstensen, "The Packers and Stockyards Act: A History of Failure to Date," *Competition Policy International*, April 29, 2010.

9. Jerry Markham, "Futures Trading," in *A Financial History of the United States* (Armonk, NY: M.E. Sharpe, 2001), 101.

10. Committee for Economic Development, "International Trade, Foreign Investment and Domestic Employment," New York, 1945, cited in Committee for Economic Development, "Trade Negotiations for a Better Free World Economy; A Statement on National Policy by the Research and Policy Committee," New York, 1964, 15.

11. "What Post-War Policies for Agriculture?" report of the USDA, Interbureau and Regional Committees on Post-War Programs, *The Farmer and the War*, no. 7 (Washington, DC: U.S. GPO, 1944).

12. Mark Ritchie and Kevin Ristau, "U.S. Farm Policy," *World Policy Journal* 4, no. 1 (1986/87): 113–34.

13. John H. Davis, "From Agriculture to Agribusiness," *Harvard Business Review*, January/February 1956, 107–15.

14. Al Krebs, "Merle Hansen, Fearless Populist Hero," *Progressive Populist* 11, no. 1 (January 2005), www.populist.com/05.01.krebs.html.

15. Interview with George Naylor, *In Motion*, June 22, 2011, www.inmotion magazine.com/ra11/g_naylor_int11.html.

16. Krebs, "Merle Hansen."

17. John Hansen, "Where Is the Outrage?" www.familyfarmer.org/sections/out rage.html.

18. E-mail exchange with author, April 11, 2012.

19. Dan Carney, "Dwayne's World," *Mother Jones*, July/August 1995.

20. Associated Press, "Senate Farm Bill Means Big Win for Big Business," *Charleston Gazette*, February 18, 1996, 5B.

21. "Big Agribusiness Enjoyed Benefits in Senate Farm Bill," *Lincoln Journal Star*, February 19, 1996.

22. Daryll Ray, "Review of US Agricultural Policy in Advance of the 2012 Farm Bill: Written Statement Extends Oral Testimony of Daryll E. Ray before the United States House of Representatives Full Agriculture Committee," May 13, 2010, www .agpolicy.org/present/2010/RayTestimonyMay2010.pdf.

23. Timothy A. Wise, "Identifying the Real Winners from U.S. Agricultural Policies," Global Development and Environment Institute Working Paper No. 05–07, Tufts University, December 2005.

24. Ibid.

25. Brad Wilson, "Rebuttal to US PIRG's Video: 'Stop Subsidizing Obesity,'" September 21, 2011, available at www.zcommunications.org/farm-subsidies-rebuttal-to -us-pirg-video-stop-subsidizing-obesity-by-brad-wilson.

26. Ibid.

27. Timothy A. Wise, "Understanding the Farm Problem: Six Common Errors in Presenting Farm Statistics," Global Development and Environment Institute Working Paper No. 05-02, Tufts University, March 2005.

28. Wise, "Still Waiting for the Farm Boom."

29. Ibid.

2. The Junk Food Pushers

1. Brody Mullins, "Koch's K Street Orbit," *Roll Call*, November 15, 2004.

2. Ibid.

3. "FTC Review (1977–84)," report prepared by a member of the FTC for the use of the Subcommittee on Oversight and Investigations of the Committee on Energy and Commerce, U.S. House of Representatives, September 1984.

4. "Written Testimony of Pamela G. Bailey, President and CEO, Grocery Manufacturers Association," Hearing on Food Safety Enhancement Act of 2009 Discussion Draft, before the Subcommittee on Health, Committee on Energy and Commerce, U.S. House of Representatives, 111th Cong., 1st sess., June 3, 2009, 1–6.

5. David Orgel, "A FMI, GMA Plan to Align Agendas, Meetings," *Supermarket News*, January 19, 2009, supermarketnews.com/retail-amp-financial/fmi-gma-plan -align-agendas-meetings#ixzz1yd21yxzJ.

6. IDEO, "Safeway Supply Chain Innovation for Kraft Foods," www.ideo.com/ work/safeway-supply-chain-innovation-for-kraft-work/featured/kraft.

7. Marion Nestle, " 'Singing Kumbaya,' GMA/FMI Displays Preemptive Label Design," *Food Politics*, January 25, 2011, www.foodpolitics.com/2011/01/singing -kumbaya-gmafmi-displays-preemptive-label-design/.

8. Ibid.

9. International Food Information Council Foundation, *2008 Food & Health Survey: Consumer Attitudes Toward Food, Nutrition & Health* (Washington, DC: IFIC Foundation, 2008), www.foodinsight.org/Resources/Detail.aspx?topic=2008_Food_ Health_Survey_Consumer_Attitudes_toward_Food_Nutrition_Health.

10. Rosa DeLauro, "DeLauro Calls for Stronger Food Labeling Regulations," press release, January 24, 2011, delauro.house.gov/index.php?option=com_content&view =article&id=848:delauro-calls-for-stronger-food-labeling-regulations&catid=7:2011 -press-releases&Itemid=23.

11. "Nutritional Info Moving to the Front of Packages," *Mass Market Retailers*, November 15, 2010, business.highbeam.com/4856/article-1G1-243278315/nutritional -info-moving-front-packages.

12. Media Awareness Network, "Marketing and Consumerism: Special Issues for Young Children," www.media-awareness.ca/english/parents/marketing/issues_kids_ marketing.cfm.

13. "Government Pulls Back on Junk Food Marketing Proposal," ABC News, October 12, 2011, available at www.wjla.com/articles/2011/10/government-pulls-back -on-junk-food-marketing-proposal-67754.html.

14. Ibid.

15. Donald Cohen, "Junk Food Companies Say Eating More Fruits and Vegetables Is a 'Job Killer,'" *HuffPost Healthy Living*, November 22, 2011.

16. Reuters, "Exclusive: Soda Makers Escalate Attacks over Obesity," July 20, 2011.

17. Tanzina Vega, "Complaint Accuses Pepsi of Deceptive Marketing," *New York Times*, October 19, 2011.

18. "The History of Pepsi-Cola," Soda Museum, web.archive.org/web/20080212 003647/www.sodamuseum.bigstep.com/generic.jhtml?pid=3.

19. "PepsiCo Lineup to Look Healthier in 10 Years: CEO," *NewsDaily*, October 17, 2011, www.newsdaily.com/stories/tre79g4zo-us-pepsico-ceo/.

20. Interview with Indra Nooyi, Fox Business *American Icon* series, February 7, 2011, embedded in Landon Hall, "CEO: Pepsi, Doritos 'Are Not Bad for You,'" *Orange County Register*, February 7, 2011, healthyliving.ocregister.com/2011/02/07/ceo-pepsi -doritos-are-not-bad-for-you/29022/.

21. "Food Processing's Top 100," 2011 rankings, FoodProcessing.com, www.food processing.com/top100/index.html.

22. Reuters, "UPDATE 6—Nestle Buys 60 Pct of Chinese Candymaker for $1.7 Bln," July 11, 2011.

23. Catherine Ferrier, "Bottled Water: Understanding a Social Phenomenon," report commissioned by the World Wildlife Fund, April 2001, 19.

24. Olga Naidenko et al., "Bottled Water Quality Investigation: 10 Major Brands, 38 Pollutants," Environmental Working Group, Washington, DC, October 2008, www .ewg.org/reports/bottledwater.

25. Ivan Penn, "The Profits on Water Are Huge, but the Raw material Is Free," *Tampa Bay Times*, March 16, 2008, www.tampabay.com/news/environment/water/ article418793.ece.

26. Elizabeth Whitman, "Bottled Water Companies Target Minorities, but So Do Soda Firms," *Madison Times*, November 23, 2011, www.themadisontimes.com/ news_details.php?news_id=1550.

27. Nestle Research, "Popularly Positioned Products: Affordable and Nutritious," www.research.nestle.com/asset-libraries/Documents/Popularly%20Positioned%20 Products.pdf.

28. Ibid.

29. "Laos: NGOs Flay Nestle's Infant Formula Strategy," *IRIN Global*, June 23, 2011, www.irinnews.org/Report/93040/LAOS-NGOs-flay-Nestlé-s-infant-formula -strategy.

30. Ibid.

31. "Davos Open Forum 2006—Water: Property or Human Right?" YouTube video, www.youtube.com/watch?v=WmFpYCtiRsw.

32. "Rainer E. Gut: Profile," *Forbes*, people.forbes.com/profile/rainer-e-gut/24010 (accessed November 19, 2011).

33. "Peter Brabeck-Letmathe, 1944–," Reference for Business, www.referencefor business.com/biography/A-E/Brabeck-Letmathe-Peter-1944.html.

34. Paul Glader, "Nestle Chairman Skeptical of Growth in Organic Food Mar- ket," *Co.Exist*, www.fastcoexist.com/1678437/nestle-chairman-skeptical-of-growth-in -organic-food-market (accessed November 19, 2011).

35. Kraft Foods, "Make Today Delicious. 2010 Fact Sheet," www.kraftfoods company.com/assets/pdf/kraft_foods_fact_sheet.pdf (accessed November 20, 2011).

36. Andrew Ross Sorkin and Chris V. Nicholson, "Kraft to Split Its Snacks and Grocery Businesses," *DealBook*, *New York Times*, August 4, 2011, dealbook.nytimes .com/2011/08/04/kraft-to-split-two-separating-snacks-and-grocery-businesses.

3. Walmarting the Food Chain

1. Barry C. Lynn and Phillip Longman, "Who Broke America's Jobs Machine?" *Washington Monthly*, March 4, 2010.

2. Phone interview with Michael Pertschuk, November 29, 2011.

3. "FTC Review (1977–84)," report prepared by a member of the FTC for the use of the Subcommittee on Oversight and Investigations of the Committee on Energy and Commerce, U.S. House of Representatives, September 1984.

4. Phone interview with Michael Pertschuk.

5. Ibid.

6. Ibid

7. "FTC Review (1977–84)."

8. Lynn and Longman, "Who Broke America's Jobs Machine?"

9. Barry C. Lynn, *Cornered: The New Monopoly Capitalism and the Economics of Destruction* (Hoboken, NJ: John Wiley & Sons, 2010).

10. Barry C. Lynn, "Breaking the Chain: The Antitrust Case Against Wal-Mart," *Harper's*, July 2006, www.harpers.org/archive/2006/07/0081115.

11. Phone interview with Bruce Von Stein, December 1, 2011.

12. Michelle Christian and Gary Gereffi, "The Marketing and Distribution of Fast Food," in *Contemporary Endocrinology: Pediatric Obesity: Etiology, Pathogenesis and Treatment*, ed. M. Freemark (New York: Springer Publishing, 2010), 439–50.

13. Ibid.

14. Wal-Mart Stores, Inc., "The Nation's First Wal-Mart Supercenter Receives Modern Design Conversion," press release, January 12, 2007; Steve W. Martinez, "The U.S. Food Marketing System: Recent Developments 1997–2006," USDA ERS Report No. 42, May 2007, 6.

15. SEC 10-K Filing, Wal-Mart Stores, Inc., 2010, 6, 10.

16. Stephanie Clifford, "Wal-Mart Tests Service for Buying Food Online," *New York Times*, April 24, 2011.

17. Martinez, "U.S. Food Marketing System," 6; Elena G. Irwin and Jill Clark, "The Local Costs and Benefits of Wal-Mart," Department of Agricultural, Environmental, and Development Economics, Ohio State University, February 23, 2006, 4.

18. Martinez, "U.S. Food Marketing System," 6–7; Christopher Leonard, "Wal-Mart Tightens Distribution to Increase Stock," *Arkansas Democrat-Gazette*, June 19, 2005.

19. Irwin and Clark, "Local Costs and Benefits of Wal-Mart," 4–5.

20. Anthony Bianco and Wendy Zellner, "Is Wal-Mart Too Powerful?" *Businessweek*, October 6, 2003.

21. Emily Schmitt, "The Profits and Perils of Supplying to Wal-Mart," *Businessweek*, July 14, 2009.

22. Ibid.; Constance L. Hays, "Big Stakes in Small Errors; Manufacturers Fight Retailer 'Discounts' in Shipping Disputes," *New York Times*, August 17, 2001.

23. Hays, "Big Stakes in Small Errors."

24. Anne D'Innocenzio, "Inventory or Invasion?; Some Fear Wal-Mart Smart Tag Plan Infringes on Privacy," *Grand Rapids Press*, July 26, 2010.

25. Ibid.; Eric Mortenson, "High-Tech Changes Face of Farmworkers," *The Oregonian*, March 5, 2008.

26. Barnaby J. Feder, "Wal-Mart Plan Could Cost Suppliers Millions," *New York Times*, November 10, 2003.

27. Steve Painter, "ID-Tag Inventory Study Finds Accuracy Up 13%; Wal-Mart Can Save Millions, Expert Says," *Arkansas Democrat-Gazette*, March 14, 2008.

28. Martinez, "U.S. Food Marketing System," 8, 9, Table 4.

29. Lynn, "Breaking the Chain."

30. Anna Lappé, "Why We Should Question Walmart's Latest PR Blitz," *Civil Eats*, January 21, 2011, civileats.com/2011/01/21/why-we-should-question-walmart%E2%80%99s-latest-pr-blitz/.

31. Stacey Roberts, "Wal-Mart Shows Its Energy-Saving Side," *Arkansas Democrat-Gazette*, August 25, 2006.

32. "Wal-Mart Says Environmental Initiatives About Money, Not Brand Image," *Environmental Leader*, April 13, 2011, www.environmentalleader.com/2010/04/13/wal-mart-says-environmental-initiatives-about-money-not-brand-image/.

33. Stefania Vitali et al., "The Network of Global Corporate Control," Swiss Federal Institute of Technology, September 19, 2011.

34. "U.S. Corporate Bond Market: 2011 Rating and Issuance Activity," Fitch Ratings, February 1, 2012.

35. Food Institute, "Merger & Acquisition Activity Is Healthy and Growing," press release, March 1, 2012.

36. Dave Fusaro, "Merger and Acquisition Activity Returns to Pre-2008 Levels," Foodprocessing.com, August 31, 2010.

37. Pauline Renaud, "Challenges and Opportunities in the Food & Drink Sector," *Financier Worldwide*, March 2010.

38. David Welch, "Kellogg to Buy Procter & Gamble's Pringles for $2.7 Billion," Bloomberg, February 15, 2012.

39. Lynn, "Breaking the Chain."

4. The Green Giant Doesn't Live in California Anymore

1. Steven Stoll, *The Fruits of Natural Advantage: Making the Industrial Countryside in California* (Berkeley: University of California Press, 1998).

2. Ibid.

3. There are various types of mandated marketing programs, depending on the crop and the jurisdiction that oversees the program—usually the federal government or state. In general, these programs are initiated by trade associations or commodity

Collaboration," press release, August 27, 2003, available at www.thefreelibrary.\
Industry+Leaders+Hain+Celestial+Group+and+Cargill+Health+%26+Foo
-a0131665463.

7. Dawn Withers, "Private Label Growing," *The Packer*, March 9, 2010.

8. Beth Kowitt, "Inside the Secret World of Trader Joe's," *Fortune*, August 23, 2010.

9. John Mackey, "The Whole Foods Alternative to Obama Care," op-ed, *Wall Street Journal*, August 11, 2009.

10. Elliot Ziebach, "Analysts Discuss Consolidation, Whole Foods," *Supermarket News*, September 19, 2011.

11. Ronnie Cummings, "The Organic Monopoly and the Myth of 'Natural' Foods: How Industry Giants Are Undermining the Organic Movement," Common Dreams, July 9, 2009, www.commondreams.org/view/2009/07/09.

12. "Top 10 Industry Deals in 2010," *Nutrition Business Journal*, January 1, 2011.

13. Kate Murphy, "Investing; Health Food Seller Is Back to Health," *New York Times*, December 31, 2000.

14. Jane Hoback, "The Future of Distribution," *Natural Food Merchandiser*, October 17, 2010.

15. Amanda Loudin, "The Right Recipe," *Food Logistics*, April 2010, www.envista corp.com/envista_case_studies/envista_case_study_NaturesBest.pdf.

16. Hoback, "Future of Distribution."

17. Nancy Luna, "Sprouts Merges with Sunflower Markets," *Orange County Register*, March 9, 2012.

18. "Sprouts, Sunflower Agree to Merge," *Supermarket News*, March 9, 2012.

19. "Private Equity Firm Scoops Up 80% Stake in Organic, Natural Food Chain or $300 Mil," *Sustainable Food News*, April 12, 2012, sustainablefoodnews.com/print tory.php?news_id=15857.

20. Ibid.

21. Melanie Warner, "Wal-Mart Eyes Organic Foods," *New York Times*, May 12, 006.

22. "Mark Retzloff from Aurora Organic Dairy," Organic Guide, www.organic ide.com/organic/people/mark-retzloff-from-aurora-organic-dairy (accessed September 14, 2011).

6. Poisoning People

1. Daniel Puzo, "Espy Calls for Overhaul of Meat Inspection System: Food: Agri- lture Secretary Tells Senators Probing Bacteria Illnesses That Contamination Can Reduced, but Not Totally Prevented," *Los Angeles Times*, February 6, 1993, articles imes.com/1993-02-06/news/mn-1076_1_meat-inspection-system.

2. Phillip Spiller, "FDA Seafood HACCP Inspection Presentation," text based on ual oral presentation, 1, nsgl.gso.uri.edu/flsgp/flsgpw93002/flsgpw93002_part1 f (accessed November 8, 2011).

3. Donald C. Smaltz, "Final Report of the Independent Counsel Re: Alphonso hael (Mike Espy)," Office of Independent Council, October 25, 2001.

groups that represent growers of a certain type of produce but are regu
government entity. They assess a fee from growers in order to develop a
campaigns that encourage consumers to buy the product. These program
criticized by smaller growers, who are compelled to participate but feel that
tising campaigns are costly and do not benefit them.

4. "Statement of C. Manly Molpus, President and CEO, Grocery Ma
of America," Agricultural Trade Negotiations, Hearings Before the Comm
riculture, U.S. House of Representatives, 108th Cong., 2nd sess., April 28 a
2004, 322–23; Robert Aldrich, "Statement of Grocery Manufacturers A
Implementation of the United States–Peru Trade Promotion Agreement,
fore the Committee on Ways and Means, U.S. House of Representatives,]
2nd sess., July 12, 2006.

5. Elliot Zwiebach, "Analysts Discuss Consolidation, Whole Foods
ket News, September 19, 2011, supermarketnews.com/retail-amp-finar
-discuss-consolidation-whole-foods.

6. Ibid.

7. Wal-Mart Stores, Inc., "Wal-Mart Releases 2012 Annual Sharehc
Materials," press release, April 16, 2012.

8. Wal-Mart Stores, Inc., "Walmart Launches Major Initiative t(
Healthier and Healthier Food More Affordable," press release, January ?

9. Wal-Mart Stores, Inc., "Illinois' Frey Farms Teams with Wal-Mar
America's Farmers," press release, September 27, 2006.

10. Frey Fruits Web site, www.freyproduce.com/profile.htm (acces
2012).

11. Julia Hanna, "HBS Cases: Negotiating with Walmart," Har
School Working Knowledge: Lessons from the Classroom, April 28, 2(
.edu/pdf/item/5903.pdf.

5. Organic Food: The Paradox

1. This chapter draws upon the excellent work of Dr. Philip H.
solidation in the North American Organic Food Processing Sector
International Journal of Sociology of Agriculture & Food 16, no. 1 (2009
Daniel Jaffe and Philip H. Howard, "Corporate Cooptation of Organ
Standards," Agriculture and Human Values 27 (2010): 387–99.

2. Diane Brady, "The Organic Myth," Bloomberg Businessweek, (

3. Howard, "Consolidation in the North American Organic
Sector."

4. Catherine Greene and Carolyn Dimitri, et al., "Emerging Issu
ganics Industry," Economic Research Service, Economic Informatio
June 2009, www.ers.usda.gov/publications/eib55/eib55.pdf.

5. Eden Organic, "Why Eden Foods Chooses Not to Use the
.edenfoods.com/articles/view.php?articles_id=78 (accessed Decem

6. Hain Celestial Group and Cargill Health & Food Techr
Leaders Hain Celestial Group and Cargill Health & Food Techr

4. USDA, "Executive Summary: Study of the Federal Meat and Poultry Inspection System," Vol. III, Washington, DC, July 1977.

5. Ibid.

6. A.M. Pearson and T.R. Dutson, eds., *HACCP in Meat, Poultry and Fish Processing* (New York: Chapman & Hall, 1995), 368, 369.

7. "Statement of Dan Glickman, Secretary of Agriculture," *USDA News* 55, no. 7 (August 1996).

8. USDA, "Glickman Announces Head of Food Safety Agency," press release, October 17, 1995, www.usda.gov/news/releases/1996/10/0570.

9. "Inspectors' Union Files Suit Against USDA Over Carcass by Carcass Rules," The-Inspector.com, April 13, 1998, www.the-inspector.com/unionsues.htm.

10. *AFL-CIO, et. al., v. U.S. Department of Agriculture, et. al.*, Appeal from the U.S. District Court for the District of Columbia, June 30, 2000.

11. Humane Society of the United States, "Rampant Animal Cruelty at California Slaughter Plant," press release, January 30, 2008, www.humanesociety.org/news/news/2008/01/undercover_investigation_013008.html.

12. Wayne Pacelle, "Case Finally Closed on 'Downers' Loophole," *A Humane Nation*, March 14, 2009, hsus.typepad.com/wayne/2009/03/obama-downers.html.

7. Animals on Drugs

1. Gardiner Harris, "New Official Named with Portfolio to Unite Agencies and Improve Food Safety," *New York Times*, September 13, 2010.

2. Elizabeth Weise and Julie Schmit, "Spinach Recall: 5 Faces. 5 Agonizing Deaths. 1 Year Later," *USA Today*, September 24, 2007.

3. Marion Nestle, "What's Up with the Hydrolyzed Vegetable Protein Recall?" *Food Politics*, March 10, 2010, www.foodpolitics.com/2010/03/whats-up-with-the-hydrolyzed-vegetable-protein-recall.

4. Tom Coburn, "Opposing View on Food Safety: Leverage the Free Market," *USA Today*, November 22, 2010.

5. François-Xavier Weill et al., "International Spread of an Epidemic Population of *Salmonella enterica* Serotype Kentucky ST198 Resistant to Ciprofloxacin," *Journal of Infectious Diseases*, August 2, 2011.

6. Phone conversation with Dr. Robert Lawrence, September 19, 2011.

7. Ralph Loglisci, "New FDA Numbers Reveal Food Animals Consume Lion's Share of Antibiotics," Center for a Livable Future blog, December 23, 2010, www.livablefutureblog.com/2010/12/new-fda-numbers-reveal-food-animals-consume-lion's-share-of-antibiotics.

8. Accelr8 Technology Corporation, "Antibiotic Resistance," www.accelr8.com/antibiotic_resistance.php (accessed September 16, 2011).

9. David Zinczenko and Matt Goulding, "Five Nastiest Things in Your Grocery Store," *Men's Health*, April 11, 2012.

10. Phone conversation with Dr. Robert Lawrence.

11. David Wallinga, "Antibiotics, Animal Agriculture and MRSA: A New Threat," Institute for Agriculture and Trade Policy, November 2009.

12. T. Khanna, R. Friendship, C. Dewey, and J.S. Weese, "Methicillin Resistant Staphylococcus Aureus Colonization in Pigs and Pig Farmers," Department of Population Medicine, Ontario Veterinary College, University of Guelph, Guelph, ON.

13. U.S. Government Accountability Office, "Antibiotic Resistance: Agencies Have Made Limited Progress Addressing Antibiotic Use in Animals," September 7, 2011, 66.

14. STOP Foodborne Illness et al., "Letter to Tom Vilsack, Secretary of U.S. Department of Agriculture," October 18, 2011, www.stopfoodborneillness.org/sites/default/files/Antibiotic%20GAO%20report%20USDA%20Group%20Sign%20On%20FINAL.pdf.

15. "Statement of Tom Vilsack, Secretary of U.S. Department of Agriculture," Future of Food Conference, Georgetown University, May 2011.

16. U.S. Government Accountability Office, "Antibiotic Resistance," Summary.

17. Konstantinos Markis et al., "Fate of Arsenic in Swine Waste from Concentrated Animal Feeding Operations," *Journal of Environmental Quality* 37, no. 4 (2008): 1626–33.

18. USDA Food Safety and Inspection Service, "National Residue Program Red Book Reports, 2000–2008"; USDA National Agricultural Statistics Service, "Poultry—Production and Value. 2009 Summary," Pou 3-1(10), April 2010; H.D. Chapman and Z.B. Johnson, "Use of Antibiotics and Roxarsone in Broiler Chickens in the USA: Analysis for the Years 1995 to 2000," *Journal of Poultry Science* 81 (March 2002): 1.

19. USDA, "FSIS National Residue Program for Cattle," Office of Inspector General, Audit Report 24601-08-KC, March 2010, 1.

20. Ibid., 1–2.

21. Bette Hileman, "Arsenic in Chicken Production. A Common Feed Additive Adds Arsenic to Human Food and Endangers Water Supplies," *Chemical and Engineering News* 85, no. 15 (April 9, 2007); Jennifer Hlad, "Poultry Farmers Resist Ban on Arsenic in Feed," *Daily Record* (Baltimore, MD), March 16, 2010.

22. Steve Schwalb, "Statement of Perdue Farms, Incorporated," Subcommittee on Conservation, Credit, Energy, and Research, Committee on Agriculture, U.S. House of Representatives, December 9, 2009, 6.

23. Hlad, "Poultry Farmers Resist Ban."

24. Chapman and Johnson, "Use of Antibiotics and Roxarsone," 1; Dave Love, "CLF Provides House Testimony on Maryland Bill 953 to Ban Arsenic from Poultry Feed," Johns Hopkins University Center for a Livable Future, March 9, 2010; Keeve Nachman, personal communication, August 10, 2010.

25. Nicholas D. Kristof, "Arsenic in Our Chicken," *New York Times*, op-ed, April 4, 2012, www.nytimes.com/2012/04/05/opinion/kristof-arsenic-in-our-chicken.html.

26. EPA, "National Primary Drinking Water Regulations; Arsenic and Clarifications to Compliance and New Source Contaminants Monitoring," Federal Register Environmental Documents, January 22, 2001; EPA, "Fact Sheet: Drinking Water Standard for Arsenic," January 2001, water.epa.gov/lawsregs/rulesregs/sdwa/arsenic/regulations_factsheet.cfm.

27. Hileman, "Arsenic in Chicken Production."

28. Don Hopey, "Chicken Feed May Present Arsenic Danger," *Pittsburgh Post-Gazette*, March 8, 2007.

29. Ellen Silbergeld and Keeve Nachman, "The Environmental and Public Health Risks Associated with Arsenical Use in Animal Feeds," *Annals of the New York Academy of Sciences* 1140 (2008): 349; FDA, "Listing of Food Additive Status Part I."

30. Zhejiang Rongyao Chemical Co. Ltd, "Chinese Manufacturer Rongyao Alleges that Pfizer Breaches Agreement by Voluntarily Removing Poultry Feed Additive '3-Nitro' from the Market," press release, October 13, 2011, Business Wire, www.businesswire.com/news/home/20111013006053/en/Chinese-Manufacturer-Rongyao-Alleges-Pfizer-Breaches-Agreement.

8. Cowboys Versus Meatpackers: The Last Roundup

1. Eric Schlosser, *Fast Food Nation: The Dark Side of the All-American Meal* (New York: HarperPerennial, 2002), 134.

2. Shane Ellis, "State of the Beef Industry 2008," *Beef Magazine*, 2008, 9.

3. Cancer Coalition, "Hormones in U.S. Beef Linked to Increased Cancer Risk," press release, World Wire, October 21, 2009.

4. James M. MacDonald and William D. McBride, "The Transformation of American Livestock Agriculture: Scale, Efficiency and Risks," USDA ERS, *Economic Information Bulletin* No. 43, January 2009, 12.

5. Ellis, "State of the Beef Industry," 11.

6. Williams Robbins, "A Meatpacker Cartel Up Ahead?" *New York Times*, May 29, 1988, 4.

7. Dale Kasler, "IPB Keeps Tight Grip on Market," *Des Moines Register*, September 24, 1988.

8. Christopher Davis and Lin Biing-Hwan Lin, "Factors Affecting U.S. Beef Consumption," Electronic Outlook Report, USDA ERS, 2005.

9. Marcy Lowe and Gary Gereffi, "A Value Chain Analysis of the U.S. Beef and Dairy Industries," Duke University Center on Globalization, Governance and Competitiveness, February 2, 2009.

10. Mike King, "JBS SA (JBSS3)—Financial and Strategic SWOT Analysis Review—A New Company Profile Report on Companiesandmarkets.com," Companiesandmarkets.com, May 28, 2012, reports.pr-inside.com/jbs-sa-jbss3-financial-and-r3201276.htm.

11. David Lewington, National Farmers Union, Ontario, Board Member, "Letter to Melanie Aitken, Commissioner of Competition at the Canadian Competition Bureau," February 19, 2009; Competition Bureau Canada, "Competition Bureau Announces Results of XL-Lakeside Merger Review," press release, February 27, 2009; Kevin Hursh, "Natural Advantage for Beef Production Squandered," *Saskatoon Star Phoenix*, March 4, 2009.

12. Cargill, "Cargill Beef Argentina," www.cargill.com/company/businesses/cargill-beef-argentina/index.jsp (accessed June 26, 2009, and on file at Food & Water Watch); Cargill, "Cargill Beef Australia," www.cargill.com.au/australia/en/home/products/beef/index.jsp (accessed June 26, 2009, and on file at Food & Water Watch).

13. Tyson Foods, Inc., Investor Relations Department, "Tyson Fact Book," 2010.

14. Ibid.

15. Wes Ishmael, "Beef Sits Down with Westley Batista," *Beef Magazine*, September 10, 2010.

16. Hoovers, JBS S.A Profile, September 15, 2011, www.hoovers.com.

17. Juan Forero, "An Industry Giant from Farm to Fork," *Washington Post*, April 15, 2011, A10.

18. Robert Pore, "JBS Swift Buys Another Huge Cattle Feeding Operation," *Aglines*, July 2, 2010, www.aglines.com/2010/07/jbs-swift-buys-another-huge-cattle-feeding-operation/.

19. Skema Business School, "JBS Company Profile," March 31, 2011, skem-a.blogspot.com/2011/03/jbs-company-profile_31.html.

20. Kate Celender, "The Impact of Feedlot Waste on Water Pollution Under the National Pollutant Discharge Elimination System (NPDES)," *William & Mary Environmental Law and Policy Review* 33, no. 3 (2009).

21. Confidential e-mail to Patrick Woodall, director of Research, Food & Water Watch, from a congressional aide, dated May 2, 2011.

22. John Howell, "Butler Battling Giants of Meat Packing Industry," *The Panolian*, September 17, 2011.

23. Ranchers-Cattlemen Action Legal Fund, "Meatpacker Apologists Engage in Deceptive Smear Campaign Against USDA Official," November 5, 2010, www.r-calfusa.com/news_releases/2010/101105-meatpacker.htm.

24. Clement E. Ward, "Feedlot and Packer Pricing Behavior: Implications for Competition Reserarch," paper presented at Western Agricultural Economics Association annual meeting, Portland, OR, July 29–August 1, 2007, 1.

25. David Domina and C. Robert Taylor, "The Debilitating Effects of Concentration in Markets Affecting Agriculture," Organization for Competitive Markets, September 2009, 46.

26. Allan Sents, testimony, Agriculture and Antitrust Enforcement Issues in Our 21st Century Economy: Livestock Industry, DOJ and USDA Public Workshops Exploring Competition Issues in Agriculture, Colorado State University, Fort Collins, CO, August 27, 2010, 81, www.justice.gov/atr/public/workshops/ag2010/colorado-agworkshop-transcript.pdf.

27. "WFU Praises Senate Bill on Competitive Livestock Markets," Wisconsin Ag Connection, September 12, 2011, www.wisconsinagconnection.com/story-state.php?Id=1094&yr=2011.

28. Ranchers-Cattlemen Action Legal Fund, "Proposed GIPSA Rule Provides a Rock-Solid Foundation for Correcting Severe Marketing Problems in the U.S. Fed Cattle Market," R-CALF USA Briefing Paper, July 26, 2010.

29. Howell, "Butler Battling Giants."

9. Hogging the Profits

1. Jeff Tietz, "Pork's Dirty Secret: The Nation's Top Hog Producer Is Also One of America's Worst Polluters," *Rolling Stone*, December 2006.

2. Nigel Key and William McBride, "The Changing Economics of U.S. Hog Production," U.S. Department of Agriculture (USDA) Economic Research Service (ERS), Economic Research Report No. 52, December 2007, 5.

3. Steve W. Martinez, "The U.S. Food Marketing System: Recent Developments 1997–2006," USDA ERS Economic Research Report No. 42, May 2007, 27.

4. Research Triangle Institute International "GIPSA Livestock and Meat Marketing Study: Volume 1: Executive Summary and Overview—Final Report," prepared for Grain Inspection, Packers and Stockyards Administration, January 2007, ES-10.

5. Land Stewardship Program, "Killing Competition with Captive Supply," April 23, 1999, www.landstewardshipproject.org/pr/newsr_042399.html.

6. James M. MacDonald and William D. McBride, "The Transformation of U.S. Livestock Agriculture: Scale, Efficiency, and Risks," USDA ERS, *Economic Information Bulletin* No. 43, January 2009, 25; Mary Hendrickson and Bill Heffernan, "Concentration of Agricultural Markets," Department of Rural Sociology, University of Missouri–Columbia, April 2007.

7. James M. MacDonald and Penni Korb, USDA ERS, "Agricultural Contracting Update, 2005," *Economic Information Bulletin* No. 35, April 2008, 17.

8. David Domina and C. Robert Taylor, "The Debilitating Effects of Concentration in Markets Affecting Agriculture," Organization for Competitive Markets, September 2009, 65.

9. "Testimony of Lynn A. Hayes, Farmers' Legal Action Group, Inc. (FLAG)," Hearing on Economic Challenges and Opportunities Facing American Agricultural Producers Today, before the U.S. Senate Committee on Agriculture, Nutrition, and Forestry, 110th Cong., 1st sess., April 18, 2007, 3.

10. David Moeller, "Livestock Production Contracts: Risks for Family Farmers," FLAG, March 22, 2003, 4.

11. "Testimony of Lynn A. Hayes," 3.

12. "Department of Agriculture, Grain Inspection, Packers and Stockyards Administration, 9 CFR Part 201 RIN 0580-AB07, Implementation of Regulations Required Under Title XI of the Food, Conservation and Energy Act of 2008; Conduct in Violation of the Act. AGENCY: Grain Inspection, Packers and Stockyards Administration, USDA. ACTION: Proposed rule," *Federal Register* 75, no. 119 (June 22, 2010).

13. Chris Petersen, testimony, Agriculture and Antitrust Enforcement Issues in Our 21st Century Economy: Livestock Industry, DOJ and USDA Public Workshops Exploring Competition Issues in Agriculture, Colorado State University, Fort Collins, CO, August 27, 2010, 66.

14. American Meat Institute, "Livestock Producers Urge President Obama to Stop GIPSA Rule," press release, August 16, 2011.

15. Informa Economic, "An Estimate of the Economic Impact of GIPSA's Proposed Rules," November 8, 2010, www.beefusa.org/uDocs/GIPSA-Executive-Summary.pdf.

16. North Carolina in the Global Economy, "Hog Farming," www.soc.duke.edu/NC_GlobalEconomy/hog/corporation.php.

17. Interview with Hugh Espey, July 15, 2011.

18. Tietz, "Pork's Dirty Secret."

19. Scott Kilman, "Restrictive Laws Dam Competition, Leave Acquisition as Best Path to Expansion," *Wall Street Journal*, August 31, 2001.

20. Smithfield Foods, Inc., "Directors & Management," www.smithfieldfoods.com/Investor/Officers.

21. Smithfield Foods, Inc., "History of Smithfield Foods," www.smithfieldfoods.com/Understand/History.

22. Office of the Iowa Attorney General, "Miller Sues Smithfield Foods to Block Acquisition of Murphy Farms in Iowa," press release, January 24, 2000.

23. Tom Miller, Iowa attorney general, "Concentration in Agriculture: Summary of Activities of the Iowa Attorney General's Office," March 12, 2010, www.state.ia.us/government/ag/working_for_farmers/farm_advisories/Ag%20Concentration%20IA%20AGO%20activity%203-12-10.pdf.

24. Office of Senator Chuck Grassley of Iowa, "Grassley Concerned About Smithfield-Premium Standard Farms Merger," press release, September 19, 2006.

25. Bob O'Brien, "Can Smithfield's Hogs Go Wild?" *Barron's*, April 11, 2012.

26. Rekha Basu, "Joe Fagan Is a Force for Activism," *Des Moines Register*, July 16, 2010.

27. Taylor Leake, "Activists Crash Iowa's Factory Farm Deregulation Hearings," Change.org, February 25, 2011, news.change.org/stories/activists-crash-iowa-s-factory-farm-deregulation-hearings.

28. Mike Augspurger, "ICCI Promotes 'Real' Family Farms," *Hawk Eye* (Burlington, IA), June 22, 2002.

29. USDA, "2007 Census of Agriculture," National Agricultural Statistics Service, 2009.

30. Cow/calf operations specialize in breeding heifers and raising calves for up to two years until they reach a weight of between four hundred and seven hundred pounds, at which time the animals go to a stock feeding operation.

30. USDA, "2007 Census of Agriculture."

31. Smithfield Foods, Inc., U.S. Securities and Exchange Commission Form 10-K, May 1, 2011, 29.

32. Ranchers-Cattlemen Action Legal Fund, "JBS Merger: Exhibit 18: Understanding the History of Smithfield," April 8, 2008, www.rcalfusa.com/industry_info/2008_JBS_merger/080409Exhibit18_HistoryofSmithfieldFoods.pdf.

33. Robbin Marks, "Cesspools of Shame: How Factory Farm Lagoons and Sprayfields Threaten Environmental and Public Health," Natural Resources Defense Council and the Clean Water Network, July 2001.

34. Tietz, "Pork's Dirty Secret."

35. Ibid.

36. Food & Water Watch analysis of USDA, NASS, "2007 Census of Agriculture."

37. S. Wing et al., "Environmental Injustice in North Carolina's Hog Industry," *Environmental Health Perspectives* 108 (2000): 225–31.

38. L. Sorg, "With Merger, the World's Number 1 Would Get Even Bigger," *North Carolina Independent Weekly*, April 4, 2007.

39. USDA, "2007 Census of Agriculture"; U.S. Census Bureau, "Annual Estimates of the Population for the United States, Regions, States and Puerto Rico," July 1, 2009.

40. USDA, "2007 Census of Agriculture."

41. Iowa Department of Natural Resources, "Manure Production Per Space of Capacity," Manure Management Plan Form, 2004, Appendix A, 2.

42. Bruce Henderson, "Hog-Waste Lawsuits Settled in Deal to Fight Water Pollution," *Charlotte Observer* (NC), January 21, 2006.

43. Estes Thompson, "Pollution Threatens Coastal Rivers," *The Herald-Sun* (Durham, NC), August 21, 1995.

44. Stuart Leavenworth, "Million Gallons of Hog Waste Spill in Jones County," *News and Observer* (Raleigh, NC), August 13, 1996.

45. Associated Press, "Report: Spill Caused by Dike Failure," April 30, 1999; James Eli Shiffer, "Waste Spill Probably Accidental, SBI Says," *News and Observer* (Raleigh, NC), April 22, 1999.

46. Greg Barnes, "North Carolina Scientists Divided over Impact of Hog Waste on the Environment," *Fayette Observer* (NC), December 13, 2003.

47. Bruce Henderson and Diane Suchetka, "Backhoes Bury Most of the Hogs Killed in Floods," *Charlotte Observer* (NC), October 1, 1999; Dennis Cauchon, "Farmers, Scientists Assess the Damage in N.C.," *USA Today*, September 27, 1999.

48. "Senate Enacts Ban on New Hog-Waste Lagoons," *News and Observer* (Raleigh, NC), April 19, 2007.

49. Neuse Riverkeeper Foundation, "Hogs and CAFOs," www.neuseriver.org/neuseissuesandfacts/hogsandcafos.html.

50. UFCW, "The Case Against Smithfield: Human and Civil Rights Violations in Tar Heel, North Carolina," www.ufcw.org/working_america/case_against_smithfield/case_against_smthfld.cfm.

51. Shafeah M'Bal and Peter Gilbert, "Smithfield Packing Struggle Mixes Black-Brown Unity, Environment and Workers' Rights," *Axis of Logic*, April 27, 2007, www.axisoflogic.com/artman/publish/article_24425.shtml.

52. UFCW, "Irresponsible Smithfield," www.ufcw.org/smithfield_justice/irresponsible_smithfield/index.cfm.

53. DOJ, "Smithfield Foods Fined 12.6 million, Largest Clean Water Act Fine Ever," August 8, 1997, www.usdoj.gov/opa/pr/1997/August97/331enr.htm.

54. Ibid.

55. Ibid.

56. Tietz, "Pork's Dirty Secret."

57. UFCW, "Workers Voices," www.smithfieldjustice.com/workersvoices.php#Main.

58. Mark Smith, "Family Farm Food Vs. Factory Farm Food," faid.convio.net/book/AmericanFamilyFarmers.pdf.

59. UFCW, "Workers Voices."

60. Research Associates of America, "Packaged with Abuse: Safety and Health Conditions at Smithfield Packing's Tar Heel Plant," report prepared for UFCW, August 2006.

61. Complaint for Damages and Equitable Relief, *Smithfield v. United Food and Commercial Workers Union International Itl.*, Civ. Action No. 3:07CV641, 26.

62. Answer for Defs. Research Assocs. of America (filed 11/20/07), 7; Answer for United Food and Commercial Workers International Union, Gene Bruskin, Joseph

Hansen, William T. McDonough, Leila McDowell, Patrick J. O'Neill, Andrew L. Stern, Tom Woodruff (filed 11/20/07), 15; *Smithfield v. United Food and Commercial Workers Union International Itl.*, Civ. Action No. 3:07CV641.

63. Memo. of Law in Support of Defs.' Jt. Mot. to Dismiss Under R. 12(b)(6). Fed. R. of Civ. Proc. (filed 11/20/07), *Smithfield v. United Food and Commercial Workers Union International Itl.*, Civ. Action No. 3:07CV641.

64. UFCW, "Case Against Smithfield."

65. Summary of Smithfield Foods, Inc., 347 National Labor Relations Board No. 109, August 31, 2006.

66. UFCW, "Case Against Smithfield."

67. Ibid.

68. Ibid.

69. Ben Wheeler, "Poland: Green Federation Gaja Fights Pollution from Industrial Feedlots," Global Greengrants Fund, January 28, 2005, www.greengrants.org/grantstories.php?news_id=72.

70. Desiree Evans, "Smithfield's Tar Heel Workers Ratify First-Ever Union Contract," Institute for Southern Studies, July 2, 2009.

71. UFCW, "Case Against Smithfield."

72. Human Rights Watch, "Blood, Sweat, and Fear: Workers' Rights in U.S. Meat and Poultry Plants," January 2005.

73. Research Associates of America, "Packaged with Abuse."

74. Ibid.

75. Ibid.

76. Doreen Carvajal and Stephen Castle, "A U.S. Hog Giant Transforms Eastern Europe," *New York Times*, May 6, 2009.

77. Wheeler, "Poland."

78. Tom Garrett, "The End of the Beginning: A Patriot Victory in the Polish Sejm," Animal Welfare Institute, *AWI Quarterly*, Spring 2005.

79. Ahmed ElAmin, "Smithfield Targets Romania for Expansion into Europe," MeatProcess.com, August 9, 2006.

80. Carvajal and Castle, "U.S. Hog Giant."

81. Mirel Bran, "Swine Plague: Romania Criticizes American Group's Attitude," *Le Monde*, August 15, 2006.

82. Ibid.

83. Ibid.

84. Ibid.

10. Modern-Day Serfs

1. Mike Weaver, e-mail communication, January 6, 2012.

2. "Top U.S. Poultry Companies 2011: Rankings," *Watt Poultry USA*, February 2011; U.S. Poultry & Egg Association, "Industry Economic Data," April 2011, www.uspoultry.org/economic_data/.

3. David Mann, "Getting Plucked: How the Poultry Industry Turns Contract Farmers into Modern-Day Sharecroppers," *Texas Observer*, March 17, 2005.

4. Jessica Chesnut, "How the Cobb 500 Changed the US Market," Poultry Site, October 21, 2008, www.thepoultrysite.com/articles/1200/how-the-cobb-500 -changed-the-us-market.

5. Oklahoma Cooperative Extension Service, "Beef, Pork, and Poultry Industry Coordination," Pig Site, www.thepigsite.com/articles/?Display=1265.

6. C. Robert Taylor, "The Many Faces of Power in the Food System," presentation at the DOJ/FTC Workshop on Merger Enforcement, FTC, Washington, DC February 17, 2004, 5.

7. C. Robert Taylor and David A. Domina, "Restoring Economic Health to Contract Poultry Production," May 13, 2010, ocm.srclabs.com/wp-content/uploads/2012/02/dominareportversion2.pdf.

8. "Testimony (Draft) of Scott Hamilton, Poultry Grower, Phil Campbell, Alabama," before the Committee on Agriculture, Nutrition, and Forestry, U.S. Senate Hearing on Economic Challenges and Opportunities Facing American Agricultural Producers Today, April 18, 2007, www.rafiusa.org/docs/hamiltontestimony.pdf.

9. Rural Advancement Foundation International-USA, "Farmers Trapped by Debt and Unfair Contracts," fact sheet, www.rafiusa.org/docs/contractagoverview.pdf.

10. Interview with Dr. Robert Taylor, August 16, 2011.

11. Ibid.

12. Ibid.

13. "Testimony (Draft) of Scott Hamilton."

14. Ibid.

15. Interview with Valerie Ruddle, August 25, 2011.

16. Ibid.

17. Ibid.

18. Ibid.

19. Valerie Ruddle, testimony, Poultry Workshop, Agriculture and Antitrust Enforcement Issues in Our 21st Century Economy: Margins, DOJ and USDA Public Workshops Exploring Competition Issues in Agriculture, USDA, Washington, DC, December 8, 2010.

20. Interview with Valerie Ruddle.

21. Ibid.

22. Ibid.

23. James M. MacDonald and William D. McBride, "The Transformation of U.S. Livestock Agriculture: Scale, Efficiency, and Risks," USDA ERS, *Economic Information Bulletin* No. 43, January 2009, 25; Mary Hendrickson and Bill Heffernan, "Concentration of Agricultural Markets," Department of Rural Sociology, University of Missouri–Columbia, April 2007, 7.

24. USDA ERS, "Consumer-Driven Agriculture," *AmberWaves*, November 2003, www.ers.usda.gov/amberwaves/november03/features/supplypushdemandpull.htm.

25. Shady Lane Poultry Farm, "A History Worth Repeating," hcfarm.com/A%20History%20Worth%20Repeating.htm.

26. Steve W. Martinez, "The Role of Changing Vertical Coordination in the Broiler and Pork Industries," USDA ERS, Agricultural Economics Report No. AER777, April 1999.

27. USDA ERS, "Consumer-Driven Agriculture."

28. John Hurdle, "Chickens Play a Big Role in Delmarva Economy, History," *DFM News*, May 1, 2012, www.delawarefirst.org/13455-chickens-play-big-role-delmarva -economy-history/.

29. Kay Doby, testimony, Roundtable Discussion on Poultry Grower Issues, Agriculture and Antitrust Enforcement Issues in Our 21st Century Economy, DOJ and USDA Public Workshops Exploring Competition in Agriculture, Alabama A&M University, Normal, AL, May 21, 2010.

30. Taylor and Domina, "Restoring Economic Health."

31. Ibid.

32. "Oklahoma Growers Win Verdict Against Tyson," *Agweek*, April 5, 2010, www .agweek.com/event/article/id16076/.

33. Associated Press, "Okla. Supreme Court Overturns $10M Tyson Verdict," March 7, 2012, ap.thecabin.net/pstories/state/ar/20120306/965234590.shtml.

34. *Feed Marketing and Distribution 2010 Reference Issue*, September 15, 2010, 4.

35. Simon Shane, "2008 Egg Industry Survey," *Watt Egg Industry* 114, no. 3 (March 2009).

36. John R. Wilke, "Federal Prosecutors Probe Food-Price Collusion," *Wall Street Journal*, September 23, 2008, online.wsj.com/article/SB122213370781365931.html.

37. Carole Morison with Polly Walker, "Organizing for Justice: DelMarVa Poultry Justice Alliance," Lecture 7, Johns Hopkins Bloomberg School of Public Health, July 2002.

38. Ibid.

39. Ibid.

40. UFCW, "Injury and Injustice—America's Poultry Industry," fact sheet, www .ufcw.org/press_room/fact_sheets_and_backgrounder/poultryindustry_.cfm?& bsuppresslayout=1.

41. Kristin Kloberdanz, "Poultry Workers," *Health Day*, March 21, 2012, consumer.healthday.com/encyclopedia/article.asp?AID=646575.

42. Jeffrey S. Passel, "Background Briefing Prepared for Task Force on Immigration and America's Future," Pew Hispanic Center, June 14, 2005, 27.

43. UFCW, "Injury and Injustice."

44. Kloberdanz, "Poultry Workers."

45. Karl Weber, ed., *Food, Inc.: A Participant Guide: How Industrial Food Is Making Us Sicker, Fatter, and Poorer—And What You Can Do About It* (New York: Public-Affairs, 2009), 61–78.

46. Emanuella Grinberg, "Humane Society Accuses Top Turkey Hatchery of Abuse," CNN, November 24, 2010, articles.cnn.com/2010-11-24/us/humane .society.hatchery.probe_1_animal-welfare-turkey-industry-animal-rights-group?_s= PM:US.

47. Ibid.

48. Kate Shatzki, "Chicken Industry Regulators Ask Congress to Raise Budget," *Baltimore Sun*, March 4, 1999.

11. Milking the System

1. Wisconsin Family Farm Defenders, "Citizens Mark the Opening Day of the World Dairy Expo with a Speak Out Against Taxpayer Subsidized Factory Farm Expansion in Wisconsin," press release, September 28, 2010, familyfarmers.org/?page_id=62.

2. Mary Hendrickson et al., "Executive Summary: Consolidation in Food Retailing and Dairy: Implications for Farmers and Consumers in a Global Food System," report to the National Farmers Union, Department of Rural Sociology, University of Missouri–Columbia, January 8, 2001. The landmark work of these researchers was an important source of information for this chapter.

3. Ibid.

4. USDA, "Economic Effects of U.S. Dairy Policy and Alternative Approaches to Milk Pricing," report to U.S. Congress, July 2004, 17–18; USDA National Agricultural Statistics Service (NASS), Agricultural Statistics Database, www.nass.usda.gov/Quick Stats (accessed August 5, 2008).

5. NASS, Agricultural Statistics Database.

6. Dennis A. Shields, "Consolidation and Concentration in the U.S. Dairy Industry," Congressional Research Service, April 27, 2010.

7. James M. MacDonald and William D. McBride, "The Transformation of U.S. Livestock Agriculture: Scale, Efficiency, and Risks," USDA ERS, *Economic Information Bulletin* No. 43. January 2009.

8. USDA, "Dairy 2007. Part 1: Reference of Dairy Cattle Health and Management Practice in the United States, 2007," October 2007, 79.

9. Lauren Etter, "Manure Raises a New Stink," *Wall Street Journal*, March 25, 2010.

10. Ibid.; Lauren Etter, "Burst Manure Bubble Causes Big Stink, but No Explosions," *Wall Street Journal*, April 1, 2010.

11. Etter, "Burst Manure Bubble."

12. Ron Cassie, "Walkersville, Farm Settle over Spill," *Frederick News Post* (MD), October 14, 2009; Jeremy Hauck, "Lawsuits Loom Large in Walkersville, Thurmont," *Maryland Gazette*, January 1, 2009.

13. Paul Walsh, "Manure Spill Closes State Park's Beach," *Minneapolis Star Tribune*, May 22, 2009; Madeleine Baran, "Dairy Fined $10K After Burst Manure Pipe Contaminates Swimming Area," Minnesota Public Radio, September 29, 2009.

14. This information is based on an analysis by Food & Water Watch of the USDA's 1997, 2002 and 2007 Censuses of Agriculture reports and data.

15. Food & Water Watch calculation comparing human and livestock waste production based on an EPA report, "Risk Assessment Evaluation for Concentrated Animal Feeding Operations," EPA/600/R-04/042, May 2004, 9. The average human produces 183 pounds of manure annually compared to 30,000 pounds for 1,000 pounds of live weight dairy cow (which is a dairy cow animal unit). Every dairy cow animal unit produces 163.9 times more manure than an average person. Food & Water Watch multiplied the number of dairy cow animal units on operations with over five hundred cows in each county by 163.9 to come up with a human sewage equivalent.

U.S. EPA reports, "A dairy CAFO with 1,000 animal units is equivalent to a city with 164,000 people," which means that one dairy animal unit is equivalent to 164 people, which matches Food & Water Watch's calculations. The human sewage equivalent was compared to the U.S. Census Bureau figures for metropolitan area population estimates. U.S. Census Bureau, "Annual Estimates of the Population of Metropolitan and Micropolitan Statistical Areas: April 1, 2000 to July 1, 2009," CBSA-EST2009-01.

16. NASS, Agricultural Statistics Database.

17. Alden C. Manchester and Don P. Blayney, "The Structure of Dairy Markets: Past, Present, Future," Commercial Agriculture Division, USDA ERS, *Agricultural Economic Report* No. 757, September 1997, 43.

18. Hendrickson, "Consolidation in Food Retailing and Dairy," 7.

19. Joel Greeno, testimony, Dairy Workshop, Agriculture and Antitrust Enforcement Issues in Our 21st Century Economy, DOJ and USDA Public Workshops Exploring Competition Issues in Agriculture, University of Wisconsin–Madison, June 25, 2011, available at armppa.webs.com/information.htm.

20. American Antitrust, "Chapter 8: Fighting Food Inflation Through Competition," Transition Report on Competition Policy, 2008, 300.

21. "USDA Report to Congress on the National Dairy Promotion and Research Program and the National Fluid Milk Processor Promotion Program," 2004, 21, www.ams.usda.gov/AMSv1.0/getfile?dDocName=STELDEV3099993.

22. Dean Foods Co. 10-Q SEC Filing, 2002, Item 2.

23. A. Cheng, "Dean Foods Cuts 2007 Forecast on Milk Price," *MarketWatch*, June 12, 2007.

24. C. Scott, "Organic Milk Goes Corporate," *Mother Jones*, April 26, 2006.

25. B. Silverstein, "Silk Soymilk: Smoooth," Brand Features: Profile, Brand channel.com, December 31, 2007.

26. Michael Ollinger et al., "Effect of Food Industry Mergers and Acquisitions on Employment and Wages," USDA ERS, *Economic Research Report* No. 13, 2005, 17.

27. "Statement of James 'Jim' W. Miller, Under Secretary for Farm and Foreign Agricultural Services, U.S. Department of Agriculture, Washington, D.C.," Hearing to Review Economic Conditions Facing the Dairy Industry, Hearings Before the Subcommittee on Livestock, Dairy, and Poultry of the Committee on Agriculture, U.S. House of Representatives, 111th Cong., 1st sess., July 14, 2009.

28. "Statement of Scott Hoese, President, Carver County Farmers Union, Dairy Farmer, Mayer, MN; On Behalf of National Farmers Union," Hearing to Review Economic Conditions Facing the Dairy Industry, Hearings Before the Subcommittee on Livestock, Dairy, and Poultry of the Committee on Agriculture, U.S. House of Representatives, 111th Cong., 1st sess., July 21, 2009, 7.

29. Letter from Warren Taylor to Friends of Snowville Creamery, undated.

30. David Domina and C. Robert Taylor, "The Debilitating Effects of Concentration in Markets Affecting Agriculture," Organization for Competitive Markets, September 2009.

31. Eric Palmer, "Dairy Co-op, Former Execs Fined $12 Million in Price Manipulation Case," *Kansas City Star*, December 16, 2008.

32. U.S. Government Accountability Office (GAO), "Spot Cheese Market: Market Oversight Has Increased, but Concerns Remain About Potential Manipulation," 2007.

33. Greeno, testimony.

34. GAO, "Spot Cheese Market."

35. John Bunting, "Dairy Farm Crisis 2009: A Look Beyond Conventional Analysis," report, March 2009, midmddairyvets.com/docs/DairyFarmCrisis2009-1.pdf.

36. Ibid.

37. "Statement of Joaquin Contente, President, California Farmers Union, Hanford, CA; On Behalf of National Farmers Union," Hearing to Review Economic Conditions Facing the Dairy Industry, Hearings Before the Subcommittee on Livestock, Dairy, and Poultry of the Committee on Agriculture, U.S. House of Representatives, 111th Cong., 1st sess., July 28, 2009.

12. Life for Sale: The Birth of Life Science Companies

1. "Herbert Boyer: Biotechnology," *Who Made America?*, PBS, www.pbs.org/wgbh/theymadeamerica/whomade/boyer_hi.html.

2. "Recombinant DNA Research at UCSF and Commercial Application at Genentech," University of California, San Francisco, Oral History Program and the Program in the History of the Biological Sciences and Biotechnology, Bancroft Library, University of California, Berkeley, 2001, archive.org/details/dnaresearchucsf00boyerich.

3. Ibid.

4. Tom Abate, "The Birth of Biotech: How the Germ of an Idea Became the Genius of Genentech," *San Francisco Chronicle*, April 1, 2001.

5. Ibid.

6. Martin Kenney, *Biotechnology: The University-Industrial Complex* (New Haven, CT: Yale University Press, 1986).

7. Andy Fell, "Nothing Ventured, Nothing Gained—Biotech Startup Illustrates Campus's Shift in Attitude," UC Davis News and Information, April 3, 2004, dateline .ucdavis.edu/dl_detail.lasso?id=7755.

8. Ibid.

9. Belinda Martineau, *First Fruit: The Creation of the Flavr Savr Tomato* (New York: McGraw-Hill, 2001).

10. Ibid.

11. Ibid.

12. Sonal Panse, "History of the Genetically Engineered Tomato," ed. Paul Arnold, Bright Hub, May 22, 2011, www.brighthub.com/science/genetics/articles/27236.aspx.

13. Bryan Bergeron and Paul Chen, *Biotech Industry: A Global, Economic, and Financing Overview* (Hoboken, NJ: John Wiley & Sons, 2004), 101.

14. "Business: More for Monsanto," *Time*, May 18, 1936.

15. Vandana Shiva, *Stolen Harvest: The Hijacking of the Global Food Supply* (Boston: South End Press, 2000), 31.

16. A.V. Krebs, "The Seed Patenters: Biotech Giants Force Farmers into Lockstep," *Progressive Populist*, July 1998.

17. Ibid.

18. Ibid.

19. Interview with Vandana Shiva, "The Role of Patents in the Rise of Globalization," New Delhi, India, August 27, 2003.

13. David Versus Goliath

1. Clive James, "ISAAA Brief 42-2010: Global Status of Commercialized Biotech/ GM Crops: 2010," International Service for the Acquisition of Agri-biotech Applications (ISAAA), 2011, 7

2. Ibid., Executive Summary, Table 1.

3. USDA, Office of Inspector General, Southwest Region, "USDA's Role in the Export of Genetically Engineered Agricultural Commodities," Audit Report No. 50601-14-Te, February 2009, 7; USDA ERS, "Adoption of Genetically Engineered Crops in the U.S.," from corn and soybean spreadsheets, updated July 1, 2011, www.ers.usda .gov/Data/BiotechCrops/ (accessed July 6, 2011).

4. Lawrence Gilbert, "Howard Schneiderman, 1927–1990," in *Biographical Memoirs*, vol. 63 (New York: National Academy of Science, 1994), 481–94.

5. Daniel Charles, *Lords of the Harvest* (New York: Basic Books, 2001), 7–10.

6. Gilbert, "Howard Schneiderman," 495–96.

7. Charles, *Lords of the Harvest*, 26–27.

8. For a copy of Rifkin's award, see foet.org/img/University%20Of%20Pennsyl vania%20Student%20Award%20of%20Mert.jpg.

9. Charles, *Lords of the Harvest*, 26–27.

10. Ibid., 28.

11. Ibid.

12. Daniel J. Kevles, "Protections, Privileges, and Patents: Intellectual Property in American Horticulture, the Late Nineteenth Century to 1930," *Proceedings of the American Philosophical Society* 152, no. 2 (June 2008): 209–12, www.amphilsoc.org/ sites/default/files/1520204.pdf.

13. Charles, *Lords of the Harvest*, 28.

14. Kurt Eichenwald, Gina Kolata, and Melody Petersen, "Biotechnology Food: From the Lab to a Debacle," *New York Times*, January 25, 2001.

15. Karen Kaplan, "New Biotech Policy Could Aid MIT," *The Tech*, March 6, 1992, tech.mit.edu/V112/N11/biotech.11n.html.

16. "Open Letter from Monsanto CEO Robert B. Shapiro to Rockefeller Foundation President Gordon Conway and Others," October 4, 1999, www.monsanto .com/newsviews/Pages/monsanto-ceo-to-rockefeller-foundation-president-gordon -conway-terminator-technology.aspx.

17. Phone interview with Rebecca Spector, October 15, 2011.

18. U.S. General Accounting Office (GAO), "Genetically Modified Foods: Experts View Regimen of Safety Tests as Adequate, but FDA's Evaluation Process Could Be Enhanced," report to Congressional Requesters, GAO-02-566, 2002, 30.

19. USDA ERS, "Adoption of Genetically Engineered Crops in the U.S."

20. Policy Statement: Foods Derived from New Plant Varieties, 57 Fed Reg. 22984 (May 29, 1992), 1.

21. Premarket Notice Concerning Bioengineered Foods, 66 Fed. Reg. 4706 (January 18, 2001).

22. Pew Initiative on Food and Biotechnology, "Guide to U.S. Regulation of Genetically Modified Food and Agricultural Biotechnology Products," September 2001, 19–20.

23. Ibid., 20.

24. FDA, GRAS Notice Inventory, accessed April 28, 2011. Data on file at Food & Water Watch and available at www.accessdata.fda.gov/scripts/fcn/fcnNavigation.cfm ?rpt=grasListing.

25. Food Additive Petitions, 21 CFR 171.1(c).

26. Determination of Food Additive Status, 21 CFR 170.38(c).

27. Premarket Notice Concerning Bioengineered Foods, 66 Fed. Reg. 4708 (January 18, 2001); Pew Initiative on Food and Biotechnology, "Guide to U.S. Regulation," 21.

28. EPA, "Concerning Dietary Exposure to CRY9C Protein Produced by Starlink: Corn and the Potential Risks Associated with Such Exposure," draft white paper, October 16, 2007, 9.

29. Emma Young, "GM Pea Causes Allergic Damage in Mice," *New Scientist*, November 21, 2005.

30. EPA, "Concerning Dietary Exposure," 7.

31. Andrew Pollack, "Crop Scientists Say Biotechnology Seed Companies Are Thwarting Research," *New York Times*, February 20, 2009.

32. Joel Spiroux de Vendomois et al., "A Comparison of the Effects of Three GM Corn Varieties on Mammalian Health," *International Journal of Biological Sciences* 5, no. 7 (2009): 716–18.

33. Manuela Malatesta et al., "Ultrastructural Morphometrical and Immunocytochemical Analyses of Hepatocyte Nuclei from Mice Fed on Genetically Modified Soybean," *Cell Structure and Function* 27, no. 5 (2002): abstract.

34. Gilles-Eric Séralini, Dominique Cellier, and Joël Spiroux de Vendomois, "New Analysis of a Rat Feeding Study with a Genetically Modified Maize Reveals Signs of Hepatorenal Toxicity," *Archives of Environmental Contamination and Toxicology* 52, no. 4 (2007): 596, 601.

35. B. Cisterna et al., "Can a Genetically-Modified Organism-Containing Diet Influence Embryo Development?: A Preliminary Study on Pre-implantation Mouse Embryos," *European Journal of Histochemistry* 52, no. 4 (2008): 263.

36. Antonella Agodi et al., "Detection of Genetically Modified DNA Sequences in Milk from the Italian Market," *International Journal of Hygiene and Environmental Health* 209, no. 1 (January 10, 2006): abstract.

37. Alejandra Paganelli et al., "Glyphosate-Based Herbicides Produce Teratogenic Effects on Vertebrates by Impairing Reinoic Acid Signaling," *Chemical Research in Toxicology* 23 no. 10 (2010): 1586.

38. Nora Benachour and Gilles-Eric Séralini, "Glyphosate Formulations Induce Apoptosis and Necrosis in Human Umbilical, Embryonic, and Placental Cells," *Chemical Research in Toxicology* 22, no. 1 (2009): 97.

39. Arnold L. Aspelin, "Pesticides Industry Sales and Usage: 1994 and 1995 Estimates," EPA, August 1997, Table 8; Arthur Grube et al., "Pesticides Industry Sales and Usage: 2006 and 2007 Market Estimates," EPA, February 2011, Table 3.6.

40. Committee on the Impact of Biotechnology on Farm-Level Economics and Sustainability, National Research Council, *The Impact of Genetically Engineered Crops on Farm Sustainability in the United States* (Washington, DC: National Academies Press, 2010), 4, 13–14. At least eight weed species in the United States (and fifteen worldwide) have been confirmed to be resistant to glyphosate, including aggressive crop weeds such as ragweed, mare's tail, and waterhemp. A 2009 Purdue University study found that glyphosate-tolerant mare's tail could "reach staggering levels of infestation in about two years after it is first detected." Even biotech company Syngenta predicts that glyphosate-resistant weeds will infest one fourth of U.S. cropland by 2013. Research shows that higher densities of glyphosate-resistant weeds reduce crop yields. Purdue scientists found that Roundup-resistant ragweed can cause 100 percent corn-crop losses.

41. Registration of Pesticides, 7 U.S.C. 136a(c)(5); EPA, "FIFRA Amendments of 1988," press release, October 26, 1988, www.epa.gov/history/topics/fifra/01.htm, and on file at Food & Water Watch.

42. Registration of Pesticides, 7 U.S.C. 136a(c)(5).

43. Pesticide Registration and Classification Procedures, 40 CFR 152.1(a).

44. Experimental Use Permits, 40 CFR 172.3(a); Tolerances and Exemptions for Pesticide Chemical Residues in Food, 40 CFR 180; Pesticide Registration and Classification Procedures, 40 CFR 152.1(a); Pew Initiative on Food and Biotechnology, "Guide to U.S. Regulation," 13–14.

45. Permits for the Introduction of a Regulated Article, 7 CFR §340.4(f)(9) (2008).

46. Permits for the Introduction of a Regulated Article, 7 CFR §340.4(g) (2008).

47. A 2010 biotechnology survey performed by the European Commission reported that 59 percent of Europeans think that GE food is unsafe for their health and that of their families, and 61 percent do not think that the development of GE food should be encouraged. These opinions are reflected in the nearly one quarter of EU member countries that have bans in effect on GE products. By September 2010 the EU had 292 regions and provinces and 4,713 local governments that are GE-free. Biotechnology regulation in the European Union is far stricter than in the United States and operates under a so-called precautionary principle, then assesses each food's safety before approving its commercialization. But, under pressure from the U.S. government and the biotech industry, the EU has approved more than thirty GE products for sale in the region, most of which is GE soy and corn (maize) in animal feed. Only two GE crops are currently approved for cultivation in the EU. EU law requires that all foods and feeds with any GE content bear notification labels, including those with more than 0.9 percent accidental biotech content, but GE products that are considered "processing aids," such as GE enzymes used to make cheese, are exempt from the labeling process.

48. Robert Wisner, "Round-Up® Ready Spring Wheat: Its Potential Short-Term Impacts on U.S. Wheat Export Markets and Prices," Iowa State University, October

2004, 1; Monsanto, "Frequently Asked Questions about Monsanto and Wheat," www .monsanto.com/products/Pages/wheat-faq.aspx (accessed February 7, 2011).

49. Monsanto Co. and KWS SAAT AG; Determination of Nonregulated Status for Sugar Beet Genetically Engineered for Tolerance to the Herbicide Glyphosate, 70 Fed. Reg. 13007-13008 (March 17, 2005).

50. "Memorandum and Order," *Geertson Seed Farms et al. v. USDA*, U.S. District Court for the Northern District of California, No. C 06-01075 CRB, February 13, 2007, 1, 3.

51. USDA, "Glyphosate-Tolerant Alfalfa Events J101 and J163: Request for Non-regulated Status," Final Environmental Impact Statement, December 2010, S-39-41.

52. Determination of Regulated Status of Alfalfa Genetically Engineered for Tolerance to the Herbicide Glyphosate; Record of Decision, 76 Fed. Reg. 5780-5781 (February 2, 2011); USDA, "Glyphosate-Tolerant Alfalfa Events J101 and J163: Request for Nonregulated Status," Record of Decision, January 27, 2011, 5, 7-8.

53. Monsanto, "2011 Technology Use Guide," 2011, 7.

54. David S. Conner, "Pesticides and Genetic Drift: Alternative Property Rights Scenario," *Choices*, First Quarter 2003, 5.

55. Union of Concerned Scientists (UCS), "Gone to Seed: Transgenic Contaminants in the Traditional Seed Supply," 2004, 28.

56. Norman Ellstrand, "Going to 'Great Lengths' to Prevent the Escape of Genes That Produce Specialty Chemicals," *Plant Physiology*, August 2003.

57. Carolyn Dimitri and Lydia Oberholtzer, "Marketing U.S. Organic Foods: Recent Trends from Farms to Consumers," USDA ERS, *Economic Information Bulletin* No. 58, September 2009, abstract.

58. William D. McBride and Catherine Greene, "A Comparison of Conventional and Organic Milk Production Systems in the U.S.," USDA ERS, prepared for presentation at the American Agricultural Economics Association Annual Meeting, Portland, OR, July 29–August 1, 2007, 13, 17; Food & Water Watch analysis of average consumer price data from the U.S. Bureau of Labor Statistics, Consumer Price Index—Average Price Data; Farmgate prices from USDA National Agricultural Statistical Service, Agricultural Prices Annual Summary.

59. Availability of an Environmental Assessment and Finding of No Significant Impact for a Biological Control Agent for *Arundo donax*, 76 Fed. Reg. 8708 (February 15, 2011).

60. USDA, "Syngenta Seeds, Inc. Alpha-Amylase Maize Event 3272, Draft Environmental Assessment," November 6, 2008, 34–35; USDA, "National Environmental Policy Act Decision and Finding of No Significant Impact, Syngenta Seeds Inc., Alpha-Amylase Maize, Event 3272," 2011, 10.

61. USDA, "Syngenta Seeds," 32–33.

62. Carey Gillam, "Monsanto Launching Its First Biotech Sweet Corn," Reuters, August 4, 2011; USDA, "2007 Census of Agriculture–United States Data," 2009, 35.

63. Jack Kaskey, "Monsanto to Sell Biotech Sweet Corn for U.S. Consumers," Bloomberg, August 4, 2011.

14. The Future of Food: Science Fiction or Nature?

1. David Ewing Duncan, "Craig Venter: The Bill Gates of Artificial Life?" *Technology Review*, June 13, 2007, www.technologyreview.com/blog/duncan/17623/.

2. "Designing Life: What's Next for J. Craig Venter?" *60 Minutes*, CBS News, June 12, 2011, www.cbsnews.com/stories/2011/06/12/60minutes/main20070141 .shtml.

3. J. Craig Venter Institute, "About: Biographies: J. Craig Venter," www.jcvi.org/ cms/about/bios/jcventer/ (accessed June 20, 2012).

4. Synthetic Genomics, "What We Do: Agricultural Products," www.synthetic genomics.com/what/agriculture.html (accessed June 20, 2012).

5. Jim Thomas, "The Sins of Syn Bio," *Slate*, February 2, 2011, www.slate.com/ articles/technology/future_tense/2011/02/the_sins_of_syn_bio.html.

6. Vítor Martins dos Santos, "Synthetic Biology in Food & Health," Systems and Synthetic Biology, Wageningen University, undated, 36.

7. Woodrow Wilson International Center for Scholars, "Trends in Synthetic Biology Research Funding in the United States and Europe," Research Brief 1, June 2010.

8. Ibid.

9. BioBricks Foundation, "About: Board of Directors," biobricks.org/about -foundation/board-of-directors/.

10. Rebecca Cathcart, "Designer Genes," *Good*, March 20, 2008, www.good.is/ post/designer_genes.

11. BioBricks Foundation, "SB 5.0: The Fifth International Meeting on Synthetic Biology," June 15–17, 2011, sb5.biobricks.org/files/sb5-program-book-v3.pdf.

12. Science Cheerleader, "About Us: Darlene Cavalier," www.sciencecheerleader .com/about-us/darlene-cavalier/.

13. Allison A. Snow, "Risks of Environmental Releases of Synthetic GEOs," invited presentation for the Presidential Commission for the Study of Bioethics, July 8, 2010, 2–3.

14. John Beddington, Government Chief Scientific Advisor, "Key Issues in Agriculture Science," Frank Parkinson Lecture, Oxford Farming Conference, January 5, 2010, www.ofc.org.uk/files/ofc/papers/2010beddingtonpaper.pdf.

15. "Biotech, Nanotech and Synthetic Biology Roles in Future Food Supply Explored," *Science Daily*, February 25, 2010, www.sciencedaily.com/releases/2010/02/ 100221143238.htm.

16. Friends of the Earth, "Out of the Laboratory and Onto Our Plates," 2008, 4.

17. International Center for Technology Assessment, "Citizen Petition for Rulemaking to the United States Environmental Protection Agency," 2008, 10.

18. Friends of the Earth, "Out of the Laboratory," 20.

19. M.C. Roco, "Broader Societal Issues of Nanotechnology," *Journal of Nanoparticle Research* 5 (2003): 182.

20. Wenjuan Yang et al., "Food Storage Material Silver Nanoparticles Interfere with DNA Replication Fidelity and Bind with DNA," *Nanotechnology*, February 2, 2009.

21. Louise Gray, "Chief Scientist Says GM and Nanotechnology Should Be Part of Modern Agriculture," *The Telegraph*, January 6, 2010, www.telegraph.co.uk/earth/earthnews/6943231/Chief-scientist-says-GM-and-nanotechnology-should-be-part-of-modern-agriculture.html.

22. National Nanotechnology Initiative, "Supplement to the President's FY 2010 Budget," 2009, 3.

23. Project on Emerging Nanotechnologies, "Consumer Products: An Inventory of Nanotechnology-based Consumer Products Currently on the Market," www.nanotechproject.org/inventories/consumer/.

24. Project on Emerging Nanotechnologies, "New Nanotech Products Hitting the Market at the Rate of 3–4 Per Week," www.nanotechproject.org//news/archive/6697/.

25. Friends of the Earth, "Out of the Laboratory," appendix A.

26. Ibid., 20.

27. Ibid., 12.

28. Ibid., 55.

29. Barnaby Feder, "Engineering Food at Level of Molecules," *New York Times*, October 10, 2006.

30. Project on Emerging Nanotechnologies, "Consumer Products."

31. Ibid.

32. Ibid.

33. Friends of the Earth. "Out of the Laboratory," 12.

34. K . Donaldson et al., "Free Radical Activity Associated with the Surface of Particles: A Unifying Factor in Determining Biological Activity?" *Toxicology Letters*, November 24, 1997, 89; Rosemary Dunford et al., "Chemical Oxidation and DNA Damage Catalyzed by Inorganic Sunscreen Ingredients," *FEBS Letters*, November 24, 1997, 89.

35. Christie Sayes et al., "Correlating Nanoscale Titania Structure with Toxicity: A Cytotoxicity and Inflammatory Response Study with Human Dermal Fibroblasts and Human Lung Epithelial Cells," *Toxicological Sciences*, April 2006, Conclusions.

36. Craig Poland et al., "Carbon Nanotubes Introduced into the Abdominal Cavity of Mice Show Asbestos-like Pathogenicity in a Pilot Stufy," *Nature Nanotechnology*, May 20, 2008.

37. National Institutes of Health, National Center for Complementary and Alternative Medicine, "Backgrounder on Colloidal Silver Products," nccam.nih.gov/health/silver/.

38. J. Wang et al., "Acute Toxicity and Biodistribution of Different Sized Titanium Dioxide Particles in Mice After Oral Administration," *Toxicology Letters*, December 2006, Conclusion.

39. "Silver Nanoparticles May Be Killing Beneficial Bacteria in Wastewater Treatment," *Nanotechnology Business Journal*, May 12, 2008.

40. L. Yang et al., "Particle Surface Characteristics May Play an Important Role in Phytotoxicity of Alumina Nanoparticles," *Toxicology Letters*, March 2005, 122–32.

41. Federicia Gillian et al., "Toxicity of Titanium Dioxide Nanoparticles to Rainbow Trout (*Oncorhynchus mykiss*): Gill Injury, Oxidative Stress, and Other Physiological Effects," *Aquatic Toxicology*, October 30, 2007, abstract.

42. Food & Water Watch, "Food and Agriculture Biotechnology Industry Spends More Than Half a Billion Dollars to Influence Congress," Issue Brief, 2010, 1.

43. Doug Gurian-Sherman, "Failure to Yield: Evaluating the Performance of Genetically Engineered Crops," Union of Concerned Scientists, 2009.

44. Veterinary Medicine Advisory Committee, "Briefing Packet: AquAdvantage Salmon," FDA Center for Veterinary Medicine, September 20, 2010 (prereleased September 3, 2010), 108.

45. Ibid., 30, 88–89.

46. Julie Nordlee et al., "Identification of a Brazil-Nut Allergen in Transgenic Soybeans," *New England Journal of Medicine*, March 14, 1996.

47. Irina Ermakova, "Genetically Modified Organisms Could Be Real Threat to the Life (Reply to ACNFP on the 'Statement on the Effect of GM Soy on Newborn Rats')," Annex 2 to ACFNP/80/8, Advisory Committee on Novel Foods and Processes, September 18, 2006, www.food.gov.uk/multimedia/pdfs/acnfp8008gm soya.pdf.

48. Manuela Malatesta et al., "Ultrastructural Morphometrical and Immunocytochemical Analyses of Hepatocyte Nuclei from Mice Fed on Genetically Modified Soybean," *Cell Structure and Function* 27, no. 5 (2002).

49. Gilles-Eric Séralini, Dominique Cellier, and Joël Spiroux de Vendomois, "New Analysis of a Rat Feeding Study with a Genetically Modified Maize Reveals Signs of Hepatorenal Toxicity," *Archives of Environmental Contamination and Toxicology* 52, no. 4 (2007).

50. Antonella Agodi et al., "Detection of Genetically Modified DNA Sequences in Milk from the Italian Market," *International Journal of Hygiene and Environmental Health* 209, no. 1 (January 10, 2006).

51. William Muir and Richard Howard, "Possible Ecological Risks of Transgenic Organism Release When Transgenes Affect Mating Success: Sexual Selection and the Trojan Gene Hypothesis," *Ecology*, November 23, 1999.

52. "100,000 Salmon Escape," Fish Site, March 12, 2010, www.thefishsite.com/fishnews/11892/100000-salmon-escape.

53. R. Naylor et al., "Fugitive Salmon: Assessing the Risks of Escaped Fish from Net Pen Aquaculture," *Bioscience*, May 2005, Introduction.

54. Ibid.

55. Dita De Boni, " 'Frankenfish' Programme Canned," *New Zealand Herald*, February 25, 2000.

56. Margot Roosevelt, "Would You Eat a Clone?" *Time*, June 13, 2005.

57. Renuka Rayasam, "Horse Is a Champ—It's in the Genes," *Austin-American Statesman*, March 31, 2006.

58. Anne Eisenberg, "Hello Kitty, Hello Clone," *New York Times*, May 28, 2005.

59. Paul Elias, "Engineered Swine Rich in Omega-3," *Monterey County Herald*, March 27, 2006.

60. H. Tamada and N. Kikyo, "Nuclear Reprogramming in Mammalian Somatic Cell Nuclear Cloning," *Cytogenetic and Genome Research* 105 (2004): 285–91.

61. G. Vatja, "Handmade Cloning—Summary," unpublished, 2004, cited in "The Science and Technology of Farm Animal Cloning: A Review of the State of the Art of the Science, the Technology, the Problems and the Possibilities," report from the project Cloning in Public, Danish Centre for Bioethics and Risk Assessment; J.P. Renard et al., "Lymphoid Hypoplasia and Somatic Cloning," *The Lancet* 353 (May 1999): 1489–91, cited in Joyce D'Silva, "Farm Animal Cloning from an Animal Welfare Perspective," Compassion in World Farming, www.ciwf.org.

62. P. Chavatte-Palmer et al., "Health Status of Cloned Cattle at Different Ages," *Cloning and Stem Cells* 6, no. 2 (June 2004): 94–100, cited in "Science and Technology of Farm Animal Cloning."

63. "Even the Few That Make It Are Abnormal," *BioWorld Today*, October 14, 2005.

64. Jeffrey Bartholet, "When Will Scientists Grow Meat in a Petri Dish?," *Scientific American*, May 17, 2011, www.scientificamerican.com/article.cfm?id=inside-the-meat-lab.

65. Ibid.

66. Michael Specter, "Test-Tube Burgers," *New Yorker*, May 23, 2011.

67. Ibid., 32.

68. Ibid., 37.

15. Eat *and* Act Your Politics

1. Shonna Dreier and Minoo Taheri, "Innovative Models: Small Grower and Retail Collaborations, Part B—Balls Food Stores' Perspective," Wallace Center of Winrock International, June 2009, 4, available at ngfn.org/resources/research-1/innovative-models/Balls%20Food%20Stores%20Innovative%20Model.pdf.

16. The Way Forward

1. Mabel Dobbs, Livestock Committee Chair for the Western Organization of Resource Councils, statement in response to USDA's final submission of Grain Inspection Packers and Stockyards Administration rule, November 4, 2011, worc.org/userfiles/file/livestock/Unde%20Preference/WORC_statement_rule_11_04_11.pdf.

2. Jim Goodman, "Occupy the Food System," *Other Words*, December 12, 2011, available at www.commondreams.org/view/2011/12/12-0.

3. Congressional Budget Office, "Trends in the Distribution of Household Income Between 1979 and 2007," Washington, DC, October 2010, cbo.gov/ftpdocs/124xx/doc12485/10-25-Householdincome.pdf.

4. Mark Winne, "Food Bank Speech—May 15, 2008—Seattle, WA: Leading the Charge, Leading the Change," June 21, 2008, www.markwinne.com/food-bank-speech-may-15-2008-seattle-wa/.

5. Mark Winne, "High Food Prices—Just Another Bad Day in the Food Line," May 11, 2008, www.markwinne.com/52.

6. Warren Buffett, "Stop Coddling the Super-Rich," op-ed, *New York Times*, August 14, 2011.

7. "Executive Summary: Food for Thought: Television Food Advertising to Children in the United States," Henry J. Kaiser Family Foundation, March 2007, www.kff.org/entmedia/upload/7618ES.pdf.

8. Marion Nestle, "The Food Industry Vs. Nutrition Standards: A First Amendment Issue?" *Food Politics*, September 6, 2011, www.foodpolitics.com/2011/09/the-food-industry-vs-nutrition-standards-a-first-amendment-issue/.

9. Jay Walljasper, "Twelve Reasons You'll Hear More About the Commons in 2012," *Huffington Post*, January 6, 2012.

10. Maude Barlow, "Our Commons Future Is Already Here," *Commons Magazine*, October 12, 2010.

11. Robert Weissman, "Pity Poor Newt Gingrich," *Huffington Post*, January 4, 2012.

BIBLIOGRAPHY

Araz, Mark, and Rick Wartzman. *The King of Cotton: J.G. Boswell and the Making of a Secret American Empire*. New York: PublicAffairs, 2003.

Barlow, Maude. *Blue Covenant: The Global Water Crisis and the Coming Battle for the Right to Water*. New York: New Press, 2007.

Bergeron, Bryan, and Paul Chen. *Biotech Industry: A Global, Economic, and Financing Overview*. Hoboken, NJ: John Wiley & Sons, 2004.

Charles, Daniel. *Lords of the Harvest: Biotech, Big Money, and the Future of Food*. New York: Basic Books, 2001.

Cochrane, Willard. *The Curse of American Agricultural Abundance: A Sustainable Solution*. Lincoln: University of Nebraska Press, 2003.

Conkin, Paul. *Revolution Down on the Farm: The Transformation of American Agriculture Since 1929*. Lexington: University Press of Kentucky, 2009.

Fromartz, Samuel. *Organic, Inc.: Natural Foods and How They Grew*. Orlando, FL: Harcourt, 2006.

Green, Dorothy. *Managing Water: Avoiding Crisis in California*. Berkeley: University of California Press, 2007.

Hauter, Wenonah. *Zapped! Irradiation and the Death of Food*. Washington, DC: Food & Water Watch, 2008.

Kaufman, Frederick. *A Short History of the American Stomach*. Orlando, FL: Houghton Mifflin Harcourt, 2008.

———. "The Food Bubble: How Wall Street Starved Millions and Got Away with It." *Harper's Magazine*, July 2010.

Kenney, Martin. *Biotechnology: The University-Industrial Complex*. New Haven, CT: Yale University, 1986.

Kimbrell, Andrew. *The Fatal Harvest Reader: The Tragedy of Industrial Agriculture*. Washington, DC: Island Press, 2002.

Krebs, Al. *The Corporate Reapers: The Book of Agribusiness*. Washington, DC: Essential Books, 1992.

Lappé, Anna. *Diet for a Small Planet: The Climate Crisis at the End of Your Fork and What You Can Do About It*. New York: Bloomsbury USA, 2010.

Lappé, Francis Moore, and Anna Lappé. *Hope's Edge: The Next Diet for a Small Planet*. New York: Jeremy P. Tarcher/Putnam, 2002.

Martineau, Belinda. *First Fruit: The Creation of the Flavr Savr Tomato and the Birth of Biotech Food*. New York: McGraw-Hill, 2001.

McCune, Wesley. *The Farm Bloc*. New York: Greenwood Press, 1968.

———. *Who's Behind Our Farm Policy?* New York: Praeger, 1956.

McMillan, Tracie. *The American Way of Eating: Undercover at Walmart, Applebee's, Farm Fields, and the Dinner Table*. New York: Scribner, 2012.

341

Nestle, Marion. *Food Politics: How the Food Industry Influences Nutrition and Health.* Berkeley: University of California Press, 2007.

Patel, Raj. *Stuffed and Starved: The Hidden Battle for the World Food System.* New York: Melville House, 2008.

Peek, George Nelson. *Equality for Agriculture.* Moline, IL: H.W. Harrington, 1922.

Ray, Daryll E., et al. "Rethinking US Agriculture Policy: Changing Course to Secure Farmers Future." Agricultural Policy Analysis Center, University of Tennessee, September 2003.

Schlosser, Eric. *Fast Food Nation: The Dark Side of the All-American Meal.* New York: HarperPerennial, 2005.

Shiva, Vandana. *Stolen Harvest: The Hijacking of the Global Food Supply.* Boston: South End Press, 2000.

Simon, Michele. *Appetite for Profit: How the Food Industry Undermines our Health and How to Fight Back.* New York: Nation Books, 2006.

Sinclair, Upton. *The Jungle.* New York: Doubleday, Page & Co., 1906.

Stoll, Steven. *The Fruits of Natural Advantage: Making the Industrial Countryside in California.* Berkeley: University of California Press, 1998.

Walljasper, Jay, and On the Commons. *All That We Share: How to Save the Economy, the Environment, the Internet, Democracy, Our Communities, and Everything Else That Belongs to All of Us.* New York: The New Press, 2011.

Weber, Karl, ed. *Food, Inc.: How Industrial Food Is Making Us Sicker, Fatter, and Poorer—and What You Can Do About It.* New York: PublicAffairs, 2009.

Winne, Mark. *Closing the Food Gap: Resetting the Table in the Land of Plenty.* Boston: Beacon Press, 2009.

Wise, Tim. "Still Waiting for the Farm Boom: Family Farmers Worse Off Despite High Crop Prices." GDAE Policy Brief 11-01, March 2011.

———. "Understanding the Farm Problem: Six Common Errors in Presenting Farm Statistics." GDAE Working Paper No. 05-02, March 2005.

———, and Alicia Harvie. "Boom for Whom? Family Farmers Saw Lower On-Farm Income Despite High Prices." GDAE Policy Brief 09-02, February 2009.

INDEX

CELEBRATING 20 YEARS OF INDEPENDENT PUBLISHING

Thank you for reading this book published by The New Press. The New Press is a nonprofit, public interest publisher celebrating its twentieth anniversary in 2012. New Press books and authors play a crucial role in sparking conversations about the key political and social issues of our day.

We hope you enjoyed this book and that you will stay in touch with The New Press. Here are a few ways to stay up to date with our books, events, and the issues we cover:

- Sign up at www.thenewpress.com/subscribe to receive updates on New Press authors and issues and to be notified about local events
- Like us on Facebook: www.facebook.com/newpressbooks
- Follow us on Twitter: www.twitter.com/thenewpress

Please consider buying New Press books for yourself; for friends and family; or to donate to schools, libraries, community centers, prison libraries, and other organizations involved with the issues our authors write about.

The New Press is a 501(c)(3) nonprofit organization. You can also support our work with a tax-deductible gift by visiting www.thenewpress.com/donate.

31901051722926